KU-098-026

333.715/TAY

Dear Pitman Publishing Customer

IMPORTANT – Read This Now!

We are delighted to announce a special free service for all of our customers.

Simply complete this form and return it to the address overleaf to receive:

A Free Customer Newsletter
B Free Information Service
C Exclusive Customer Offers – which have included free software, videos and relevant products
D Opportunity to take part in product development sessions
E The chance for you to write about your own business experience and become one of our respected authors

Fill this in now and return it to us (no stamp needed in the UK) to join our customer information service.

Name: ______________________ Position: ______________________

Company/Organisation: ______________________

Address (including postcode): ______________________

____

______________________ Country: ______________________

Telephone: ______________________ Fax: ______________________

Nature of business: ______________________

Title of book purchased: ______________________

Comments: ______________________

____

---------- **Fold Here Then Staple** ----------

We would be very grateful if you could answer these questions to help us with market research.

1 Where/How did you hear of this book?

- ☐ in a bookshop
- ☐ in a magazine/newspaper (please state which):

____

- ☐ information through the post
- ☐ recommendation from a colleague
- ☐ other (please state which):

____

2 Which newspaper(s)/magazine(s) do you read regularly?:

____

____

____

3 When buying a business book which factors influence you most?
(Please rank in order)

- ☐ recommendation from a colleague
- ☐ price
- ☐ content
- ☐ recommendation in a bookshop
- ☐ author
- ☐ publisher
- ☐ title
- ☐ other(s):

____

4 Is this book a

- ☐ personal purchase?
- ☐ company purchase?

5 Would you be prepared to spend a few minutes talking to our customer services staff to help with product development? YES/NO

The Business Publisher

Written for managers competing in today's tough business world, our books will help you get the edge on competitors by showing you how to:

- increase quality, efficiency and productivity throughout your organisation
- use both proven and innovative management techniques
- improve the management skills of you and your staff
- implement winning customer strategies

In short they provide concise, practical information that you can use every day to improve the success of your business.

FINANCIAL TIMES
PITMAN PUBLISHING

No stamp necessary in the UK

Free Information Service
Pitman Professional Publishing
FREEPOST
128 Long Acre
LONDON
WC2E 9BR, UK

The Environmental Management Handbook

■

The Institute of Management (IM) is at the forefront of management development and best management practice. The Institute embraces all levels of management from students to chief executives. It provides a unique portfolio of services for all managers, enabling them to develop skills and achieve management excellence. If you would like to hear more about the benefits of membership, please write to Department P, Institute of Management, Cottingham Road, Corby NN17 1TT. This series is commissioned by the Institute of Management Foundation.

The Environmental Management Handbook

■

BERNARD TAYLOR
COLIN HUTCHINSON
SUZANNE POLLACK
RICHARD TAPPER

PITMAN PUBLISHING
128 Long Acre, London WC2E 9AN

A Division of Longman Group Limited

First published in Great Britain 1994

A CIP catalogue record for this book can be obtained
from the British Library.

ISBN 0 273 60185 7

1 3 5 7 9 10 8 6 4 2

Photoset by PanTek Arts, Maidstone
Printed and bound in Great Britain by
Biddles Ltd, Guildford and King's Lynn

The Publishers' policy is to use paper manufactured from sustainable forests

Contents

■

PART II
DEVELOPING BUSINESS OPPORTUNITIES

PART III
DEVELOPING POLICIES AND SYSTEMS FOR ENVIRONMENTAL MANAGEMENT

PART V
ORGANISING THE CHANGE

Foreword

■

JAN-OLAF WILLUMS, D.SC.
Executive Director, World Industry Council for the Environment

The United Nations Conference on Environment and Development (UNCED), held in Rio de Janeiro in 1992, drew worldwide attention to the challenges of managing our environment properly and effectively. Governments were asked to commit to a wide scope of actions and initiatives, and industry was called on – not as a culprit to be scolded for past sins, but as a key instrument in moving sustainable development from idea to action. Rio gave a focus and a timetable for addressing important issues.

UNCED was conceived to mobilise the political will of nations to take concrete actions. It also mobilised the involvement of the non-government groups, and the 'independent sector', including the corporate sector. Business was actively involved in the UNCED process from the very beginning. The participation of business leaders from many countries helped policy-makers to understand more clearly what role industry does play today, and what role it could play in the future, given the right policy framework.

The first task is to convince our fellow industrialists that sustainable development makes good business sense – if done the right way. This handbook is therefore an important instrument for conveying the message that sustainable development makes sound business sense.

Just before the Rio conference, I contacted many of the (now over 1,200) corporations from all continents that had already supported the ICC Business Charter for Sustainable Development. We wanted to know how they had begun to implement some of the charter's sixteen principles, and which were the most important challenges they saw in such a commitment. Within weeks we received several hundred letters and reports from both small and large corporations, from northern Canada to southern India, filled with examples and concrete suggestions for fellow businessmen. We compiled the most interesting examples in our book *From Ideas into Action,*[1] which the ICC released at the Rio conference.

The examples illustrated what benefits a company which had pursued a proactive attitude to the environment could achieve.

[1] Willums, J-O. and Goluke, U. *From Ideas to Action: Business and Sustainable Development*, ICC report on the Greening of Enterprise. ICC Publishing, Oslo (1992).

- Henkel, the German chemical manufacturer, learned that the best and the brightest young people preferred to join the companies with the best long-term environmental strategy – and quality of manpower is for Henkel a guarantee to stay competitive in the long run.
- Xerox and AT&T each substantially decreased the use of production resources by implementing their very focused 'design for environment' philosophy. To do this properly, they considered the total life cycle of the product, and evaluated the product from 'cradle-to-grave'.
- Volvo, the Swedish car manufacturer, tried to look at the sum of all environmental effects from extraction of raw materials, through manufacturing, to the use of the product and its final scrapping.

It is important that such experiences, and the understanding of the usefulness and present limitations of the new management tools, is shared by many others. A widely applicable handbook that goes into depth in these issues is therefore very timely.

How do we implement the message? Treating the environment as an asset, rather than as a free good, leads to major changes in corporate management. The need for change must be recognised among the highest corporate priorities. The first condition for success is a clearly stated top management commitment. Once the commitment is clear the business opportunities will emerge. And once everyone knows that the environmental performance of his or her work is a key factor for the long-term stability of the whole corporation, the entire organisation can move more swiftly towards a new environmental goal.

The challenges and tasks that face managers across industry are to establish new industrial infrastructures for, among other things:

- recycling and re-manufacture;
- measuring the economic value of natural resources;
- looking at product costs over their life cycle;
- assessing environmental risks;
- calculating production costs in terms of energy, resource use, waste, and environmental impacts; and
- greater corporate openness and reporting to add information to the environmental debate.

These will involve new relationships between businesses and their suppliers, and also between businesses and their customers and end-users. The relationships will be achieved through use of the tools for environmental management and organisational development which are described in this handbook.

This handbook may not only be useful for senior managers in public and private organisations; its practical ideas and suggestions should also inspire specialists, be they in health, safety and the environment, or in design, manufacturing and sales. By getting involved in discussing not what sustainable development is or is not, but how we can strive towards a more sustainable way of producing and consuming, they lend their weight to the worldwide campaign on sustainable development. And the more people get involved in these issues, the easier it will be to implement new ideas and actions. To navigate a new and difficult course we need a determined captain, a clear vision, and the right maps to steer by. This book may serve as such a map.

Introduction

■

A thousand organisations worldwide have signed the Business Charter for Sustainable Development[1] and 150 companies in the UK have joined the Environment Business Forum[2] and submitted their environmental action plans as part of their 'agenda for voluntary action'. The number is growing around the world.

There are already many books about the environment but few of these deal with the response of business and even fewer provide practical information in some depth on what management is doing to deal with environmental issues. The *Environmental Management Handbook* fills this gap. Each chapter is written by a senior executive who has first-hand practical experience of what has been done in a particular organisation. The chapters represent *global best practice* and explain the company cases with appropriate illustrations and diagrams. The book is organised into five Parts:

I Monitoring the Pressures for Change
II Developing Business Opportunities
III Developing Policies and Systems for Environmental Management
IV Management Approaches and Techniques
V Organising the Change

We recognise that executives are at very different places on the learning curve. For some the first Part will be very important; for others much of the information will be well known. Also, companies will be at different stages in implementing their environmental strategies. Readers are encouraged to select the Parts which are most relevant for themselves and their businesses.

The book is written for directors and senior executives working on four different aspects of the environmental challenge. They are those who are:

- formulating a corporate strategy;
- working on corporate development projects, product development, new ventures, acquisitions and investments;
- managing organisational change involving restructuring, human resource management, corporate culture and process re-engineering;

[1] Published by the International Chamber of Commerce, 38 Cours Albert ler, 75008 – Paris, France. Tel: (33) (1) 49-53-28-28; Fax (33) (1) 42-25-86-63. Reproduced as Appendix 1.

[2] Environment Business Forum, Confederation of British Industry, Environment Management Unit, Centre Point, London WC1A 1DU. Tel: 071-379 7400.

- responsible for Health, Safety and Environment.

We believe that managing the environment and striving for sustainability is emerging as a new function of management and is still being defined. This handbook represents a snapshot of an evolving area of accountability which is developing fast. Its philosophy is sustainability, its processes include environmental auditing and life cycle analysis, its market contains new opportunities for processes, products, packaging and services, but it still requires better measurement of performance and closer integration with financial management for more effective resource allocation. Even now organisations are having to discover their own ways to feed audit findings into strategic planning and management practices, and most environmental managers work alone or in small units.

Several potential authors whom we approached declined to contribute because they had been reassigned to other work, indicating that environmental management is seen as a transient responsibility. We believe that this will change as the area becomes better defined.

Those who teach management need the frameworks, concepts, skills and case studies and we believe that this *Handbook* will fulfil a growing need in business schools, universities and management colleges.

Environmental management: the three stages

As editors we have learned a great deal about the subjects addressed in this book. We have come to realise that the first steps, which are often reactions to the environmental challenge, do not end there. The reactive response soon has to move on to a broad-based strategic approach which often requires fundamental re-thinking about the business.

There are already signs that this too will be insufficient and executives will need to re-examine their own inner world and the fundamental values which lie at the heart of the business enterprise. This third level is only just being appreciated. For example, Volvo stresses the importance of training and development of their staff and Rank Xerox, with their cross-functional committees, show the value of getting people in different departments working out solutions together. We believe that new approaches to *managing in ecological terms* will emerge during the next decade and this will become a fundamental issue for managers at every level.

During this process it is inevitable that disagreement and conflict will emerge. Opinions will differ, facts will be challenged and priorities will be hard to establish. This has, of course, arisen over many issues in addition to the environmental challenge. However, we belive that learning about better ways to resolve differences and manage conflict is a crucially important aspect of the search for environmental solutions. We hope that in the future companies will be willing to describe this sensitive area of their work because the creative ways in which conflict management skills are developed in business are likely to contain lessons which have application in other spheres. This will be increasingly important as businesses move from reacting to the environment, through a proactive phase and on to the bigger challenge of becoming a sustainable business. The three stages are:

- the reactive firm;
- the proactive firm; and
- the sustainable organisation.

We can already see signs that reporting and public accounting are receiving more attention, that risk management relating to environmental issues is becoming more important, that loss control is emerging as a discipline in its own right and that insurance risks and the resulting premiums are becoming harder to assess.

Our beliefs

As a result of our study of the environment as it effects business, aided by the contributions of the authors of this *Handbook,* we believe that:

1 Environmental management should be as *professional* and as well organised as any other aspect of the business such as Finance and Accounting, Marketing and Sales, Production and Operations or Purchasing and Procurement.
2 Environmental management is a *core* activity of management, not an option or frill undertaken for public relations reasons.
3 Environmental management is *financially important* – it is about managing losses and reducing risks to health and safety of staff and the community.
4 Environmental management also represents a new area of *competitive advantage* and includes *opportunities* as well as threats.
5 Environmental management is the responsibility of *the board of directors* and the *executive management.* It is a fundamental and personal responsibility of all *line managers and all staff* – not a matter which can be delegated to a specialist staff function.

The diffusion of environmental management

Environmental management is a new and vital aspect of management which has emerged in the 1980s and will be adopted by an increasing number of firms in the 1990s.

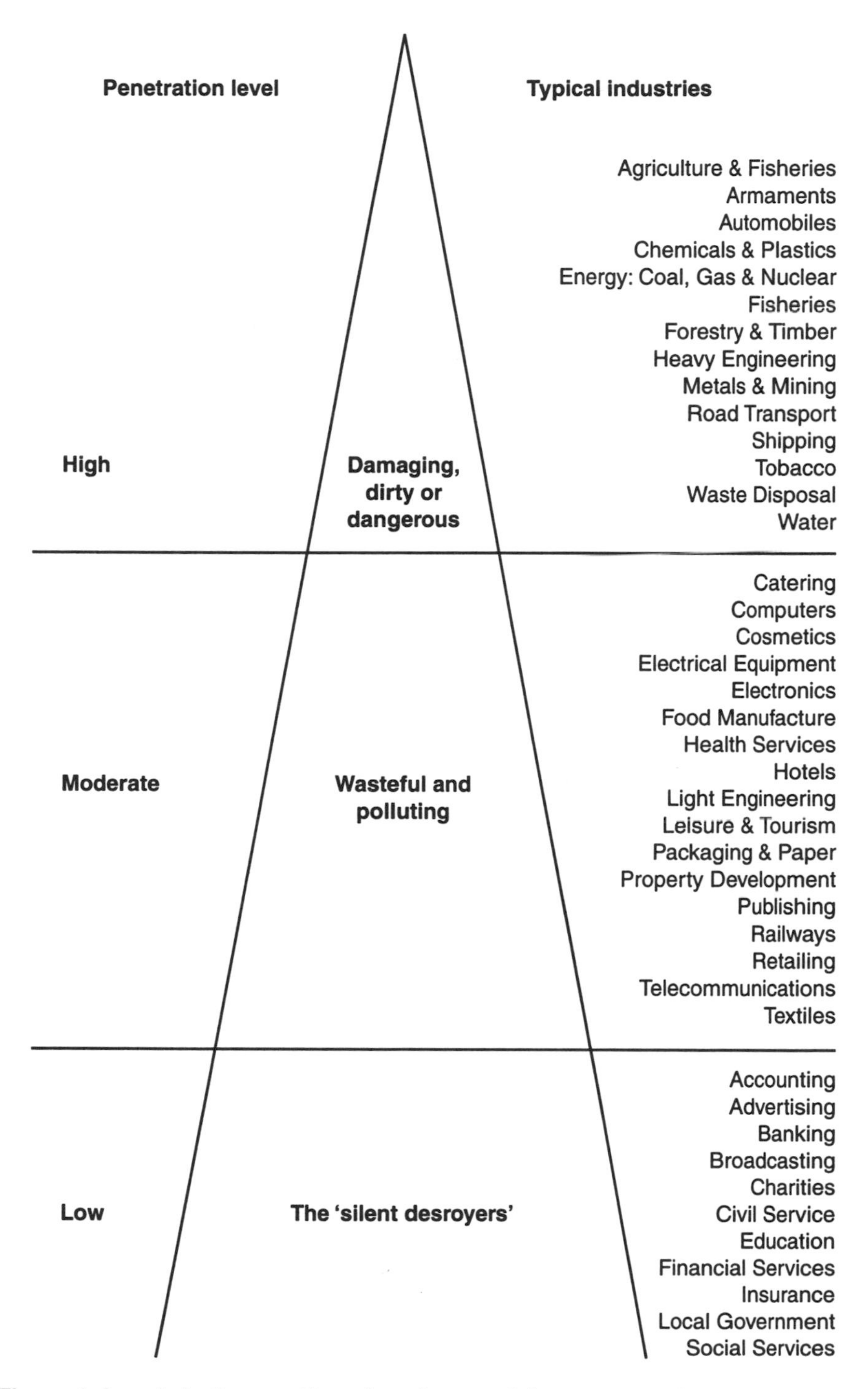

Figure 1 Levels in the practice of environmental management
(Source: Bernard Taylor, Henley Management College)

LEVEL 1

Inevitably, the pioneers of environmental management are the industries which directly damage the environment and/or handle *damaging, dirty,* or *dangerous* materials, To these industries environmental management is a matter of survival.

LEVEL 2

Industries in a second category, while not producing materials which are extremely damaging, dirty or hazardous, are still engaged in destroying the environment – incidentally – because they make excessive use of valuable resources such as trees, minerals, energy, or land; or they may damage the environment by disposing of waste or through air and water pollution. We describe these as *wasteful and polluting*.

LEVEL 3

In the category of service industries and government, activities are neither as dirty and dangerous nor as heavily polluting and wasteful as individual businesses. However, because of their large number, these organisations and their employees account for a significant proportion of the pollution of the environment (for example, through their cars and commercial vehicles), the majority of the waste (through their use of paper, packaging, office furniture, office machinery, central heating, etc.), and much of the environment's destruction (because they need roads, office buildings, car parks, shops, hospitals, services, theatres and cinemas).

These are the *Silent Destroyers*, the people who are damaging the environment without knowing it because they are the 'Quiet Polluters'. These Silent Destroyers work mainly in 'clean' businesses which produce no products at all, and apparently no waste.

Environmental management is already practised in the *Dirty or Dangerous* businesses (Level 1).

Environmental management has made a moderate penetration in the *Wasteful and Polluting* industries (Level 2).

However, environmental management appears to have made a low impact amoung the *Silent Destroyers* (Level 3).

In this *Handbook* we have selected illustrations of environmental management in all three types of organisation in roughly equal proportions because we believe that interest in Level 2 is increasing fast and interest in Level 3 will do so during the 1990s as the realisation dawns that we are all involved. Level 3 businesses which provide services to Levels 1 and 2 will need to learn fast because their approach is likely to get out of date quickly if they are unaware of the challenges faced by their clients.

Colin Hutchinson, on behalf of the Editors,
January 1994.

Biographical notes

■

Allen H. Aspengren graduated from Marquette University, in Milwaukee, Wisconsin, in Mechanical Engineering and began his career with 3M in 1968. With over twenty-five years' experience in the environmental area he is the author of many publications on environmental matters in Europe and the USA. After holding a number of environmental jobs in St Paul, USA, he took up the position of Manager, Environment, Health and Safety, Europe, in January 1991.

He works closely with 3M's various operations in Europe to make sure that they comply with Corporate Policies and Regulatory Requirements of the local government and the EU. He also helps solve technical problems in the environment, health and safety areas.

He is Chairman of the American Chamber of Commerce, Safety and Health Committee, and has been involved in the Environmental Management and Audit Scheme developed by the EU. He has conducted two pilot audits for 3M in Europe, in addition to several audits when working in the USA, and has given many speeches on 3M's 3P, Auditing and Environment Programs.

David Ballard was educated at Jesus College, Cambridge and the London Business School. His business career includes a wide variety of strategic, operational and change management roles with Esso Europe Inc. and with Thorn EMI Rental (UK) Ltd. In 1990 he became involved in the latter company's environmental programme, as Quality and Environment Adviser from 1991 to 1993.

He is now Environmental Consultant, Aspinwall and Company, and is also Organisational Consultant with the Bath Consulting Group. He is also the Director of Swindon and Malborough Hospitals NHS Trust.

Victor Bruns is a First Vice President with Corporate Banking Division (Head Office) at Deutsche Bank. In addition to his activities in international business, he is also responsible for the development of strategies and for supporting branches in their business with local authorities and for customer-related environmental protection.

After completing his bank training, and studying law in Germany and the USA, he joined Deutsche Bank where he was initially employed in the International Department at Head Office. Subsequently he played a part in setting up Deutsche Bank (Suisse) SA where he became General Manager.

Richard Dalley is an Associate Consultant with Aspinwall & Company Ltd, one of the UK's oldest independent environmental consulting practices.

He joined Aspinwall after eight years' experience in the chemicals industry with International Paint plc, where he was responsible for many environmental and regulatory aspects of the company's worldwide operations. In recent years he has led a number of major projects to assist corporate organisations establish environmental management systems, and has undertaken a wide range of audits in many different business sectors. Before joining industry, Richard Dalley spent several years undertaking pure research into various aspects of marine biology – an area in which he continues to maintain an active interest.

A. J. (John) Forte was born in Italy and studied modern languages at Victoria College in Jersey, Channel Islands. He joined Forte's of London in 1964 as a 'Trainee Executive' where he gained experience of all aspects of hotel and catering management. He joined Gardner Merchant in 1969, became General Manager of the first 'Cook-Freeze' plant in the UK, and developed it into the central frozen food production centre for the Group.

In 1975 he was appointed Technical Director and was responsible for a new Laboratory and Evaluation Kitchen; energy conservation and environmental responsibilities were added to the portfolio. With the continued expansion of the Group and its reorganisation he became Executive Director of Environmental Services with the new laboratory facilities built at Gardner Merchant International Management Centre at Kenley.

In January 1993, following the Gardner Merchant management buy-out, he was invited back to the Forte Head Office in London to continue work on Environmental Management.

Bradford S. Gentry is managing partner of Morrison & Foerster's London office and is the partner in charge of the firm's Land Use and Environmental Law Group for Europe. He advises companies on how best to manage the environmental requirements, concerns and opportunities affecting their international operations. He has worked on projects in the United States, throughout Western and Central Europe, as well as in the former Soviet Union, the Middle East, Southeast Asia and Latin America. He has spoken and written frequently on international environmental issues, and is a visiting lecturer on comparative environmental law at the University of London.

J. A. (Tony) Hill is a Manager in Environmental Quality Co-ordination, Europe, at Procter & Gamble's Technical Center in Brussels. One of his main responsibilities is to work with suppliers, ensuring P&G's environmental quality requirements are met. He has spent his whole career with P&G, initially joining their Manufacturing Division at the Manchester factory and later working in a variety of positions in the UK, Cincinnati and Brussels.

He graduated in chemistry at London University (BSc and PhD) and holds the Certified Diploma in Accounting and Finance of the Association of Certified Accountants.

Martin Houldin is a Fellow of the Chartered Institute of Management Accountants and an independent consultant specialising in the environmental management field. At the time of writing he was a partner of KPMG Peat Marwick's National Environment Unit team. He was responsible for environmental management services and R and D and advised clients on environmental strategies policies and management systems. Areas of specialism included integrating the environment with accounting and finance, purchasing and supply, information management and technology, performance measurement and environmental reporting.

He led KPMG's work with Business in the Environment and is an active working member with a number of organisations, including the American Chamber of Commerce, the Environmental Auditors Registration Association (EARA), the Association of Environmental Consultancies and the environmental task force of the Féderation des Experts Comptables Européens (FEE).

Andrea Jacobs, KPMG's National Environment Unit's expert on matters related to accounting, risk and liability, assisted in writing the chapter.

Colin Hutchinson is an independent Environmental Change Consultant specialising in creating awareness of the environmental challenge, the business opportunities that arise from it, and how to develop strategies for managing individual and organisational change.

His early career was with a multinational company where he gained sales and marketing experience in Africa, Malaysia and Europe. He became their management and organisation development adviser with a worldwide remit before moving into consultancy. He worked with Sheppard Moscow, consultants in organisation development, from 1976, and was Chairman from 1986 to 1991.

He has had a life-long interest in environmental matters and during his twenty years' experience in a voluntary capacity, he has been involved with campaigning, environmental education and in setting up and directing new initiatives. He is a Director of GAP UK Ltd, the UK arm of the Global Action Plan, and the Centre for Sustainable Industry.

Stanley Johnson, an international environmental expert with over 25 years' experience of environment and development issues, is a former head of the European Commission's prevention of pollution division and special adviser to DGXI, the environment directorate.

He was a member of the European Parliament from 1979–1984, and Vice Chairman of the Parliament's Committee on Environment, Public Health and Consumer Protection, and later became Director of Energy Policy at the Commission. He won the 1984 Greenpeace prize for outstanding services to conservation. He was recently Director, International Policy Services at Environmental Resources Management (ERM). He has since joined Coopers & Lybrand as Special Adviser on the Environment.

Robert J. Jones is a consultant on energy policy with Energy Policy Studies (EPS). At the time of writing he was responsible for co-ordinating the external energy conservation activities of British Gas, relating to and supporting government initiatives and customer services. He was also responsible for the management of the company's own-use energy spend of around £80 million p.a. In 1991 he joined the government's Advisory Committee on Business and the Environment (ACBE) and served on its Global Warming Working Party and on groups concerned with energy in buildings.

From 1978 to 1980 he was a member of the Advisory Council on Energy Conservation (ACEC). He was Technical Author of ACEC's Energy Papers 25 and 48. He was also a member of the CIC Energy and Environment Group and the CBI Energy Efficiency Working Party. He is a non-executive director of the Energy Action Grants Agency (EAGA) and a Trustee of the EAGA Trust.

Karl Kummer gained a diploma in Communications Engineering, Dpl. Ing. Nachrichtentechnik, from the Staatliche Ingenieurschule für Maschinenwesen, Dortmund, Germany. He held several senior positions in Quality Control with leading German companies before joining Rank Xerox in 1974.

During twenty years with Rank Xerox he gained experience in manufacturing, service, personnel, marketing and general management, and held leading management positions or directorships in Holland, Germany, India and the UK. He was Director of Business Operations in India where Rank Xerox has a joint venture company with the Modi Group. He was appointed Environment Director, Europe, in January, 1992 and is responsible for the realisation of the Rank Xerox Leadership goals.

During four years spent in India he gained valuable insights into life in the third world and their environmental problems.

John Lawrence is a Senior Honorary Fellow and Visiting Professor at the Monitoring and Assessment Research Centre at King's College, London.

For more than 12 years he was Director of the ICI Group Environmental Laboratory at Brixham, Devon. The Laboratory, now part of Zeneca, helps units of the company to understand their environmental commitments and to carry out the technical work to meet them. In the year before retiring from ICI at the end of 1992, John Lawrence was responsible for developing policies and practise for managing the rapidly emerging issue of potential land contamination.

Rolf Marstrander is Senior Vice President and member of the Management Group of Hydro Aluminium a.s, a fully owned subsidiary of Norsk Hydro a.s, with responsibility for technology and ecology. Previously he was Senior Vice President Health, Safety and Environment of Norsk Hydro.

He specialised in Industrial Engineering and held the appointment of Systems Manager in two different companies before moving on to the Norwegian Scientific and Industrial Research Council where he was appointed Assistant Director in 1972.

He was general Director of the Norwegian State Pollution Control Agency from 1978 until he left for Norsk Hydro in 1982 and has held several honorary posts related to environmental research in Norway. He is at present Chairman of the Ecology Committee in the European Aluminium Association.

Tim Mohin is a Congressional Fellow with the US Senate Committee on Environment and Public Works. Working for Chairman Max Baucus of Montana, Tim Mohin has been primarily responsible for environmental technology and has coordinated the introduction of Senator Baucus's Environmental Technology Legislation – the National Environmental Technology Act of 1993.

Prior to his fellowship in the Senate, he was a Section Chief in EPA's Office of Air Quality Planning and Standards, responsible for implementation of provisions of the air toxics title of the Clean Air Act amendments. In this capacity he led the EPA team dedicated to the development of Life Cycle Assessment Guidelines. He also served on the EPA team that helped to negotiate passage of the Clean Air Act amendments in 1990.

Tim Mohin has a Bachelor of Science degree from the State University of New York at Cortland, and a Master's degree of Environmental Management from Duke University.

Suzanne Pollack joined Henley Management College in 1987 having completed a doctoral thesis in the Management of Change. Her work with corporate clients, such as Shell UK Ltd, and Grand Metropolitan plc, has enabled her to develop her interest in environmental management. This has included the development of specialist workshops, conference organisation and speaking, also the writing of a long-distance learning MBA module about environmental management.

Nicholas L. Reding joined Monsanto Company in 1956 after graduating from Iowa State University with a BS in chemical engineering. He held various technical, product, marketing and management positions before becoming President of Monsanto's Agricultural Division in 1986. He became Vice Chairman of Monsanto Company in 1993 with responsibility for the company's environmental, manufacturing, human resources and administrative functions.

He is a member of the Board of Directors of Monsanto Company, and two of their subsidiaries, The Nutrasweet Company and Searle, and serves on the Advisory Committees for the Agricultural Group and the Chemical Group. He is also a member of he board of directors of Multifoods Corp., Meredith Corp., CPI Corp., The Keystone Centre, the United Way of Greater St Louis, and the St Louis Zoo.

In 1982 Nicholas Reding received the National Agri-Marketing Association's Robert S. Kunkel Award for Excellence in Agricultural Marketing, and was named the St Louis Agri-Business Club Leader of the Year in 1984. He was also elected an honorary member of Alpha Zeta Fraternity, the US national honour society of agriculture.

Philip K. Rees is Group Director of Environmental Affairs with John Laing plc, based in London.

He has been responsible for the development and implementation of environmental strategy and policy since 1989, and prior to that he was the Divisional Chairman responsible for the Group's environmental businesses. His career has been centred on commercial and general management within the construction, property development and housing industries. He represents the construction industry on a number of UK and European environmental committees and work groups, has provided input into the work of Business in the Environment, and acted as a Liaison Member to his Chairman on the Business Council for Sustainable Development. He is a Fellow of the Royal Institution of Chartered Surveyors and a Fellow of the Institute of Management.

Werner B. Rothweiler holds a PhD degree in chemistry from the University of Basel, Switzerland. He joined Ciba-Geigy in 1966 and has held various positions in chemical development and in production (inter alia: head of Pigment Division's worldwide production, factory manager). He is now with Ciba-Geigy's Corporate Safety and Environment Unit, where he is responsible for safety and environmental auditing, and liaisons.

William G. Seddon-Brown was educated at Cambridge University and the CEI Genera. He is Director, European Government Affairs, Waste Management International, Brussels.

Previously he worked in manufacturing, marketing, government affairs and, general management in eight countries and speaks nine languages – five fluently. He opened Monsanto offices in Rumania and Yugoslavia, and ran Monsanto Agricultural Company operations in China (Peoples' Republic). He has published various articles and organised a major photographic exhibition in Hong Kong, sponsored by Kodak Far East, on closed areas of China.

He was for seven years The Chairman of The American Chamber of Commerce, Health and Safety Sub-committee in Brussels, and is currently Chairman of the EU Committee's Waste Working Group, and the Steering Committee of the European Energy from Waste Coalition (EEWC).

Mike Seymour is the Executive Director in Burson-Marsteller, London, with responsibility for crisis management and planning in the UK, and he co-ordinates these specialised areas across Europe.

He joined Burson-Marsteller, the largest PR agency in the world, in 1988 as a senior counsellor specialising in issues management, working on a wide range of business and industrial incidents. After six months he was made a director and took over a series of corporate accounts including the Industrial Development Board of Northern Ireland. He facilitates corporate strategy sessions and trains senior managers, using simulation exercises, and counselling them on issues which impact nationally and internationally. Since 1989 he has developed awareness and incident-handling programmes for companies operating world wide. He manages issues and handles crisis management for major UK-based and international companies operating out of London, involving all work on industrial and environmental incidents, including Piper Alpha. In 1990 he launched a package which brings together, for the first time, public relations, insurance, re-insurance and legal expertise in support of corporate management.

Richard Tapper World Wide Fund for Nature (WWF)

Richard Tapper is Head of Industry Policy at WWF UK. He was a member of WWF's delegation to the UN Conference on Environment and Development (UNCED) in 1992, and is a member of WWF's team on UNCED follow-up. His work focuses on promoting action by industry on sustainable development, strategic environmental management, and the links between industry and biodiversity in implementation of the Biodiversity Convention. He has previously worked in the power generation industry on the integration of environmental policy with corporate strategy, and for the BBC where he edited a multimedia, interactive database on the environment. Following research into symbiosis for his doctorate from the University of Bristol, he lectured in ecology from 1977 to 1979 at the University of Durham.

Richard Tapper has initiated programmes of work within WWF on the tourism industry, environmental management, and review of industry impacts on the environment, as well as contributing to WWF's work on trade and environment, energy, and resource consumption and pollution as a whole. He has spoken at various international meetings, including the UNEP/Norway Expert Conference on the Biodiversity Convention in Norway, examining follow up to UNCED, and has participated in seminars organised by UNEP, CEST, SETAC (on LCA) and the Anglo-German Foundation amongst others

Bernard Taylor is Professor of Business Policy at Henley Management College, Oxfordshire.

Before joining the College he held responsible positions in marketing with Procter & Gamble, and education and training with Rank Xerox and the Chartered Institute of Marketing, also the University of Bradford Management Centre.

He has been closely associated with the development of strategic planning in Britain and has produced twenty books on business strategy and strategic management. He is Editor of the *Long-Range Planning Journal*, Founding Editor of the *Journal of General Management* and Director of the European Council on Corporate Strategy.

Bernard Taylor is a consultant to business and government internationally and has been a visiting professor at the Universities of Western Australia, Ottawa, Cape Town, Witwatersrand, Giessen, and at the Hubert Humphrey Institute, Minnesota. He has also worked as a UNDP Consultant with the National Productivity Council in New Delhi.

He has had an interest in Environmental Management for many years, and has written articles on the Strategic Management of Resources. He has also worked as a consultant with the Canadian government and organised workshops on Global Energy Strategy.

Tessa Tennant graduated from King's College, London, with a degree in Human Environmental Sciences. In 1988 she co-founded the UK's first green fund and is now Head of Green and Ethical Investments at NPI, London..

She was Special Projects Director for the Green Alliance from 1983 to 1987 and her duties included parliamentary liaison and developing the industry and environment programme. In 1987 she worked in the USA with Franklin Research and Development Corporation, a leading social investment management company, to develop criteria for the assessment of corporate environmental performance.

In 1992 she became an Honorary Fellow of the University of Dundee in recognition of her work in green investment. She is Chair of the UK Social Investment Forum, a member of the finance working group of the UK government's Advisory Committee on Business and the Environment, a member of the Environment Committee of the RSA, and a member of the Advisory Committee to HRH The Prince of Wales's Business in the Environment initiative.

Dennis Vaughn is a graduate of Iowa State University with a Bachelor of Science degree in chemical engineering and has 23 years experience in environmental engineering, environmental safety and environmental management. He worked initially for a state regulatory agency before assuming management responsibilities for environmental engineering for the North American operations of a chemical company, where he designed waste water treatment systems, directed remediation activities and developed regulatory compliance programs.

In 1986 Dennis Vaughn relocated to a pharmaceutical manufacturing company to develop environmental programs for chemical production facilities on a worldwide basis, working in the USA, England, Belgium and India. Activities focused on prevention of accidental chemical releases and designs for pollution control systems.

He started his career at Grand Metropolitan in 1990. His responsibilities cover the worldwide food processing sector and include integrating consistent environmental programs, designing compliance systems, implementing risk avoidance initiatives and developing standards for world-wide application.

He is a member of The Conference Board's Advisory Counsel of Environmental Affairs in the USA and Europe, the Water Environment Federation, the American Institute of Chemical Engineers, and is a registered professional engineer in five states.

Axel Wenblad graduated with an MSc in limnology and chemistry from the University of Uppsala, Sweden in 1974. He is Corporate Manager, Environmental Auditing, and Head of Environmental Affairs at Volvo Group Headquarters.

Before joining Volvo in 1990, he was responsible for the co-ordination of local and regional activities at the Swedish Environmental Protection Agency. He has worked in the Swedish Ministry of the Environment and the United Nations' Food and Agriculture Organisation. He has extensive experience in the field of environmental protection and is the author of many articles on various aspects of the subject. He has also served as Chairman of the Swedish Association of Environmental Auditors .

David Wheeler is General Manager, Ethical Audit, The Body Shop, and is responsible for their Environmental Audits. Their third audit statement was published in June 1994, while their first, in 1992, was widely acclaimed.

He began his career in the water industry and specialised in pollution control of bathing water. As an academic he became a leading critic of the declining standards of water quality in the UK. He also became a frequent consultant to the World Health Organisation (WHO) as well as to a wide variety of development agencies working on water and sanitation projects in developing countries.

Before joining The Body Shop he was full-time adviser to the Shadow Secretary of State for the Environment.

Robert M. Worcester is Chairman of MORI (Market & Opinion Research International) and Visiting Professor of Government at the London School of Economics and Political Science, and at the Graduate Centre for Journalism at City University, London. He is a Vice President of the International Social Science Council/UNESCO, and a member of its Standing Committee on Human Dimensions of Global Environmental Change. He is also a Trustee of the World Wide Fund for Nature (WWF UK) and of the Natural History Museum Development Trust. His most recent book (co-authored with Samuel H. Barnes) is *Dynamics of Societal Learning about Global Environmental Change.*

Acknowledgements

Like most handbooks, this one is the result of a team effort. In thanking the team, we particularly wish to thank the authors who have contributed chapters. They are all people who have direct experience in working to improve environmental performance. It is through their efforts, and those of others like them, that businesses are moving forwards on the road to sustainable development.

We are similary grateful to Jan-Olaf Willums, Exectutive Director of the World Industry Council for the Environment, for the Foreword that he has written.

All of us also recognise our debt to many colleagues behind the scenes. Their support has been of the greatest value in preparing and editing this handbook.

It is impossible to acknowledge everyone by name, but we particularly wish to thank Susanne Walton at the Henley Management College, who has been at the centre of the project from the start, and whose superb organisation has kept the handbook on track. Thanks also to Peter Denton, Rachel Hall, Sally Nicholson and Gill Witter at WWF UK.

Bernard Taylor
Colin Hutchinson
Suzanne Pollack
Richard Tapper

Part I

■

MONITORING THE PRESSURES FOR CHANGE

'An image as an environmentally responsible company is becoming an essential part of a competetive strategy.'

Business Week, 18 June 1990

'Only when the last tree has died and the last river been poisoned and the last fish been caught will we realise that we cannot eat money.'

19th Century Cree Indian

Introduction to Part I

■

Commerce has a huge impact on the environment within which we all live. Its impact is so vast that the scientific findings explored by Richard Tapper in Chapter 2 of this section suggest that changes to our environment, such as the loss of biodiversity or climate change, are happening at rates never experienced on earth before.

This has caused many to stop and think. In boardrooms round the world, directors discuss the actions they should take in their businesses. Governments, financial institutions, insurers, consumers and shareholders, among others, have changed their expectations of business. This section explores how these expectations have changed and the resulting pressures for improved corporate environmental performance affecting businesses in Europe and other parts of the world.

In Chapter 1, Robert Worcester charts the trends that MORI have found in the public perception of environmental issues in the UK.

Then Richard Tapper presents a comprehensive picture of the environmental problems created by business practices which are often wasteful, harmful and linear – linear meaning to use raw materials once only, so that once a product is finished with by consumers it becomes rubbish. Attempts are not made to re-use, recycle or repair in any part of the product's life cycle.

Stanley Johnson, Chapter 3, then shows how environmental concerns have been tackled by the EC. Johnson outlines the current EC environment programme which is called 'Towards Sustainability' and how this affects various business sectors.

In the fourth chapter, Bradford Gentry sketches some of the key features of the US regulatory system and the lessons that should be learnt for other regulatory systems, particularly those within Europe.

The final chapter by Tessa Tennant provides a wealth of information on the rise in popularity of ethical and environmental investments funds. Tennant points to the need for fuller and more systematic environmental disclosure in business.

Executive summaries

■

1 Public opinion on environmental issues

The public is concerned about the state of the environment, confused as to whom to believe, disenchanted with governmental attempts at all levels to deal with the problems of pollution and ecological degradation, and ready for radical action. What's more, the public's view is shared by elite opinion as well, from captains of industry, the City, the media and, significantly, legislators both at Westminster and at the European Parliament. The impression let in the aftermath of the 1992 British general election and in discussions among the chattering classes, is that the environment is yesterday's issue. Recent survey evidence indicates that nothing could be further from the truth.

Over the past five years MORI has conducted a suite of surveys in its Communications Research Programme that has tracked the concern of various publics, general and elite, to the issues surrounding the environment, conservation, pollution and sanctions relating to corporate environmental behaviour.

More recently MORI has extended its survey evidence across Europe, and looked at attitude and behavioural data in North and South America and in Australia and New Zealand as well. These findings are supported by other survey evidence, especially that carried out by the Eurobarometer, the semi-annual sounding of European public opinion across the 12 EC member states.

Chapter 1 outlines what has been found in these surveys, why these findings are important, what their relevance is to corporate decision-makers, and what the public and legislative reaction is likely to be in the light of public and elite opinion both in Britain and abroad.

Facing up to limits: the challenge of sustainable development

Increasing acceptance of the concept of sustainable development means that attention is now focused on the way in which both the activities and management of business and industry contribute to this goal. In the European Union (EU), this goal is set out in the Maastricht Treaty, which has as an overall aim 'to promote a harmonious and balanced development of economic activities, (and) sustainable and non-inflationary growth respecting the environment . . .'

Global and regional targets for environmental goals already provide a framework for policy on industry. For example, combating global warming by reduction of CO_2 emissions requires action by all industrial sectors. The EU and Japan have set targets for this. Governments have also agreed targets to phase out various ozone-depleting chemicals under the Montreal Protocol, and to reduce pollution from acid rain emissions and other pollutants under a range of national and international agreements.

The environment is therefore a central business issue. Management is becoming more aware of the scale of environmental problems and of the impact which business has on the environment. Business leaders have declared the need for corporations 'to assume more social, economic and environmental responsibility in defining their roles'.[1] They went on to say: 'We must expand our concept of those who have a stake in our operations to include not only employees and shareholders but also suppliers, customers, neighbours, citizens' groups, and others.' Over 1,200 companies have now signed the ICC's Business Charter for Sustainable Development (Appendix 1) which incorporates a commitment 'to establish policies, programmes and practices for conducting operations in an environmentally-sound manner'. This figure includes over a quarter of the Fortune-500 companies.

This chapter looks at the science and knowledge-base that underlies our understanding of environmental issues, and of ways of handling them. It describes some of the key environmental issues for business.

The science of environment is not particularly new. What is new is scientific observation of the whole Earth, and the use of scientific models as a means to integrate diverse measurements into forecasts of global environmental change. This has enabled scientific studies of environmental problems to be translated into global and regional policy. But while policies may change, the fundamental environmental problems will remain.

3 The philosophy and approach to modern environmental regulation: the experience of the European Community

This chapter discusses the approach to environmental policy and regulation adopted by the European Community over the twenty years 1973–1993.

It covers the early emphasis on pollution control legislation, particularly in the fields of air, water and waste; and the way in which EC environmental policy evolved with successive Environment Action Pro-

[1] Schmidheiny, Stephan, with the Business Council for Sustainable Development. *Changing Course*, MIT Press, Cambridge, Massachusetts (1992).

grammes, first towards the preventive (as opposed to reactive) approach and, latterly, towards the integration of environmental considerations in all major sectors of EC activity.

A substantial part of the chapter is devoted to the presentation of the main elements of the EC's Fifth Environmental Action programme, 'Towards Sustainability', and to the implications for the European Community of Agenda 21, the action plan for integrating environment and development which was agreed at the United Nations Conference on Environment and Development (UNCED) held in Rio de Janeiro, Brazil, June 1992.

4 US regulatory approaches – the implications for Europe

Chapter 4 outlines the differences that currently exist between the regulatory approaches to environmental issues in Europe and the US. It begins with a discussion of the different priorities assigned to environmental issues in Europe and the US. It then considers the specific techniques which are used in the US to control or respond to environmental problems, and examines the governmental structure under which the US laws are enacted and implemented. In particular, it highlights the problems created by this structure.

Finally, the major implications of these factors for efforts by European businesses to manage the environmental risks facing their product sales or manufacturing operations in the US and elsewhere, are presented.

5 The growth in environmentally-responsive investment

Environmentally responsible investment funds are growing: they represent a practical way of achieving sustainable development, as defined by Agenda 21; they offer a value-added dimension to money management; they fit the ethos of the coming century and provide a key mechanism by which environmental accountancy will be practised and refined. The message for industry is 'get to know your environmental and social bottom-line before the ecofinancier does it for you'.

In March 1993, newspapers reported that 'green' and ethical funds had grown dramatically in Europe over the last five years with total funds under management of over £730 million, and more than 70 products for the investor to choose from. These findings support other evidence which shows that green and ethical investment is a permanent fixture of the financial markets, and continues to grow in size and influence.

[1] M. Campanale *et al, Green and Ethical Funds in Continential Europe*, Merlin Research Unit, London (1993).

Chapter 5 defines environmentally responsible investment; it describes the origins of the green investment industry, the types of products which are available and the principal driving forces behind its growth. The implications for manufacturing industry are then considered with a review of the evaluation techniques employed by green investors.

1

Public opinion on environmental issues

ROBERT WORCESTER

Chairman,

MORI (Market and Opinion Research International), London

- The British public is concerned about the state of the environment, disenchanted with government attempts to deal with environmental degradation, and ready for radical action. This view is shared by the City, the media and parliamentarians.
- In a corporate social responsibility survey, 31% of respondents view protection of the environment as an essential corporate responsibility, 16% cite looking after employees' welfare, 12% say serving the community, and 11%, listening to the consumer and giving good service.
- Four out of ten captains of industry agree that 'British companies do not pay enough attention to their treatment of the environment'. Nearly half believe that the most effective way of dealing with polluters would be 'legal penalties for companies which damage the environment'.
- Seventy per cent of MPs believe that 'the penalties imposed on companies causing pollution in Britain are not severe enough'.
- In the City, 56% of institutional investors and 42% of analysts agree that 'British companies do not pay enough attention to their treatment of the environment'.
- Since 1986 the number of people in EC member states who describe their concern about protecting the environment and fighting pollution as an immedi-

The view of the British public

Since its founding in 1969, MORI has tracked the view of the British public towards environmental matters. From 1969-1973, an annual Attitudes to Air and Water Pollution was carried out as a multi-client study for such companies as ICI, Shell, BP, Esso, British Airways and other forward-looking corporations doing business in Britain. Even then, from a third to half of the British public said they were 'concerned' about air and water pollution, with noise pollution and with pollution of the seashores and coastal waters.

More recently, and as part of a Communications Research Programme that surveys not only the views of the public but also of captains of industry, the City, the media and other elite groups, MORI has carried out a pair of related studies among the general public. The first, 'Business and the Environment', has been conducted in July of each year since 1989. The most recent survey was carried out in 1992 among 1,923 members of the general public aged 15+ in 146 constituency sampling points throughout Great Britain; the other, 'Corporate Social Responsibility', has been conducted annually since 1990, and in the 1992 study 1,846 adults nationwide were interviewed in late August and early September.

These studies found a bewildered British public, with **nearly four in ten (39%) agreeing with the statement 'I don't fully understand environmental issues'.** Hardly surprising, but more worrying, nearly as many, **37%, were in agreement that 'even the scientists don't really know what they are talking about when it comes to the environment'.** In 1989, 36% agreed that scientists aren't to be trusted about their knowledge on environmental matters, and almost identical findings have been recorded in any recent year.

In fact, fewer than half of the British say they have a 'great deal' or even a 'fair amount' of trust in what scientists working either in industry (47%) or for the government (48%) say about environmental issues. In contrast, **more than eight in ten, 82%, say they trust what scientists working for environmental groups have to say about environmental issues**. While only 8% of ABs (the one-in-six households headed by senior managerial/professional people) say they trust industry's scientists, and only 9% trust government scientists 'a great deal', 45% of ABs say they trust environmental groups' scientists a 'great deal' on what they have to say on environmental matters.

Over the same period, 1989–1992, there has been an increasing percentage of the public who say they disagree that 'There isn't much that ordinary people can do to help protect the environment', with 29% in 1989 declining steadily to just one in five, 20%, this past year. **Nearly seven in ten (69%) of the British public think that pollution and environmental damage are things that affect them in their day-to-day life**, and only one in ten (11%) believe that 'Too much fuss is made about the environment nowadays'.

No one issue dominates people's minds in their environmental considerations. When asked: *What issues to do with the environment and conservation, if any, most most concern you these days?*, pollution of rivers and streams at 16% and destruction of the ozone layer, also at 16%, shared the top place recently. Four years before, the ozone layer scored 27%, well ahead of the 15% who then volunteered water pollution as their main environmental concern.

When asked in the Corporate Social Responsibility Study: *It has been said that companies have two main kinds of responsibilities – commercial responsibilities (that is, running their business successfully), and social responsibilities (that is, their role in society and the community). What kinds of responsibilities, if any, do you think companies have?*, the most frequently mentioned responsibility, by 31%, was 'to protect the environment', followed by 16% who said looking after employees' welfare, 12% said serving the community and 11%, listening to the consumer and giving good service.

Following up that question was a specific question relating to corporate priorities, asking how important it is for companies to do various things, using a five-point importance scale running from extremely important to not at all important. Once again, the environmental question ran ahead of all others, with 45% saying that companies taking positive steps to prevent or reduce environmental damage was extremely important. These figures compare with 16% for supporting government training schemes, down from a quarter two years ago, 19% for supporting education in schools/colleges, 11% for helping small businesses to get started and donating their products or services to needy causes, 3% sponsoring national or international sporting events and 2% who thought sponsorship of cultural activities and the arts was 'extremely important'.

For several years MORI has used behavioural rather than attitudinal measures to define Green Consumers (those people who say they have selected one product over another because of its environmentally-friendly packaging, formulation or advertising) to identify psychologically Green Consumers. The test is minimal, asking people only to recall a single, conscious consumer choice over a lengthy, intentionally imprecise, time period. The figures for Britain (Table 1.1) and for a number of other countries (Table 1.6) indicate how discriminating this simple typology has turned out to be. Even in Spain and Italy, with Environmental Activists below 10%, around four in ten of their populations have 'acted green', even if they have not, consciously, 'thought green' to any significant degree. 'Environmental Activism' is a much tougher test, asking consumers to combine passive green behaviour with active campaigning, petitioning and lobbying, as well as joining green organisations and contributing financially to them, in order to qualify for inclusion in the Green Activist typology, qualification gained by doing five or more of the behavioural items from the list of ten.

The 'Environmental Activist' and the 'Green Consumer'

For many years MORI clients have had access to the important and powerful 'Socio-Political Activist' typology, the 10% or so of the British public who are the 'movers and shakers' of British society. In an adaptation of that concept in 1988, I introduced the 'Environmental Activist' typology described above, using a behavioural scale of items indicating a degree of interest in environmental matters (see Table 1.1). It has proved powerfully predictive, and has enabled us to track the environmental movement in this country over the past five years (see Fig. 1.1). In 1991 we extended the use of the typology internationally for the World Wide Fund for Nature (WWF), as a part of the continuing monitoring of their attitudes and behaviour of people in countries where WWF either now has, or intends to have, an important presence (see International Perspective, below).

As indicated in Fig. 1.1 and Table 1.1, there has been a sharp rise in the proportion of the British public who have done various green activities: they are slight in such sedentary activities as walking in the countryside or watching environmental programmes on TV; more spectacular in the case of purchasing green (double over the five years), giving money to conservation charities and requesting green information. Over the period, the proportion of Environmental Activists (defined as those who

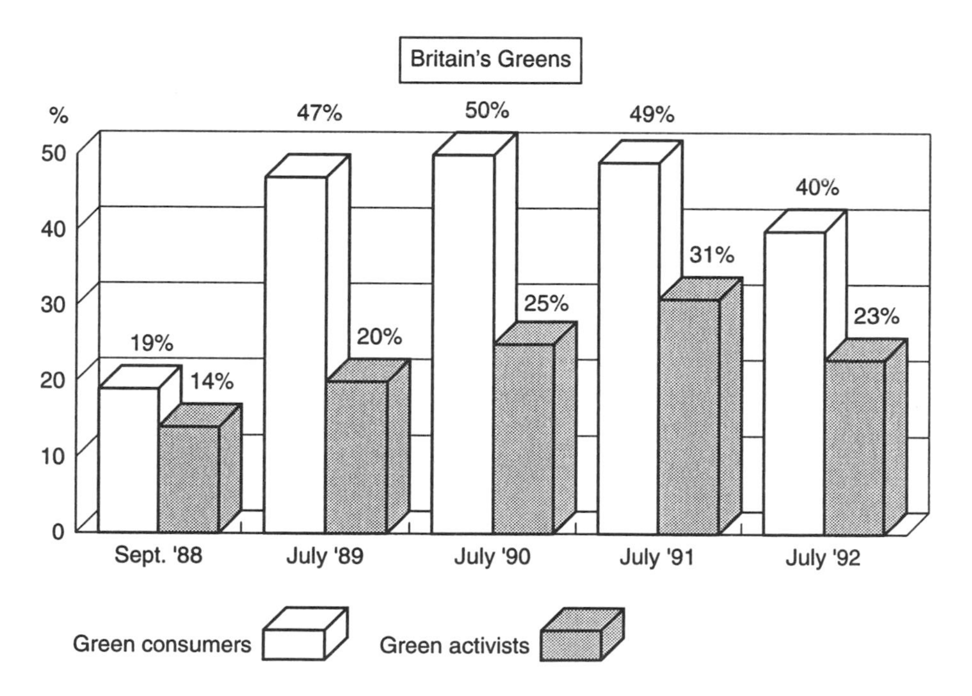

Figure 1.1 The environmental movement of Britain's 'Greens' 1988–92
(Source: MORI)

have taken five or more of the listed activities) has more than doubled, from 14% of the adult population in 1988 to 31% in 1991, before falling back to 23% in 1992, although it recovered to 28% in 1993.

Table 1.1 British 'Green' activism

Q. 'Which, if any, of the following things have you done in the last year or two?'

Date of Fieldwork	*1988* %	*1989* %	*1990* %	*1991* %	*1992* %	*1993* %
Read/Watched TV about wildlife/ conservation/natural resources/ Third World	79	80	84	87	85	83
Walked in the countryside/ along the coast	72	75	77	81	75	77
Given money to or raised money for wildlife/conservation or Third World charities	28	45	48	57	49	54
Selected on product over another because of its environmentally friendly packaging, formulation or advertising (Green Consumers)	19	47	50	49	40	44
Requested information from an organisation dealing with wildlife/ conservation/natural resources/ Third World	7	14	15	15	12	13
Subscribed to a magazine concerned with wildlife/conservation/natural resources or the Third World	8	14	13	15	10	13
Been a member of an environmental group/charity (even if you joined more than two years ago)	6	8	9	13	8	12
Visited/written a letter to an MP/ councillor about wildlife/conservation/ natural resources or the Third World	5	5	5	5	4	5
Campaigned about an environmental issue	4	4	6	5	3	5
Written a letter for publication about wildlife/conservation/natural resources or the Third World	2	2	2	3	2	2
Total	**230**	**294**	**309**	**330**	**288**	**308**
Environmental Activists (5+ Activities)	**14**	**20**	**25**	**31**	**23**	**28**
USE LEAD-FREE PETROL IN YOUR CAR	**12**	**22**	**30**	**35**	**37**	42

Source: MORI (c. 2,000 British adults.)

The use of unleaded petrol in a car was one of the early indicators of Environmental Activism, but when international tests were begun, this proved not to be an effective inclusion, as in some countries, e.g., Germany, all petrol was unleaded, and this therefore gave a misleading ingredient in the scale for purposes of international comparison. Still, it is of interest in Britain, so we include it in the list, but no longer use the findings to calculate the Activism scale.

Several points from the statistical analyses we have carried out on these data are worth commenting on.[1] The first is that there is a low correlation between Green Activism or Green Consumerism on the one hand, and writing letters or speaking to MPs or local councillors, about green matters on the other. It seems that people who present a case to MPs or local councillors, even on green issues, are no more likely than the average person to be 'green' in any other way, despite efforts made by the green movement to mobilise their members and supporters to bombard MPs with letters and personal pleas. If MPs' postbags are filling with letters about the environment and conservation, never mind animal welfare, the perennial top of the pile of MPs' volume of correspondence, think what a mobilised Green movement would do to the House of Commons' postbags! (See Fig. 1.2.)

The second point is that there is only a two-thirds correlation between Green Activists and Green Consumers. In other words, a third of those who act green in the shops do so with few of the other green activities.

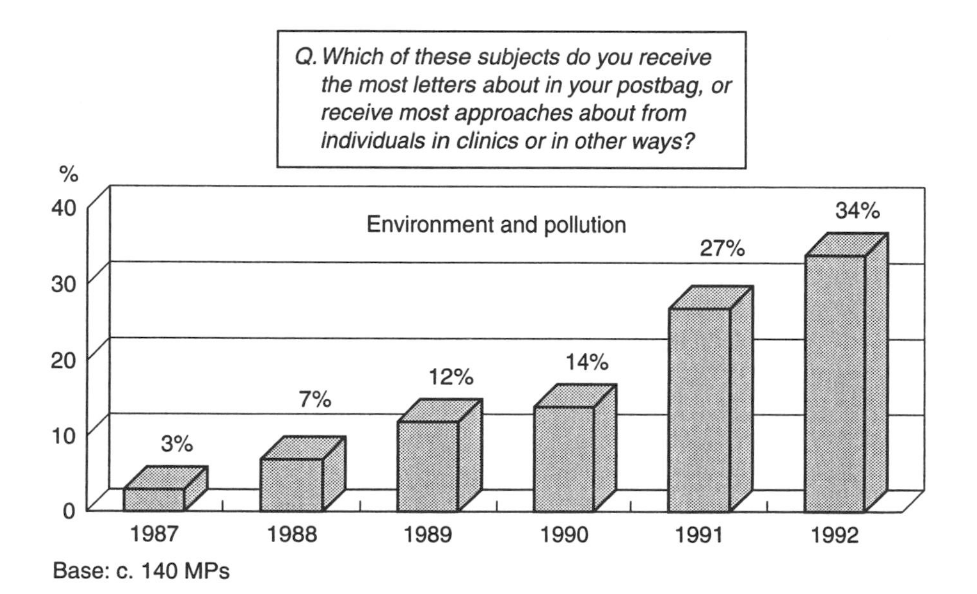

Figure 1.2 Increased interest in environmental matters as shown in MPs' postbags 1987–92
(Source: MORI)

As can been seen in Table 1.2, Greens as we have defined them are neither youngsters dedicated to saving our planet before it is too late, nor are they middle-class do-gooders. Green Activists, just under a quarter of the adult population (23%), are slightly more likely to be women than men, to be more middle-aged than young, but, significantly, are unlikely to be drawing their pension. They are more middle-class, with 62%, middle-class and 38% working-class. However, as there is still about a 60/40% skew to working-class people in Britain, this means that there are almost equal numbers of middle-class and working-class people in Britain today who qualify under this typology as Green Activists. They are more likely to be Liberal Democrats, and more likely to live in the South of England; but still, a majority of Green Activists live in the Midlands (including Wales) and the North (including Scotland).

Table 1.2 Profile of the 'Green' public

	General Public	*Environ. Activists*[1]	*Green Cons.*[2]	*Green Cons. Activists*[3]
Base:	(1,923)	(433)	(766)	(572)
	%	%	%	%
All	100	100	100	100
Men	48	44	41	42
Women	52	56	59	58
15–24	20	18	20	16
25–34	18	22	24	23
35–44	15	22	19	21
45–54	14	17	16	15
55–64	14	13	12	13
65+	19	7	10	11
AB	18	31	26	26
C_1	23	31	29	30
C_2	28	19	23	23
DE	31	19	22	21
North	35	28	31	28
Midlands	25	26	27	26
South	40	46	42	46
Cons.	33	33	34	32
Labour	37	36	35	34
LibDem.	12	17	15	18
E.A.	23	100	49	54
G.C.s	40	87	100	75
G.C.A.s	30	71	56	100

[1] E.A. =Environmental Activist (Table 1.1)
[2] G.C. =Green Consumer (Table 1.1)
[3] G.C.A. =Green Consumer Activists (4+ at Table 1.4)

Source: MORI, 1992

Table 1.3 Britain's 'Green' consumers

Q. Which, if any, of these do you do, or have you done, in the last 12 months as a result of concern for the environment?

	GENERAL PUBLIC				*GREEN CONSUMERS*			
	1990	*1991*	*1992*	*1993*	*1990*	*1991*	*1992*	*1993*
Base: %	*100*	*100*	*100*	*100*	*50*	*49*	*40*	*44*
	%	*%*	*%*	*%*	*%*	*%*	*%*	*%*
Buy 'ozone friendly' aerosols	73	71	65	71	92	91	88	90
Buy products which come in recycled packaging	41	55	52	50	58	74	77	71
Buy products made from recycled material	40	52	51	54	58	72	75	75
Buy household, domestic, or toiletry products that have not been tested on animals	43	51	47	51	59	70	70	72
Buy free-range eggs or chickens	44	46	44	45	55	58	58	56
Regularly use a bottle bank	39	39	43	46	53	58	55	57
Keep down the amount of electricity and fuel your household uses	44	44	42	51	49	55	52	58
Send your own waste paper to be recycled	31	36	36	38	51	43	47	47
Buy 'environmentally friendly' phosphate-free detergents or household cleaners	38	37	35	36	37	56	58	55
Avoid using chemical fertilisers or pesticides in your garden	41	38	31	40	55	49	45	53
Buy products which come in biodegradable packaging	26	34	29	36	52	53	43	57
Buy food products which are organically grown	25	28	24	27	43	41	36	41
Avoid using the services or products of a company which you consider has a poor environmental record	23	19	16	22	33	31	30	38
Keep down the amount you use your car	19	19	13	16	23	25	21	21
Avoid buying chlorine bleached nappies	13	10	7	7	19	15	10	10
Have a catalytic converter fitted to your car	9	7	6	9	12	9	7	12
Loft insulation	–	–	–	9	–	–	–	36
TOTAL	**549**	**586**	**541**	**599**	**749**	**800**	**772**	**813**
AVERAGE	**36.6**	**39.1**	**36.1**	**37.4**	**46.8**	**50.0**	**48.2**	**50.8**
INDEX (Based on 1990)	**100**	**106.7**	**99.5**	**110.8**	**100**	**106.8**	**103.0**	**108.5**

Source: MORI

Green consumers, as loosely defined here, make up four in ten of adult British people. Most are women (56%), middle-class (55%) and younger (63% under 45 v. 53% of the total adult population). But what do they do? Are Green Consumers really more consistent in their behaviour? If they are, then the Green Consumer typology has some meaning (see Table 1.3).

In 1990, the first year that their activity level was tested across a wide spectrum of green behavioural activities, Green Consumers were on average 12.5 percentage points more likely to take green action; this widened to 13.4 in 1991, and to 14.4 in 1992. Thus, while the percentage of the adult population represented by Green Consumers declined over the past three years, the gap between the general public and those who take green consumer action increased substantially (see Table 1.4).

Table 1.4 'Green' activity

Q. In the last 12 months, which, if any, of these things have you or your household done?

Q. And which, if any, of these things have you or your household done in the last 12 months out of concern for the environment?

	Done at all %	Environment-motivated action	Done for environment %
Avoid using aerosols/buy ozone-friendly aerosols	69	80%	55
Avoid buying products tested on animals	53	62%	33
Buy products made from recycled materials	47	68%	32
Regularly use a bottle bank	46	80%	37
Use lead-free petrol in your car	41	73%	30
Conserve water	41	56%	23
Buy products in recycled/biodegradable packaging	40	68%	27
Reduce the amount of fuel and electricity ... used	39	36%	14
Send your waste paper/cardboard to be recycled	38	79%	30
Bought a different ... product ... less damaging	33	76%	25
Avoid ... chemical fertilisers for pesticides in ... garden	29	66%	19
Use energy-efficient appliances	19	47%	9
Compost household refuse	17	53%	9
Avoid using peat in your garden	17	59%	10
Joined/been a member of an environmental group	4	75%	3
Done none of these	4		
Done 1–3	27		
Done 4+	69		

Base: 1,064 British adults, 15+, in 105 Constituencies throughout Great Britain, 26 June–10 July 1992.

Source: MORI/UK Ecolabelling Board, Department of the Environment, HMG

There are a number of behavioural measures being used by various researchers around the globe, including my own which is based on the premise that *doing* is a better test of 'greenness' that talking about it.

Another measure that we have developed has been used in work carried out for the UK's Ecolabelling Board (see Table 1.5). A third measure has been developed by SCPR (Witherspoon 1992), and a fourth has recently been tested by the Eurobarometer (Table 1.8).

The added value in the Ecolabelling Board work is the dual test that differentiates between those activities that are done for economic and other motives, and those done consciously out of concern for the environment. This shows that, at a minimum, one action in five, 20%, which might be described by others as 'green', is not thought of by the respondent as green, but is done for other, chiefly economic, reasons.

In the case of household energy conservation, as many as two-thirds of consumers who claim to have reduced the amount of fuel and electricity used in their household did so for other than 'green' reasons. Using a four-plus test (four or more items ticked), this scale offers a 'Green Activity' typology which equates to 42% of the British public at a time when the more established Green Activist typology currently defines only 23% of the British public.

In the 'Green Activity' test, women, the key purchasers of day-to-day household goods, show a much higher level of 'green' purchasing than men, particularly when it comes to avoiding products tested on animals and buying products made from recycled or biodegradable packaging. Generally, the 25–44 age group is most conscious of the environment in its purchasing behaviour, and ABs, the one-in-six of the British public whose household is headed by someone in the upper managerial/professional class, are some way ahead of those in other social classes in acting green.

What people want to know

During the Ecolabelling study, we asked respondents about the ways in which they had heard about the environment in the last year and found, predictably, that television, at 92%, ranked first, and that 78% said they found television most valuable in giving them information about the environment. The press was nominated as a source by three-quarters of the British public (76%), but only 25% felt it most valuable, while leaflets and family/friends were a source of environmental information for half (50% and 49% respectively) and of 'most value' to only 8% each.

Women tended to find the leaflets of greater value than men, as did younger people; middle-class respondents gave higher marks to the press, no doubt reflecting greater coverage of the subject in quality newspapers than in the tabloid press.

A detailed analysis is contained in Table 1.5.

Table 1.5 'Green' information sources

Q. In which of these ways, if any, have you heard about the environment in the last year?

Q. And which ways do you feel were most valuable in giving you information about the environment?

	Heard %	Valuable %
TV documentaries	83	64
TV news programmes	76	28
Newspaper articles	68	16
Magazine articles	52	12
Radio programmes	47	9
Posters	36	3
Manufacturers' ads.	32	2
Leaflets delivered to your home	32	5
Friends	31	2
Leaflets in supermarkets	28	3
Children at school	23	5
Other family members	21	2
Leaflets in libraries	11	1
Organisation membership	10	3
(Others below 10% omitted)		
None	2	3
Don't know	3	4

Base: 1,064 British adults, 15+, in 105 Constituencies throughout Great Britain, 26 June–10 July 1992

Source: MORI/UK Ecolabelling Board, Department of the Environment, HMG

THE VIEW OF BRITISH CAPTAINS OF INDUSTRY AND OTHER ELITE PUBLICS

Four captains of industry in ten agree with the proposition that 'British companies do not pay enough attention to their treatment of the environment'. Nearly half, 47%, believe that the most effective way of dealing with polluters would be 'legal penalties for companies which damage the environment'. These are the findings from 1992's MORI 'Captains of Industry Survey', a key element in the MORI Corporate Communications suite of research studies that monitors the current thinking of publics of importance to companies operating in Great Britain.

In a more general question, and one which has profound implications for anyone who cares about what these top businessmen think (the sample was drawn from the 'Times 500' largest companies – 74% of respondents were chairmen, MDs and/or CEOs), the way companies treat their environmental responsibilities was thought by a third (34%) of Captains to be one of the most important factors they take into account when making their judgements about other companies, ranking it above the company's growth performance and balance sheet and on a par with the implementation of its strategy, and just below productivity/efficiency.

WHO IS LEADING THE PARADE?

There is a question as to who is leading – the leaders or the led? – in the matter of environmental action. In the case of lead in petrol, there is no question but that it was the actions of two men, Godfrey Bradman providing the resources, and Des Wilson providing the campaigning, that forced the combined opposition of the petrol companies, the car companies and the government, to reverse their foot-dragging over the lead in petrol issue and, first, provide the fuel, and then the economic incentive in the form of tax savings, for motorists who used unleaded petrol in their cars. As shown earlier, over the past few years the environment as an issue has grown steadily in that most sensitive of public opinion barometers, MPs' postbags. Few MPs are immune to the clarion call of the sincere message, often hand written, from constituents who feel strongly about an issue. The environment has been just such an issue, as shown in the graph (Fig. 1.2) from the annual MORI Survey of the Attitudes of MPs, first begun in 1976 and conducted now on a semi-annual basis.

Even more worrying about what MPs think, is their reaction to the statement that 'British companies do not pay enough attention to their treatment of the environment', as a majority (53%) agree, and seven in ten (69%), believe that 'The penalties imposed on companies for causing pollution in Britain aren't severe enough', according to a survey (June-July 1992) of the attitudes of a representative sample of 158 MPs (64% response rate). Work done by the consultancy WBMG among MPs, shows that new MPs are on nearly every issue substantially more 'green' than the House as a whole.

While 26% of the entire sample of MPs gave 'utmost priority', scoring a maximum of ten points, to 'reducing acid rain emission at power stations', 37% of newly elected MPs gave this the top score, and while 29% of all Members gave maximum points to the 'goal of a 20% reduction from 1990 levels in the release of carbon dioxide by the year 2005, through the introduction of programmes of energy efficiency', 45% of the new MPs gave ten out of ten priority rating to this goal.

Journalists too think that it's time to get tough over environmental matters. Another of the MORI Corporate Communications Programme special groups interviewed annually are journalists. In the case of industrial journalists, who cover major companies' activities, no fewer than 83% interviewed in May 1992 agreed that 'British companies do not pay enough attention to their treatment of the environment'. The City also agreed in the MORI Summer 1992 City Study, when 56% of the 173 institutional investors interviewed face to face (73% response rate) and 42% of the 287 analysts interviewed by telephone (again, a high 74% response rate), agreed that 'British companies do not pay enough attention to their treatment of the environment'.

The international perspective

Protecting the environment also topped the poll in parallel studies carried out in June 1992 in Great Britain (by MORI, with a sample of 2,400 British adults), France (by Institut Francais de Démoscopie, among 1,000 respondents) and in Spain (Démoscopia, 1,200 respondents). Using a somewhat different questioning technique (Q. *Here is a list of areas of concern to business and industry. Which three or four do you think companies should pay particular attention to over the next few years?*), comparable results were obtained. In Britain, 49% selected 'protecting the environment', followed closely by 'keeping price rises to a reasonable level' (47%) and 38% each choosing 'training workers' and 'providing good quality products and services' and 35% said 'providing more jobs'.

In France however, more chose 'protecting the environment' (62%), but this ran second to the 64% who said that 'providing more jobs' was among the three or four things companies should pay particular attention to over the next few years. In Spain, jobs was chosen by 62%, and in second place was 'protecting the environment', at 45%.

In a telephone survey, also conducted in June 1992, for *The European* newspaper, Britain, with 72% of the British public in agreement, was among the countries which companies' protection of the environment highest, with Portugal topping the poll at 80%, followed closely by the

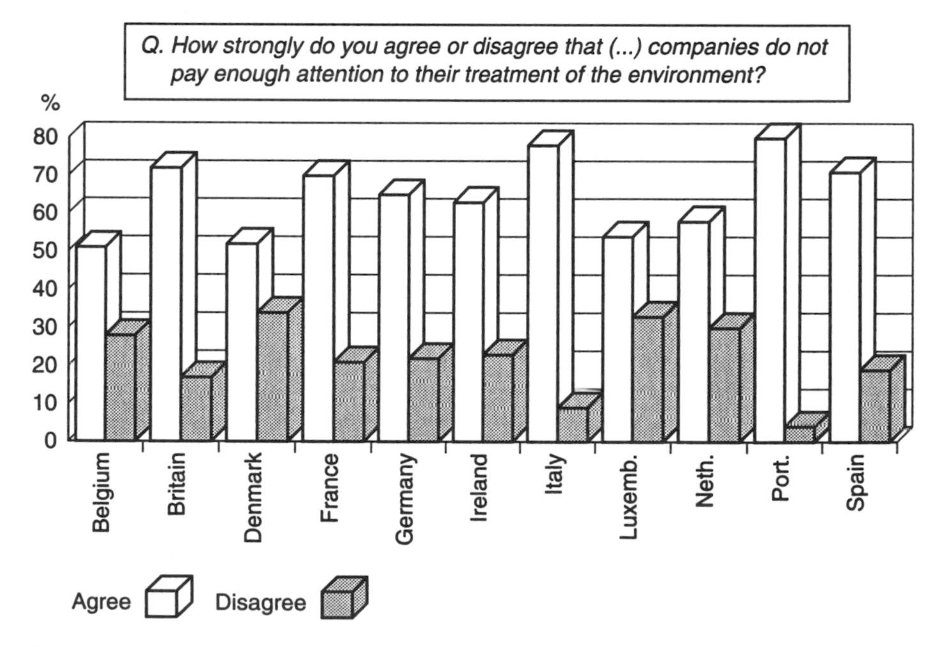

Figure 1.3 Rating of European countries of companies' attitudes towards the environment

(Source: MORI/*The European*, 1992)

Table 1.6 International 'Green' activism

Q. Which, if any, of the following things have you done in the last year or two?

	Aust'la	New Z.	Austria	Canada	Britain	France	Germ	Italy	Neth	Spain	Sweden	Switz	AVG
Base: *	1,200	771	785	2,004	2,028	1,000	500	969	508	1,007	1,057	886	1060
	%	%	%	%	%	%	%	%	%	%	%	%	%
Read/TV	87	93	54	80	86	87	93	62	96	69	93	90	83
Walked	88	90	77	77	81	91	87	85	95	86	89	89	87
Given	46	62	28	33	49	21	50	15	73	12	47	47	40
Gr Cons	**67**	**64**	**50**	**64**	**54**	**46**	**84**	**42**	**81**	**38**	**66**	**64**	**60**
Requested	10	15	9	13	11	5	19	6	17	3	8	11	11
Subscribe	13	26	7	20	11	13	12	5	27	4	16	15	14
Member	11	29	6	14	10	6	10	4	53	4	21	16	15
Visited	7	8	5	5	4	1	5	2	5	2	2	1	4
Campaign	7	9	16	7	4	7	21	7	10	6	7	16	10
Written	2	2	5	3	2	1	3	1	2	2	1	1	2
Environ Activists	**28**	**48**	**17**	**20**	**27**	**15**	**29**	**8**	**64**	**5**	**35**	**23**	**27**
Lead-free	36	56	35	▲	37	23	▲	9	48	8	54	57	43

Fieldwork in Austria, New Zealand, Spain, Switzerland in 1992, all others in 1991.

* The questions used in this survey are exactly the same as shown in Table 1.1. They are shortened here for reasons of space.

▲ Not asked, use mandatory.

Source: MORI/WWF International

Table 1.7 European Parliament and the environment

Q. For each of the following, could you tell me if you think the European Parliament plays an important or unimportant part in 'Environmental protection policy'?

	Belgium	Denmark	France	Germany	Greece	Ireland	Italy	Lux.	Neth.	Portugal	Spain	UK	EC 12
	%	%	%	%	%	%	%	%	%	%	%	%	%
Important	60	53	61	64	57	67	65	68	63	68	61	62	63
Unimportant	20	29	24	22	8	9	13	18	26	9	10	13	17
Net	40	24	37	42	49	58	52	50	37	59	51	49	46

EUROPEAN COMMISSION AND THE ENVIRONMENT

Q. Some people believe that certain areas of policy should be decided by the (national government) while other areas of policy should be decided jointly within the European Community. Which do you think it should be with respect to protection of the environment?

	Belgium	Denmark	France	Germany	Greece	Ireland	Italy	Lux.	Neth.	Portugal	Spain	UK	EC 12
	%	%	%	%	%	%	%	%	%	%	%	%	%
National Gov't	27	39	28	22	35	35	21	31	13	25	21	28	24
European Comm.	66	60	70	76	60	61	75	64	86	69	69	71	72
Difference	39	21	42	54	25	26	54	33	73	44	48	43	48

THE EUROPEAN PUBLIC AND THE ENVIRONMENT

Q. Many people are concerned about protecting the environment and fighting pollution. In your opinion is this . . .

	Belgium	Denmark	France	Germany	Greece	Ireland	Italy	Lux.	Neth.	Portugal	Spain	UK	EC 12
	%	%	%	%	%	%	%	%	%	%	%	%	%
An immediate and urgent problem?	85	87	80	88	97	70	91	83	84	73	82	82	85
More a problem for the future?	11	10	17	10	3	19	7	12	11	14	12	10	11
Not a problem?	2	3	2	1	0	8	1	1	3	1	1	3	2
Increase since '87	**12**	**5**	**21**	**4**	**15**	**7**	**6**	**–1**	**19**	**9**	**8**	**15**	**11**
Base:	1,036	1,000	1,005	1,065	1,000	1,001	1,046	496	1,002	1,000	1,000	1,319	13,082

Source: '*Eurobarometer:* Europeans and the Environment in 1992', Spring 1992

Italians (77%) and the Spaniards (71%) and the French (70%). The study also showed that there is general agreement across Europe that companies do not pay enough attention to their treatment of the environment (see Fig. 1.3).

The 1985 data from the six countries participating in the International Social Survey Program (ISSP) indicated wide variation in concern for the environment, as shown by the answers to the question eliciting the degree to which respondents (n = 677 in the USA to 1,580 in Italy) wished to see more government spending for the environment. Germany led, with 82% of those interviewed saying they would like the German Government to spend more on the environment, followed by its neighbours Austria (73%) and Italy (62%) while fewer than half of the English-speaking countries, USA (43%), Great Britain (37%) and Australia (32%) wanted such spending.[2]

A 1990 Community Attitudes Survey in Australia found that worry or concern about the environment, at 56%, topped the poll from a list of nine national issues, followed closely by education (55%), interest rates (54%) and unemployment (54%). In terms of priority, unemployment came first, with 24%, followed by interest rates (19%), and in third place, the environment, with 17% saying that it gave them most worry.[3]

The European Community's Eurobarometer has been monitoring Europeans' attitudes to the environment for a decade, although the 1982 study is described as 'limited' and 'exploratory'. In virtually every European country, there has been a sharp increase in the number of citizens since 1986 who describe their concern about 'protecting the environment and fighting pollution' as 'an immediate and urgent problem': Belgium (from 72% to 85%, up 13), Denmark (up 10, to 85%), France (up 24, to 80%), Germany (up 8, to 88%), Greece (up 13, to 97%), Ireland (up 14, to 70%), Italy (up 6, to 91%), Luxembourg (no change, at 83%), Netherlands (up 21, to 84%), Portugal (up 2, to 73%), Spain (up 10, to 82%), and United Kingdom (up 18, to 82%).[4]

The Eurobarometer study in 1992 attempted to determine the degree to which EC citizens were ready to balance economic realities to the natural desire to improve the environment, and found that while seven in ten (69%) across Europe believed that 'economic development must be ensured but the environment protected at the same time', of those who chose between protecting the environment and economic development, 22% said that 'concerns about the environment would get higher priority than economic development' and only 4% the reverse.

There was a remarkable consistency across demographic sub-groups on the answers to this question, but vast differences by country, from a 5:4 ratio in Ireland to 5:1 in the UK up to 17:1 in Denmark and 15:1 in The Netherlands. Concern about the effect of development on the environment focused on industry (69%), energy (45%) and transport (33%), more than agriculture (21%) or tourism (7%).

What does it all add up to?

Whether or not companies in Britain do live up to their environmental responsibilities is hardly the point. We can bring two types of findings to our clients: the first is when a genuine problem is perceived by the public, and companies have to deal with that problem or live with its consequences. The second type of finding is when something is perceived as being negative when, in reality, it is not. In this case, a properly constructed programme of communicating the reality, instead of leaving the misconception intact, can correct a false perception. Without that, it is likely that the European Commission or the European Parliament will step in. The balance of British public opinion is, surprisingly, on the side of Brussels: **when asked *'Who would you trust to make the right decisions about the environment?'*, a plurality of the British, by 43% to 37%, said they would trust the EC rather than the British Government. The gap is an even wider 51% to 34% among ABs, and nearly two to one 55% to 30%, among Environmental Activists.**

It is not that the British public has great faith in the European Community to make the right decisions about the environment. The recent Eurobarometer asked which level of government can act efficiently to protect the environment, and found all wanting, but, relative to other levels, from local to world wide, the EC came off least badly. But when, in the Eurobarometer survey the question was asked: *'Some people believe that certain areas of policy should be decided by the British Government, while other areas of policy should be decided jointly within the European community. Which (when it comes to protection of the environment) should decide?'* the replies indicated that **by 71% to 28%, joint EC decision-making was preferred to a purely British Government decision**. This was, interestingly, spot-on the EC average response. **And when asked again by the Eurobarometer to choose from a list which was the most trusted source of information on the state of the environment, environmental protection organisations received 11 times the level of trust that public authorities did, and five times that of the media**. People in Britain seem to want more information on the environmental risks posed by everyday products (87%),ways of disposing of waste (86%), the way public authorities spend the money intended for protecting the environment (86%), and the potential risks of nuclear radiation (84%).

When MEPs were asked whether or not they supported the principle that companies should be fully responsible for the cost of clearing up any environmental damage which they cause, 87% said they supported this in MORI's 1991 MEPs' study; this rose to 93% in the 1992 study. And 72% of MEPs (including 69% of British MEPs) support the view that the

Table 1.8 Actions taken to protect the environment

Q. Which, if any, of these things have you ever done?

	Belgium	Denmark	France	Germany	Greece	Ireland	Italy	Lux.	Neth.	Portugal	Spain	UK	EC 12
	%	%	%	%	%	%	%	%	%	%	%	%	%
A.	85	86	90	88	90	85	89	84	79	88	86	88	88
B.	62	77	87	74	55	55	54	55	65	58	54	73	65
C.	58	71	58	87	12	33	55	75	82	34	31	54	60
D.	50	55	54	67	58	38	57	55	49	71	70	46	58
E.	49	43	65	54	78	40	59	53	39	69	67	52	55
F.	41	54	45	54	28	42	43	66	52	25	30	57	46
G.	35	36	41	50	42	22	39	41	43	43	34	38	41
H.	15	18	7	46	11	10	5	27	22	6	5	17	19
I.	32	13	32	26	25	6	30	27	13	32	15	11	23
J.	11	7	8	16	8	11	10	21	5	6	5	9	10
K.	12	5	13	12	10	8	10	18	4	7	7	5	9
L.	15	28	5	13	12	15	7	37	22	5	4	14	10
M.	10	16	5	7	2	5	6	18	20	2	4	8	7
Index	**37**	**39**	**39**	**46**	**33**	**28**	**36**	**44**	**38**	**34**	**32**	**36**	**38**

Source: *Eurobarometer:* Europeans and the Environment in 1992, Spring 1992

A. Avoid dropping papers or other waste on the ground
B. Save energy, for example, by using less hot water, by closing doors and windows to save heat
C. Sort out certain types of household waste (glass, paper, oil, batteries) for recycling
D. Save tap water
E. Not make too much noise
F. Buy an environmentally-friendly product even if it is more expensive
G. Use less polluting means of transport (walking, bicycle, public transport) than your car
H. Have your car fitted with equipment to limit the pollution such as, for example, a catalytic converter
I. Go on a type of holiday that is less harmful to the environment
J. Take part in a local environmental initiative, for example, cleaning a beach or park
K. Demonstrate against a project that could harm the environment
L. Financially support an association for the protection of the environment
M. Be a member of an association for the protection of the environment

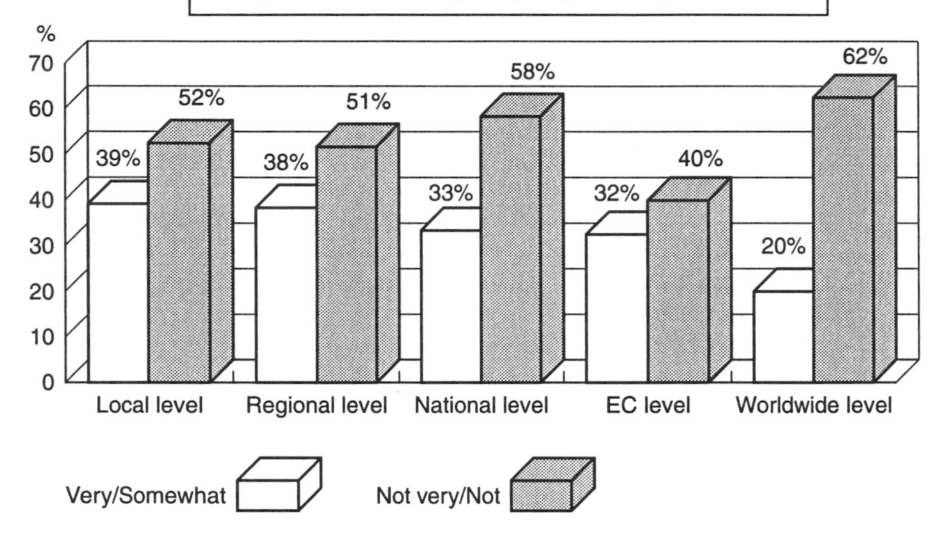

Figure 1.4 Rating in EC countries of the efficiency of public bodies in protecting the environment, 1992
(Source: Eurobarometer)

recently established European Environment Agency should assume full responsibility for the enforcement of agreed European standards on the environment.

So if action is needed by British companies to get their act together either to correct what they are in fact doing about the environment, or, alternatively, to correct the misconception that they are not doing enough in their treatment of the environment, they should get moving. Otherwise, the British Government or the European Environment Agency, acting for the EC, or the European Parliament, will do it for them, and legal penalties may well bc high.

References

1 Worcester, Robert M. 'Business and the Environment', *Admap*, January 1993.
2 Skrentny, John D. 'Concern for the Environment: A Cross-National Perspective', *International Journal of Public Opinion Research*, vol. 5, no. 4, (1993).
3 McAllister, I. and Studlar, D. 'Trends in Public Opinion on Environmental Issues in Australia', *International Journal of Public Opinion Research*, vol. 5, no 4, (1993).
4 Martier, E. 'Europeans and the Environment in 1992', Eurobarometer, no. 37, August 1992.

Recommended reading

McIntosh, Andrew. 'The Impact of Environmental Issues on marketing and politics in the 1990s', *Journal of the Market Research Society*, vol. 33, no. 3, July 1992.

Witherspoon, Sharon and Martin, Jean. 'What Do we Mean by Green?', chapter in *British Social Attitudes – 9th Report* (SCPR), Jowell *et al* (eds), Dartmouth Publishing Company, Aldershot (1992).

Worcester, Robert M. and Corrado, Michèle. 'Attitudes to the Environment: A North-South Analysis', chapter in *Environment and Development: Problems and Prospects for Sustainable Development*, Benachenhou, A. (ed.) UNESCO, Paris (1991).

2

Facing up to limits: the challenge of sustainable development

RICHARD TAPPER

Head of Industry Policy,
WWF UK (World Wide Fund for Nature), Surrey

■ The environment is a central business issue. Attention is focused on the way in which the activities and management of business contribute to sustainable development and improve their environmental performances.

■ Sustainable development is 'development that meets the needs of the present without compromising the ability of future generations to meet their own needs'. Sustainable rates of use of resources mean living off the interest rather than the biological and environmental capital of the Earth.

■ The resources of the Earth are finite. However, the Earth's stock of renewable resources is being used at rates far in excess of their capacity to regenerate; and the Earth's ability to mop up wastes and pollution is even more severely limited.

■ Biodiversity is declining fast, and an estimated 25% of all species are likely to become extinct over the next 50 years – a rate of extinction between 1,000 and 10,000 times of that recorded for past geological periods.

■ The dynamics and functioning of living ecosystems cannot be mimicked by any existing technologies, nor can biotechnology and genetic engineering create new genes or genetic diversity.

■ Economists estimate that at least 3–5% of GDP in industrialised countries is spent on cleaning up some of the damage caused by pollution and the effects of economic activity on the environment. Many economic and business activities do not pay for the costs of the environmental damage that they cause.

■ Business and industry can reduce environmental damage by achieving absolute reductions in use of resources and energy, and in their generation of pollution and wastes. Action to prevent pollution or overuse of resources from the outset often costs far less than paying for damage once it has been caused.

■ Measures which increase the efficiency of resource use through regulation, taxation, or incentives, also encourage the efficient allocation of resources by consumers and businesses, and so improve overall economic performance.

■ The structure and function of ecosystems provide a model for sustainable industry. We need to change industrial process so that they will become:

- cyclical;
- more efficient at conserving resources; and
- avoid the release of wastes, especially toxic wastes.

■ Long-term wealth creation and industrial development will not be achieved without sustainable development. Companies which look ahead and direct their R and D and capital investment programmes to meet the needs of sustainable development will be the successful companies of the future.

Sustainable development

In 1992, the United Nations Conference on Environment and Development (UNCED), held in Rio de Janeiro, Brazil, put the spotlight on environmental issues and on the need to achieve 'sustainable development'. A summit meeting attended by 120 Heads of Government, demonstrated the global realisation that healthy economies depend on healthy environments.

Sustainable development is about altering the way we use resources – about changing both our production and our consumption to patterns that are more environmentally sound. It has been defined in the Brundtland Report as 'development that meets the needs of the present without compromising the ability of future generations to meet their own needs.'[1]

Sustainable development involves four key components:

- Integrating the environment and the economy – recognising that the economy and the environment interact, and that economic gains in the short term which are bought at the expense of the environment cannot be economically sound.
- Long-term planning – recognising that a focus on short-term economic planning inevitably undermines the long-term delivery of a decent environment, since it places little value on long-term outcomes; and that to deliver a healthy environment requires long-term planning which respects environmental time-scales.
- Equitable sharing of resources throughout the world – recognising that without a fairer sharing of resources within and between countries, those who have the most are liable to over-consume and to use resources in ways that are environmentally unsound, while those who have least are forced into short-term, environmentally damaging activities in order to survive.
- Protecting environmental assets – recognising the need to retain environmental assets also known as 'critical environmental capital' (including the quality of air and water, soil fertility and biodiversity)

above minimum thresholds, and that this imposes limits on the use of physical space, the management and consumption of resources, and emissions of pollution and waste.

Five major documents were agreed at Rio, which tried to capture the global consensus on the environment. There were two international Conventions aimed at halting climate change, and at protecting biological diversity, which were both signed by over 150 governments present, along with a statement on Forest Principles, and Agenda 21, which sets out a programme for sustainable development. Finally, the Rio Declaration, which reiterates commitments by Governments to the Precautionary Principle, the Polluter Pays Principle, Environmental Assessment, and the use of economic instruments to internalise (make consumers, companies and governments pay for) environmental and social costs.

Table 2.1 UNCED – Agenda 21: Chapter titles on environment and related topics

Many of the chapters are inter-linked. Pollution reduction, protection and sustainable use of renewable resources, are emphasised throughout.
Human health
Human settlements
Protection of the atmosphere Stratospheric ozone depletion Transboundary atmospheric pollution
Planning and management of land resources
Deforestation
Desertification and drought Soil conservation
Mountain ecosystems
Sustainable agriculture
Conservation of biological diversity
Biotechnology
Protection of oceans, seas, coastal areas, and their living resources Fish stocks Marine pollution
Protection of freshwater resources Freshwater pollution
Toxic chemicals
Hazardous wastes
Solid wastes and sewage-related wastes
Radioactive wastes

The chapter headings of Agenda 21 indicate the range of environmental issues that are of concern at a global level (see Table 2.1). Other chapters dealt with population and consumption issues; with the responsibilities of different sectors of society, including business and industry; and with technology transfer, information provision, and finance. This list of issues touches on all aspects of economic activity.

The environmental issues considered at UNCED all interact and overlap with each other. Emissions of CO_2 contribute to global warming and climate change. They also are linked to the extraction and combustion of fossil fuels. The extraction of fossil fuels itself creates environmental impacts, whether from open-cast mining, or from oil or gas extraction. Coal and oil, the world's primary sources of fossil fuels, contain sulphur and traces of heavy metals. These are released on combustion, and have caused intense regional pollution across North America, Europe, China, parts of South America, and in industrial regions around the world.

Climate change, regional pollution, and the local impact of fuel extraction, all damage the environment. In different ways, and on different time scales, they impact on biological diversity.

Furthermore, the direct damage to biological diversity that results from deforestation also reduces the amount of carbon stored in standing vegetation. Loss of forest cover leads to the release of large amounts of carbon dioxide from the oxidation of soil organic matter and plants which are left to rot. Destruction of peatlands, wetlands and other natural ecosystems similarly leads to large releases of CO_2. These emissions of CO_2 also contribute to global warming and climate change.

Climate change has global effects, and it can only be tackled effectively through international agreement. Other environmental issues, such as the disposal of growing volumes of waste in landfill or by incineration, have more local effects in the first instance, and they can be dealt with at the national level. However, because of their cumulative impact around the world, they also require international action. Action is certainly required where international trade shifts impacts around the world – for example, through trade in hazardous and other wastes.

Science and the global environment

Over the past twenty-five years, new scientific methods have been developed for the study of the global environment, and of the way in which the different components of the environment interact with human activities:

1. *Environmental monitoring*

Remote-sensing technologies, such as Earth Observation Satellites (EOS) combined with information technology, now provide a global observatory for the measurement and observation of environmental change.

These technologies make it possible to observe changes as they happen. For example, it is possible to monitor crop growth or the spread of agricultural pests, to detect the dispersal of pollutants, or to follow weather systems across the oceans.

2. *Scientific understanding of environmental effects*
Scientific understanding of the links between the various factors affecting different components of the environment, and between those environmental components, has advanced enormously. The effect of CFC propellants sprayed from aerosols on ozone depletion in the stratosphere above Antarctica is one example. Understanding of the links which keep biogeochemical cycles turning has also increased greatly.

3. *Forecasting environmental effects*
Approaches have also been developed for the analysis of environmental data and forecasting on a global scale, using techniques such as systems analysis. Although they are more complex, global environmental models are like the forecasting models used in business strategy development, or for macroeconomic planning. These models are being further developed to improve the way in which they handle the effects of oceans on the atmosphere, and forecasts at a regional level.

In addition to their scientific role, global environmental models provide a link between science and government policy, enabling a translation to be made between the languages of science, government policy-making and business.

The combination of these scientific methods has made it possible to integrate information, from many sources, about different aspects of the environment. Local observations from the ground can be combined in global environmental models with information from satellites covering an entire hemisphere. In this way physical measurements of CO_2 in the atmosphere and the CO_2 dissolved in the oceans, can be linked, through environmental models, to CO_2 absorption by the world's forests, and other elements of the global carbon cycle.

Global models can also be used to understand the impacts of pollution, energy use, or deforestation. Forecasts from these models can be fed into models of the world economy, of food production, population growth, or resource availability.

In short, such models enable the linkages between the environment, the economy, and human activities, to be explored and investigated in a way never before possible.

Economics and the environment

Economics is intimately entwined with ecology and the environment. Many economic and business activities do not pay for the costs of the environmental damage that they cause. The costs of energy, waste disposal and transport are widely acknowledged by business leaders and economists to be far cheaper than the cost to society as a whole of the environmental impact of these activities. Conversely, people are willing to pay more to live in attractive environments that are unpolluted, as studies of housing prices in the US demonstrate.

Economic activity, like it or not, is part of the Earth's ecological system. All economic activity affects the environment, whether or not the resulting effects are accounted for in the balance sheets.

PAYING FOR POLLUTION

'Externalities' are what economists call effects that are not paid for by those who cause them.

The world is full of examples of externalities where economic activity leads to costs that are paid for by others. Whoever pays for them, the costs still exist and they affect the national and local economy directly.

Economists estimate that at least 3–5% of GDP in industrialised countries is spent on cleaning up some of the damage caused by pollution and the effects of economic activity on the environment. This figure does not account for the economic benefits that flow from clean environments, but which are foregone in polluted ones. For example, it does not account for reduced crop yields, or declining fish catches, caused by pollution.

Pollution certainly causes damage, but that damage is presently paid for in the form of increased health care costs and time lost at work through pollution-related illnesses, expenditure to restore damage to buildings, losses from reduction in productivity of natural and managed ecosystems, and so on.

In the same way, overharvesting of fish stocks, or over-intensive use of agricultural land, may increase income in the short term, but do so at the expense of declining incomes in the future, as the ability of the environment to regenerate those resources that are harvested excessively is continually diminished. The recovery of overharvested stocks takes longer the more they have been diminished. And if they have fallen below critical thresholds, recovery may become impossible.

Experience tells us that the costs of taking action to prevent pollution or overharvesting in the first place are often far less than the costs to society of paying for damage once it has been caused, and the costs of restoring past damage.

The problem is that current economic practice generally takes only a short-term view. In business most investment is expected to be written off over a maximum of five years, and many business people regard rates of return on investment of less than 10% as not worth bothering with. By contrast the environment operates on much longer time-scales. Even when the use of CFCs stops completely in the year 2000 under the terms of the Montreal Protocol, it will take a further 100 years for the existing pool of CFCs in the upper atmosphere to disappear; during this time these CFCs will continue to cause ozone depletion in the stratosphere.

And ozone depletion is likely to result in a far higher incidence of certain cancers, leading to increased health care costs, as well as adverse effects on agriculture, and even increased rates in the decay of certain materials as higher levels of ultra violet radiation reach the Earth's surface. In other words, environmental damage results in economic costs – the only question about these costs is who pays them, and when.

Limits to growth

The resources of the Earth are finite. Depletion of fish stocks, loss of forest cover, overabstraction of freshwater from rivers and groundwater, and soil erosion, all demonstrate how the Earth's stocks of renewable resources are being used at rates far in excess of their capacity to regenerate.

But even more, the Earth's ability to mop up the consequences of resource use is far more severely limited, and the environmental sinks for pollution are overstretched – even crumbling – in the face of ever growing burdens of pollution and wastes.[1]

Non-renewable resource stocks, too, are being fast depleted. Between 1970 and 1990, world energy consumption almost doubled; so did the number of registered automobiles. The use of minerals and raw materials also rose dramatically, matched by increased levels of consumption. World population rose from 3.6 billion to 5.3 billion. It is projected to reach 8 billion by 2025.[2]

It might be thought that such increases in resource use and pollution just represent rises due to population expansion, plus a modest reduction in poverty, worldwide. But the reality is that while population increases in the developing world have lead to absolute increases in energy use in those regions, per capita energy consumption has changed little there. It is in the industrialised world, where population levels have changed little, that per capita energy consumption has rocketed, along with consumption of other materials.

Developed countries contain just 24% of the world's population, yet their share in global consumption of various commodities ranges from around 50% to 90% or more. This quarter of the population consumes 48% of cereals, 64% of meat, 60% of fertilisers, 75% of total energy, 80% of iron and steel, 81% of paper, and 92% of cars.[3]

Increases in the production of goods and services, and the consumption of raw materials and energy, and population, have all followed exponential patterns of growth. For how long can these exponential growth patterns continue before reaching limits – for example, exhaustion of oil or gas reserves, or near-extinction of major species of edible fish stocks?

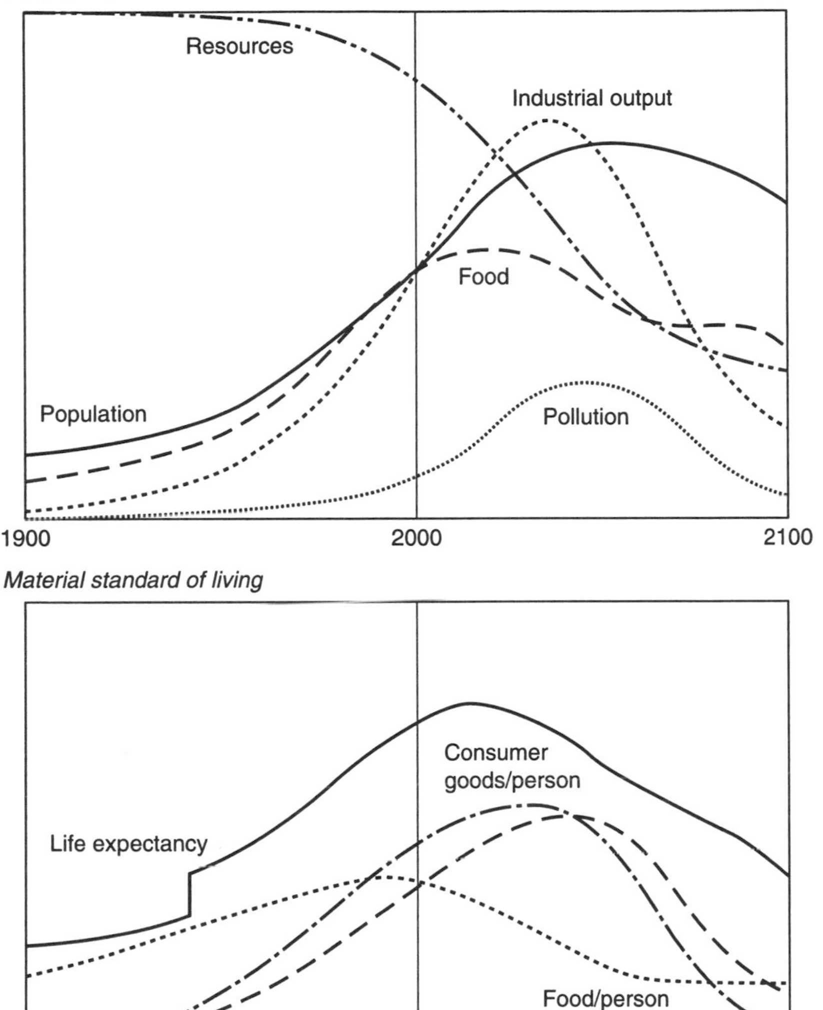

Figure 2.1 Technology, markets and overshoot: results shown in scenario from World3 model

The recognition of tough and inevitable resource constraints on growing consumption and growing population levels, led the Club of Rome to commission the 'Limits to Growth' study, published in 1972. The follow-up study *Beyond the Limits* revises and extends the original Limits to

Growth models, which predict that growing population levels and rising consumption levels will eventually overshoot the Earth's resources and regenerative capacities, and that this will be followed by a collapse in population numbers, as food availability declines.

Fig. 2.1 shows the results from one of the scenarios considered in *Beyond the Limits* using the 'World3' model. The features of increasing population, industrial output, and food, alongside decreasing resources are common, in broad outline, to all the scenarios considered. So too is the crash that follows from overshooting the Earth's capacity to support an ever-growing population, and to cope with continual declines in resource availability and ever-increasing pollution.

The scenario displayed in Fig. 2.1 is more optimistic than the base scenario. It assumes that the Earth's resources are double those that are now known, and that increasingly effective pollution control technology is available, which will reduce the amount of pollution generated per unit of industrial output by 3% per year. Nevertheless, pollution rises high enough to produce a crisis in agriculture that draws capital from the economy into the agricultural sector and eventually stops industrial growth.[2] The stresses engendered by food and resource scarcity, and by environmental pollution, which lead to a fall in population and life-expectancy, could well create great social instability.

The authors of *Beyond the Limits* conclude that:

> One lesson from these runs is that in a complex, finite world if you remove or raise one limit and go on growing, you encounter another limit. Especially if the growth is exponential, the next limit will show up surprisingly soon. There are *layers of limits*. World3 contains only a few. The 'real world' contains many more. Most of them are distinct, specific, and locally variable. Only a few limits, such as the ozone layer or the greenhouse gases in the atmosphere, are truly global.
>
> We would expect different parts of the 'real world', if they keep on growing, to run into different limits in a different order at different times. But the experience of successive and multiple limits in any one place, we think, would unfold much the way it does in World3. *And in an increasingly linked world economy, a society under stress anywhere sends out waves that are felt everywhere. Free trade enhances the likelihood that those parts of the world included in the free trade zone will reach limits simultaneously*. [Author's italics.]
>
> A second lesson is that the more successfully society puts off its limits through economic and technical adaptations, the more likely it is in the future to run into several of them at the same time. In most World3 runs, including many we have not shown here, the world system does not run out of land or food or resources or pollution-absorption capability, it *runs out of the ability to cope*.

Industrial and ecological systems

The industrial system uses water, energy and resources to produce goods and services. Wastes are generated in the production process and, once the goods and services are used up, they, too, are disposed of, and become wastes. The industrial system is powered predominantly by fossil fuels. It has high reliance on non-renewable resources, and on use of renewable resources, including freshwater supplies, at rates in excess of their capacity to regenerate. A low proportion of wastes is recycled; pollutants and most wastes are produced at rates in excess of environmental capacities to detoxify them, and accumulate in the environment – land, air, and water. Many are inherently toxic.

By contrast, ecological systems – ecosystems – are powered by renewable energy ultimately derived, through photosynthesis, from the sun. Ecosystems rely predominantly on the recycling of limited stocks of essential elements; ecosystem growth is resource-limited and does not exceed rate of nutrient recycling. In the most mature ecosystems a large proportion of the nutrient stock is incorporated into the biomass of organisms. Dead organic matter is decomposed by organisms forming detritus ecosystems. These take wastes as their resources and energy source; the capacity of detritus ecosystems matches the rate of inflow of wastes and dead organisms.

Healthy ecosystems provide services, such as binding soil in place, or providing shelter from wind and weather, as well as products, such as food.

As industrial systems and ecosystems interact, changing land-use and overuse of biological resources reduces the overall level of biological capital and of ecosystem functioning. Intensive harvesting of particular species changes ecosystem composition and functioning, and introduces instabilities. Industrial wastes entering ecosystems are an additional burden and exceed ecosystem waste handling capacity; many of the wastes are toxic, and further impair waste handling, as well as the regenerative capacity of ecosystems. Where industrial wastes are also ecosystem nutrients, growth of certain species is promoted and pushed out of balance with the rest of the ecosystem; the waste products from this accelerated growth overwhelm waste handling capacity and reduce available resources to the rest of the ecosystem – this positive feedback loop continues until ecosystem function is severely impaired, and ecosystem composition is changed.

The behaviour of either system is determined by the ways in which the different components within each system interact. For example, successful company management juggles factors of production and marketing to ensure that production neither exceeds overall sales, nor falls below thresholds that force major changes, and that its capital is efficiently employed in the long run. Within these bounds, the choices and outcomes of corrective action are fairly predictable, but once they are crossed, corrective action becomes highly unpredictable, and the possibility of catastrophic change is large.

Table 2.2

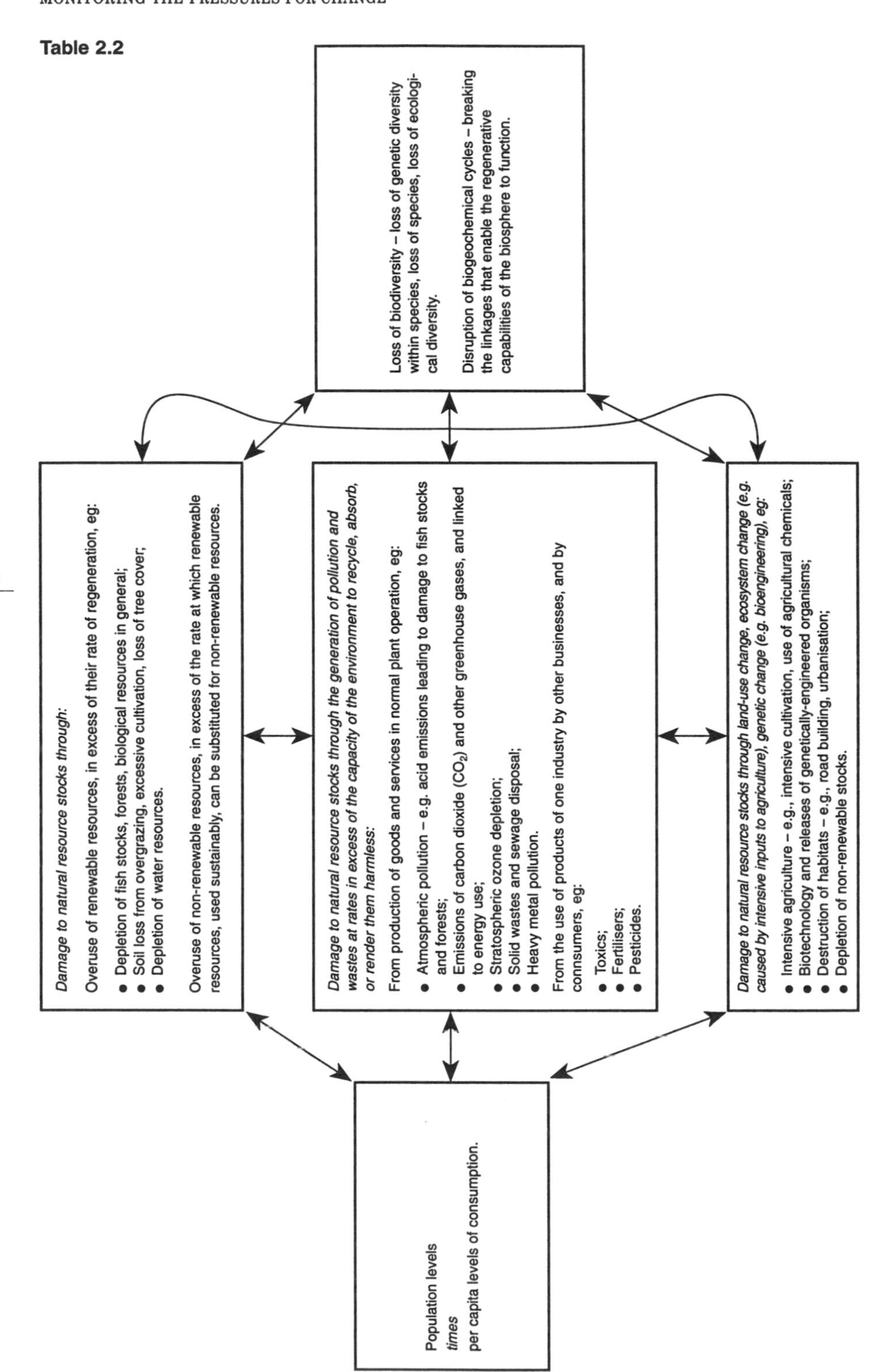
Loss of biodiversity – loss of genetic diversity within species, loss of species, loss of ecological diversity.
Disruption of biogeochemical cycles – breaking the linkages that enable the regenerative capabilities of the biosphere to function.
Damage to natural resource stocks through:
Overuse of renewable resources, in excess of their rate of regeneration, eg:
Depletion of fish stocks, forests, biological resources in general;
Soil loss from overgrazing, excessive cultivation, loss of tree cover;
Depletion of water resources.
Overuse of non-renewable resources, in excess of the rate at which renewable resources, used sustainably, can be substituted for non-renewable resources.
Damage to natural resource stocks through the generation of pollution and wastes at rates in excess of the capacity of the environment to recycle, absorb, or render them harmless:
From production of goods and services in normal plant operation, eg:
Atmospheric pollution – e.g. acid emissions leading to damage to fish stocks and forests;
Emissions of carbon dioxide (CO_2) and other greenhouse gases, and linked to energy use;
Stratospheric ozone depletion;
Solid wastes and sewage disposal;
Heavy metal pollution.
From the use of products of one industry by other businesses, and by connsumers, eg:
Toxics;
Fertilisers;
Pesticides.
Damage to natural resource stocks through land-use change, ecosystem change (e.g. caused by intensive inputs to agriculture), genetic change (e.g. bioengineering), eg:
Intensive agriculture – e.g., intensive cultivation, use of agricultural chemicals;
Biotechnology and releases of genetically-engineered organisms;
Destruction of habitats – e.g., road building, urbanisation;
Depletion of non-renewable stocks.
Population levels
times
per capita levels of consumption.

The same is true of ecosystems. Once critical thresholds are crossed, outcomes become highly unpredictable, and the possibility of dramatic and catastrophic change is large.

The essence of this is that the structure and function of ecosystems provide a model for sustainable industry. We need to change industrial systems as well as processes so that they will become:

- cyclical;
- more efficient at conserving resources; and
- avoid the release of wastes, especially toxic wastes.

This will require the redesign of industrial processes and products so as to reduce the consumption of resources per unit output, and to reduce absolute levels of consumption of resources and energy. It also requires a move away from those industries that are least sustainable, and towards those that have a sustainable long-term basis.

The impacts of human activities

In broad terms, the impacts of human activity on the environment are proportional to population levels *times* per capita levels of consumption. Per capita consumption rises with affluence, and the affluent populations of Northern industrial societies each on average consume more than sixteen times the resources consumed by a counterpart in the South. Population levels combined with consumption give rise to three groups of impacts (see Table 2.2), that cause damage to natural resource stocks through:

- overuse;
- the generation of pollution and wastes; and
- alteration, as a result of land-use change, ecosystem change (e.g., caused by intensive inputs of agrochemicals), or genetic change (e.g., bioengineering).

All of these environmental impacts also affect human health and the built environment.

Technology strongly influences human impacts on the environment. It is clearly helpful that new technologies can be applied to clean up some of the environmental damage that has been created in the past, and to reduce impacts from production of specific goods and services. In poorer countries and communities, the introduction of even the most basic of technologies can greatly reduce impacts on the environment.

But on the other side, the development of technology, and its associated infrastructure, has led to high levels of consumption per capita in industrial countries and continues to do so. And, while technological development is also leading to increased efficiency of resource use per unit of output, and consequent reductions in the generation of pollution and wastes, the rate of increase of consumption in industrialised countries, and of world population growth, outstrip these efficiency gains and impose growing burdens on the environment.

The consequences of damage to natural resource stocks, whether from overuse, from pollution and wastes, or from intensive inputs or land-use changes, are:

- loss of biodiversity – loss of genetic diversity within species, loss of species, loss of ecological diversity; and
- disruption of biogeochemical cycles – breaking the linkages that enable the regenerative capabilities of the biosphere to function.

To make sense out of this complexity, and to define sustainable, long-term, limits to use, Herman Daly, one of the World Bank's senior economists, has proposed three simple rules:[4]

- for a renewable resource, the sustainable rate of use should not exceed the rate of regeneration of that resource (this is equivalent to the concepts of 'carrying capacity' or 'sustainable yield');

- for a non-renewable resource, the sustainable rate of use should not exceed the rate at which a renewable resource, used sustainably, can be substituted for it;
- for a pollutant, the sustainable rate of emission should not exceed the rate at which that pollutant can be recycled, absorbed, or rendered harmless by the environment (this rate is sometimes referred to as a 'critical load').

In other words, sustainable rates of use of resources mean living off the interest rather than the biological and environmental capital of the Earth, and maintaining environmental assets above the minimum levels critical for their continued survival. And there are limits to the sustainable rates of use of renewable and non-renewable resources, and of the sinks for pollution and wastes.

Being sustainable – priorities for business and industry

Business and industry have a major role to play in moving national and global economies onto a sustainable development footing, that does not exceed the sustainable rates of use of resources outlined by Herman Daly. At the end of the 1980s manufacturing output was about a third of the GDP of OECD countries, and represented on average for these countries:

- 15% of water consumption (excluding water used for cooling);
- 25% of nitrogen oxide emissions;
- 35% of final energy use;
- 40–50% of sulphur dioxide emissions;

- 50% of contributions to global warming, through emissions of CO_2, methane, nitrous oxide, and CFCs;
- 60% of biological oxygen demand and particulate discharges to water ('conventional' water pollution);
- 75% of non-inert waste; and
- 90% of toxic substances discharged into water.[5]

These figures exclude agriculture, transport, the service sector, or housing and construction development; taking these sectors along with manufacturing, business and industry account for as much as 80% of GDP.

Clearly, business and industry have a pervasive influence within economies and, consequently, on decisions about the allocation of resources. This includes the allocation of resources to address key environmental problems, and to take action to reduce, and ultimately prevent, them from occurring. In addition, the finance sector, along with the insurance industry, underpins all the other sectors. The way in which banks appraise loans and demand security, or insurers appraise risks, and charge premiums, has a strong influence on patterns of industrial development and environmental impact. Furthermore, the way in which banks and insurers assess, and protect against, environmental risk, is a crucial factor in improving environmental performance of companies and especially new projects. By harnessing the financial and technological resources of business and industry to sustainable development, the prospects of securing this objective will be enormously enhanced.

Table 2.1 sets out the range of environmental issues of concern at a global level that are part of the consensus reached at Rio. Among these areas of environmental concern, what are the priorities that require most urgent attention by business and industry? In 1990, the US Environmental Protection Agency (US EPA) published a report that aimed to answer this question.[6] Building on an earlier report entitled *Unfinished Business*[7], the US EPA's 1990 report set out eleven priorities based on ecological and health impacts. These were:

Ecological priorities

Global climatic change
Stratospheric ozone depletion
Habitat alteration
Species extinction and biodiversity loss

Health priorities

Criteria air pollutants (e.g., smog)
Toxic air pollutants (e.g., benzene)
Radon
Indoor air pollution
Drinking water contamination

Occupational exposure to chemicals
Application of pesticides
Stratospheric ozone depletion

The US EPA acknowledges that this is not an inclusive list. Some key issues like acid rain or solid wastes are not placed as top priorities since they are already subject to strong control measures in the US. The factors taken into account in setting these priorities include the risks and the costs associated with their effect on:

- human health;
- ecology – resulting from habitat modification and environmental pollution; and
- welfare (including damage to property, goods and services, to which a monetary value can often be assigned) for example, damage to natural resources (e.g., crops, forests, and fisheries), recreation, materials, aesthetic values, and various public and commercial activities.[7]

Another study undertaken by the Centre for the Exploitation of Science and Technology[8] in the UK has looked at priorities for environmental action in relation to the proportion of manufacturing GDP that will need to take action in different countries (see Fig. 2.2). Reducing emissions of CO_2, energy use and waste production require action by all sectors of industry, while in other cases a few industries are responsible for considerable amounts of raw material consumption and pollution. These are the 'Dirty and Dangerous Industries' mentioned in the Introduction and shown there in Fig. 1. In Canada it has been estimated that six industries – agro-foodstuffs, metal extraction and processing, cement works, pulp and paper, oil refining and chemicals – account for more that two-thirds of all industrial pollution, although they represent just one-fifth of the value added.[5]

Of all the priorities for action for business and industry, two need action by all enterprises, in all sectors:

- reducing energy demand, and the environmental impacts associated with energy demand, from extraction to pollution and global warming; and
- reducing the generation of wastes and the use of toxic materials.

Reducing energy demand and its impacts on climate and the environment

The inevitable consequences of fossil fuel use are release of CO_2, and of pollutants such as SOx, NOx, and traces of heavy metals. Governments

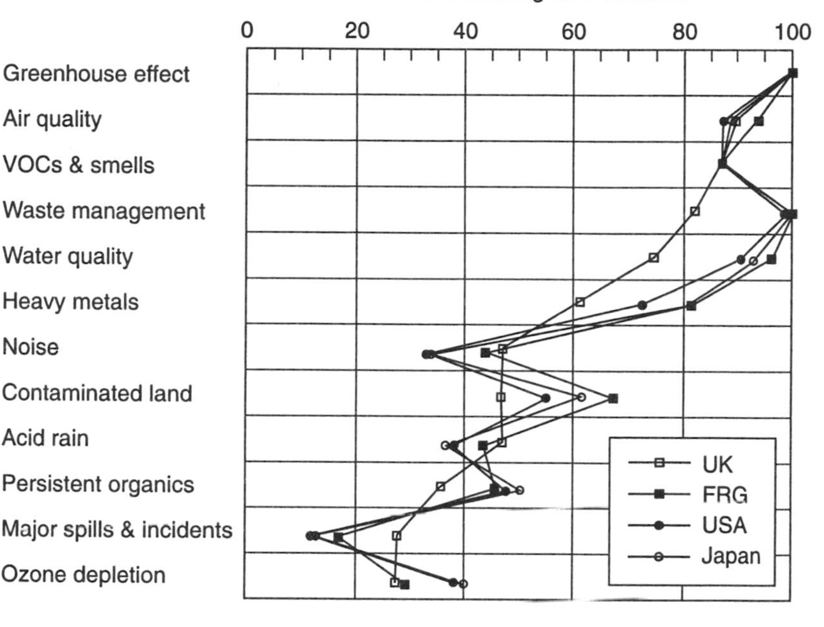

Figure 2.2 The impact of the environmental problems in the UK, former FRG, USA and Japan

in North America and Europe have already taken action to control and reduce emissions of SO_2 and NOx – the major pollutants that cause acid rain. The US Clean Air Act Amendment (1990), the UNECE Convention on Long-range Transport of Air Pollutants, and the EC Large Combustion Plant Directive all set out vigorous measures to reduce emissions of SO_2 and NOx – in the US, the aim is to reduce major point source emissions of SO_2 by 40% in 2010 compared to 1980 levels. The EC's Large Combustion Plant Directive requires cuts in SO_2 and NOx emissions from existing plant of 40% and 30% respectively by 1998 against 1980 levels, while all new plant must incorporate SO_2 and NOx emissions abatement technology. Ultimately, these measures will be extended to bring emissions in Europe down to levels that do not exceed 'critical loads' defined for habitats across Europe.

At the Rio Earth Summit in 1992, 154 governments signed the UN Framework Convention on Climate Change. The objective of the Convention is the 'stabilisation of greenhouse gas concentrations in the atmosphere at a level which would prevent dangerous anthropogenic interference with the climate system'. This is to be achieved '. . .within a timeframe sufficient to allow ecosystems to adapt naturally to climate change, to ensure food production is not threatened and to enable economic development to proceed in a sustainable manner.'

The UN Framework Convention on Climate Change is based on the broad consensus on climate change set out by the Intergovernmental Panel on Climate Change (IPCC). Established in 1988 by the World Meteorological Organisation and United Nations' Development Programme, the IPCC has investigated effects of the changing composition of carbon dioxide and other atmospheric trace gases on world climate. The IPCC involved more than 300 scientists from over 40 countries, and combined scientific data and analysis from many different groups into a unified report that summarised the extent of scientific knowledge about climate change and its impacts on the Earth, as well as the development of response strategies.

In its 1990 report,[9] the IPCC concluded that emissions from human activities are substantially increasing the atmospheric concentrations of the greenhouse gases: carbon dioxide, methane, CFCs and nitrous oxide. CO_2, thought to be the major anthropogenic contributor to perturbation of the Earth's radiative balance, has increased from 280 ppm before the Industrial Revolution to 353 ppm today, and its concentration in the atmosphere continues to rise as fossil fuel use accelerates. Continuing emissions of greenhouse gases on present trends may result in a rate of increase of global mean temperature during the next century of 0.3°C per decade – a rate greater than that seen over the past 10,000 years – according to the IPCC's projections. This would result in a likely increase in global mean temperature of about 1°C above the present value by 2025 and 3°C before the end of the next century. By 2030, global mean sea level may have risen by 20 cm, mainly due to thermal expansion of the oceans and the melting of some land ice. Regional climate changes would differ from the global mean. For example, temperature increases in Southern Europe and cental North America are predicted to be higher than the global mean, accompanied on average by reduced rainfall and soil moisture.

Such changes will affect agriculture. While some regional variations may lead to local conditions more favourable to crop growth, a high proportion of the main food-producing regions in both North and South hemispheres are likely to be affected adversely. Increases in climatic variability will increase the risk of low agricultural yields and potential for crop failures.

As more than 70% of the world's population live on coastal plains, the impacts of sea-level rise are particularly critical. Virtually all the land of the Maldives, in the Indian Ocean, is less than two metres above sea level and the entire nation is at risk from inundation from sea-level rise. Large parts of Bangladesh are also at risk, as are densely populated and fertile coastal strips around the world. Even where the risk of flooding is less serious, the ingress of sea water into groundwater supplies used for irrigation and drinking is a major threat.

Table 2.3 Stabilisation of atmospheric gases

	Atmospheric Concentrations			
	Pre-industrial	*Present day*	*% contribution from human-made GHGs to the change in radiative forcing from 1980 to 1990*	*% reduction in human-made emissions to stabilise GHG concentrations at present day levels*
Carbon dioxide	280 ppmv	353 ppmv	55%	>60%
Methane	0.8 ppmv	1.72 ppmv	15%	15–20%
Nitrous oxide	28.8 ppbv	310 ppbv	6%	70–80%
CFCs*			24%	70–85% (Montreal Protocol – 100%)
CFC–11	0 pptv	280 pptv	–	–
CFC–12	0 pptv	484 pptv	–	–
HCFC–22	0 pptv		–	40–50%

ppmv – parts per million by volume
ppbv – parts per billion by volume
pptv – parts per trillion by volume

*Measures are already in place to phase out production of CFCs, which contribute to global warming, by the end of 1998, owing to their effect on stratospheric ozone. However, HCFCs and HFCs, which have been developed as replacements with lower ozone depletion potentials than CFCs, are themselves still potent greenhouse gases.

Sources: Houghton, J.T., Jenkins, G.J., Ephraums, J.J. *Climate Change: The IPCC Assessment*, Cambridge University Press (1990).

A recent study by the US Environmental Protection Agency estimates that the gross cost to the US of protecting strategically-significant sites – such as coastal industrial sites and urban areas – from a one-metre rise in sea-level, caused by global warming, would be as much as $400 billion. Such a sea-level rise would also involve other costs, including costs resulting from salt-water ingress into groundwater, and impacts from increased storm frequency. These costs directly, or indirectly affect industry. Hurricane Andrew devastated the headquarters of Grand Metropolitan's Burger King subsidiary, causing $10 million of damage, while 300 out of 700 employees there lost their homes. In the immediate aftermath of Hurricane Andrew and Cyclone Iniki in August 1992, no fewer than nine US insurance companies went out of business.[10] Disruption of agriculture would impose further costs, as would changes to the availability of freshwater supplies.

As Table 2.3 illustrates, stabilisation of atmospheric concentrations of greenhouse gases will require major reductions in emissions of these gases arising as a result of human activity. For example, a 60% reduction in CO_2 emissions is needed to achieve atmospheric stabilisation – the aim of the United Nations Framework Convention on Climate Change.

The most critical measures needed to attain the objectives of the UN Framework Convention on Climate Change remain dramatic reductions in CO_2 emissions which can only be achieved by reducing reliance on fossil fuels, and by a decline in the global demand for energy. In 1990, world consumption of energy reached an estimated 8.7 billion tonnes oil equivalent (btoe) with CO_2 emissions from fossil fuels totalling 5.5 billion tonnes carbon (btC).[11] On present trends, the World Energy Council projects that by 2020, world energy demand will probably have increased to between 13 btoe/year to 16 btoe/year. The bulk of this will be met by combustion of fossil fuels, and once adjustments are made for changing fuel mixes, the result will be a rise in CO_2 emissions from fossil fuels to between 5.8 btC/year and 10.6 btC/year.

Unless these trends reversed, and annual CO_2 emissions reduced from 1990 levels by at least 60%, CO_2 concentrations in the atmosphere will continue to rise. The longer the delay in achieving such a reduction, the higher will be the final atmospheric concentration of CO_2 once stabilisation is reached – and the greater the consequent effects on global mean temperatures and climate.

The trend of rapidly growing energy demand is based on conventional energy economics, which fail to take into account the economic costs of coping with the impacts of climate change if CO_2 emissions are not reduced, or of the costs of other forms of pollution and environmental damage resulting from energy production and use. The World Energy Council concludes however that CO_2 emissions could be levelled out if serious attention is given to energy policies to control demand and introduce new technologies.

The alternative view to conventional energy economics is put most strongly by Amory Lovins, a widely recognised US expert on energy efficiency, who calculates that as much as three-quarters of US electricity can be saved at costs of one cent per kilowatt hour or less, and that three-quarters of US oil use can be saved at a cost of less than three dollars per barrel equivalent, using existing technologies.[10]

These costs are far less than the costs of supplying the equivalent amounts of energy and, on this basis, Amory Lovins estimates that world energy demand could be satisfied by 2020 with just 50% of the energy used today, with concomitant reductions in CO_2 emissions. Add to this the costs saved by averting climate change and energy-related pollution, and there is enormous potential to fund new technologies for the low-energy-use society of the future, including photovoltaics, decentralised energy networks, and public transport and communications systems.

Lessons from Japan, which has reduced the amount of energy per unit of GDP by 34% over the period 1973 to 1990, demonstrate the enormous potential of energy efficiency. This has been achieved through clear and long-term Government policies, which have been implemented through a package of measures including financial incentives, regulations, standards and education. Regularly updated energy efficiency standards for appliances and buildings, set by the Minister of International Trade and Industry (MITI) in conjunction with industry representatives, have led to dramatic efficiency improvements. Indeed, straightforward energy efficiency measures have played a major role in ensuring the high degree of Japan's international competitiveness.[18]

Inefficient and profligate energy use reflects past failures to value energy sufficiently, and to price energy use at levels that account for its full environmental and social costs. The energy market is consequently distorted in favour of energy use rather than energy conservation. According to the World Bank, eliminating subsidies for energy consumption would raise more than $230 billion worldwide.

Initial measures to reduce CO_2 emissions are the removal of such market distortions, and the introduction of energy efficiency and conservation measures. These include regulation, economic incentives and taxation. Such measures are fully compatible with the precautionary principle and 'no-regrets' policies. In the face of uncertainties about the extent of climate change and its impacts, these measures keep national and international options open, and they are also desirable in their own right.

Energy efficiency is about the more efficient use of resources, and consequently it reduces costs. But, while the efficient use of resources is an important factor, the environment responds to absolute rather than relative improvements. Although energy use per unit of GDP has declined in most industrialised countries for the past 20 years, absolute levels of energy use and CO_2 emissions have increased. The improvements in the fuel efficiency of car engines have been far outstripped by the increased numbers of cars in use, and by the increased overall mileage driven. Clearly, such efficiency alone is not enough, and more fundamental measures are necessary. Economic measures, such as environmental taxes, are a beginning, but in the future, it will be increasingly necessary to restructure and relocate markets in such a way that they provide incentives to industry to achieve absolute reductions in resource and energy use, waste generation and pollution.

Capital investment is of course necessary to achieve cost reductions from energy efficiency, but the pay-back is quick, often as little as one or two years, even for major investment. This has been demonstrated repeatedly by organisations such as 3M. And some measures have zero outlay costs. Likewise in national economies, measures which increase the efficiency of energy use, whether through regulation, taxation, or incentives, also encourage the efficient allocation of resources by consumers and businesses. Consequently they improve overall economic performance; for example, through a reduction in energy imports. This frees up the portion of GDP that would otherwise be used to control or restore the damage done by pollution, so that it can then be applied to more productive uses.

Of course, some might point to nuclear power, but reserves of uranium would only be sufficient for 60 years should nuclear energy production rise to 20% of global energy supplies.[12] The energy required to extract purified uranium from existing grades of ore is already large, and would increase even further as lower and lower grades of ore are exploited. But even before viable uranium supplies yielding a net energy gain are exhausted, nuclear technology may itself be limited by supplies of strategic metals such as chromium and nickel. And nuclear fusion which also requires supplies of strategic metals does not fare any better. Nuclear power is additionally unattractive because of serious doubts on its economics, and concern over long-lived wastes, and security risks.

Limits to energy use are just some of the constraints imposed by the physical limits of the Earth's resources to supply energy on a sustainable basis, and to mop up CO_2 and pollutant emissions from energy consumption. No doubt further reserves will be discovered, which will extend the time horizons of reserves a little further – indeed proven reserves of fossil fuels have kept pace with rate of energy use in recent years. However, two things are certain: the Earth is not supplied with an infinite stock of non-renewable resources and, even if more reserves are developed, the Earth's capacity to absorb CO_2 and pollutant emissions is already saturated.

Reducing the generation of wastes and the use of toxic materials

The current mode of operation of industrial activities leads to the accumulation of wastes in the environment. No part of the Earth is immune – Even the Arctic icecap contains traces of dry cleaning fluids that have worked their way northwards from North America and Europe. Unless they are degradable, wastes never go away. Even degradable wastes accumulate in the environment if conditions prevent their breakdown. In other cases, the amounts of waste produced far exceed the ability of the environment to break it down.

At the international level, concern over wastes has lead to the adoption and implementation of the Basle Convention on the Transboundary Movement and Disposal of Hazardous Wastes.

In healthy ecosystems, virtually the only wastes that are produced are dead organic matter, the mineral nutrients that this contains, and any soil that may be washed away. But these wastes are the basic foodstuffs and nutrients for detritus ecosystems and soil organisms, including worms, fungi and bacteria, that decompose and rot dead material. Detritus ecosystems handle most of these wastes right where they are produced. The vast majority of wastes in healthy ecosystems are recycled either directly into living organisms, or into water, carbon dioxide and mineral nutrients released back into the biogeochemical cycles of the environment.

Industrial systems recycle very little. What is recycled has first to be concentrated into separate types of material at centralised waste handling depots. The organisation of collection and waste handling systems is presently quite rudimentary. Less than 30% of aluminium, where recycling is well-developed, comes from recycled scrap. For paper and glass between 30–35% comes from recycling, but for most other materials, the rate of recycling is far below 5%.[5, 13]

In 1990, OECD countries together produced close to 1,500 million tonnes of industrial wastes (including over 300 million tonnes of hazardous wastes) – a figure that has risen from 1,000 million tonnes per annum in the early 1980s. A further 7,000 million tonnes of other wastes included residues from the production of energy, agricultural waste, mining spoil, demolition debris, dredge spoil and sewage sludge. Municipal wastes added another 420 million tonnes, bringing the total to nearly 9,000 million tonnes of wastes to be managed by OECD economies during 1990. Industrial wastes produced outside the OECD amounted to just over 600 million tonnes, including 35 million tonnes of hazardous wastes.[5]

The problems, actual and potential, that waste disposal can cause are well-known and include risks of groundwater contamination and damage to health. Every country has its roll-call of waste disposal disasters.

Love Canal in the US is possibly the most notorious, but it is just one of over 21,000 hazardous or potentially hazardous landfill sites identified in the US, at least 1,750 of which require urgent remedial action. European countries each have several thousand potentially hazardous sites. The scale of contamination in central and eastern Europe has yet to be accurately assessed, but it is likely to be considerable.

Apart from the sheer volumes of waste produced each year, wastes present enormous problems of environmental persistence and toxicity. Many synthetic chemicals have been developed specifically to be inert and unreactive to other materials with which they might come in contact in industrial applications. In consequence, once they enter the environment, either accidentally or at the end of their life in a product, they cannot be broken down. CFCs, for example, persist in the upper atmosphere for around 100 years. PCBs, developed as insulating and coolant liquids for use in transformers and other electrical equipment, also persist in the environment where they find their way into the food chain, and accumulate in the fatty tissues of fish and mammals. Heavy metals too, such as chromium, cadmium, lead, mercury, nickel and zinc, which are discharged in large quantities, accumulate in the environment and also enter the food chain.

Organisms are now exposed to a cocktail of artificial chemicals that have accumulated in the environment. There is increasing evidence that implicates exposure to these chemicals with cancers, and with interference to reproductive fertility, in humans as well as many animals. Chemicals as diverse as PCBs, some pesticides and nonylphenols (used to manufacture plastics) disrupt the functioning of the reproductive and other hormone systems. General environmental exposure of the population to these chemicals, collectively termed 'endocrine disruptors' appears to be linked to a twofold fall in average human sperm counts, and a similar increase in testicular cancer, since the 1930s.

Parallel reductions in reproductive fertility are also being detected in a number of other animal species, and may well be linked with environmental exposure to a cocktail of artificial chemicals.[19] Many of these chemicals are released daily into our environment to combat insects, weeds, and mould. Industrial chemicals such as cadmium, lead, mercury, chlorinated biphenyls, penta to nonylphenols, phthalates, and styrenes, also disrupt hormone messages.

Over 100,000 chemicals are available commercially, although only 1,500 of them account for 95% of world production. A further 1,000 new substances are put on the market each year. Synthetic organic and inorganic chemicals are used in a growing number of applications, from pigments for paints, lubricants, fertilisers, food additives, and solvents, to medicines.[5] It is estimated that an average television set contains over 4,000 different chemicals. Even if all these could potentially be recycled, the task of separating them from an old TV would be almost impossible.

There are fortunately many practical steps that industry can take to reduce the production of wastes, and levels of toxicity, as well as to improve waste management and recycling. These are dealt with in Part II, Chapter 7, on business opportunities in waste management. Legal and policy aspects of waste in Europe and the US are also covered in Part III, Chapters 14 and 15. Governments also have an important role to play in setting the stage for the development of high quality waste disposal businesses.

Reducing the amounts of toxic materials used in industrial processes and products, and their levels of toxicity, are key actions that are vital in moving towards sustainable development, and to cyclical rather than linear industrial systems. The preparation of inventories – such as Toxic Release Inventories (TRIs) – at company and national level, of chemicals in use as well as those being emitted to the environment (combined with a re-evaluation of the need for many of these chemicals) are also central tasks in moves to achieve sustainable development.

Biodiversity, loss and disruption of biogeochemical cycles

The diversity of living organisms and ecosystems is the basis of the health of the Earth's life support systems. A significant reduction in biodiversity has profound implications. Human beings depend on biological resources and on the diversity that underpins the continuing survival of these resources. Directly, we rely on living organisms for food, medicines, clothing, construction, and many of the basic inputs of raw materials into industrial processes. Trade in agricultural products alone amounts to more than $3 trillion.

Just as importantly, we rely on the Earth's biogeochemical cycles to purify the atmosphere and water, maintain fertile soils, recycle wastes, and maintain the balance of natural processes. Without living organisms, these cycles cease to function.[14] The dynamics and functioning of living ecosystems cannot be mimicked by any existing technologies, nor can biotechnology and genetic engineering create new genes or genetic diversity.

The importance of protecting biodiversity is recognised in the Convention on Biological Diversity, signed along with the Framework Convention on Climate Change by over 150 Governments in Rio in 1992. The aim of the Biodiversity Convention is to stem losses in biodiversity that are occurring throughout the world, through measures for the conservation and sustainable use of biological resources and biodiversity.

The main sources of the impacts on business and industry on biodiversity are overuse, generation of pollution and wastes, land use and ecosystem, and genetic changes resulting from industrial activity (see Table 2.2). Financial decisions that enable such activities to take place and consumption of the goods and services which they produce, are contributory factors to these impacts.

The direct value of biodiversity is considerable. About 4.5% of the US GDP is attributable to the economic benefits from wild species. Genes found in low-yielding or wild relatives of crop plants are essential to breeding programmes developing commercial varieties. For example, plants from a wild wheat population in Turkey were used to give disease resistance (worth $50 million annually to the US alone) to commercial wheat varieties, and resistance bred in from an Ethiopian barley now protects California's $160 million annual barley crop from yellow dwarf virus.[13]

The indirect, and irreplaceable values of biodiversity are more difficult to measure, but are likely to be even greater. For example, environmental assets are the mainstay of many economies, particularly in developing countries.

Currently biodiversity is declining fast, and an estimated 25% of all species are likely to become extinct over the next 50 years – a rate of extinction between 1,000 and 10,000 times higher than recorded in the geological record of previous ages[16]. Conversion of land from natural and semi-natural ecosystems, to intensively managed agricultural systems, or putting it under concrete, leads to a simplification of ecosystems, and a reduction in ecosystem diversity. Furthermore, the genetic diversity of crops and animals in intensive agriculture monocultures, is orders of magnitude less than that in traditional, low-intensity agriculture, or among the wild relatives of domesticated crops and livestock.

In temperate regions, forests are being felled and replaced with single species plantations. Intensive agriculture, linked to inputs of fertilisers and pesticides, is based on monocultures of just a few dozen varieties of any crop. The clearance of land for agriculture, forestry, mineral extraction, urban expansion, or new roads and infrastructure, has created swathes of much simpler, managed systems which do not provide the environmental services of the diverse ecosystems which they have replaced.

Extinction of species and other forms of biodiversity loss are essentially irreversible, and can lead to economic losses and instabilities that are often immediate. This is well illustrated by commercial whaling which has driven stocks of the Blue and Northern Right whales close to extinction. Overfishing of the Blue Fin Tuna has so reduced its numbers in the West Atlantic that it is now endangered. Other commercially-fished species, such as cod in the North-West Atlantic and in the North Sea, have been so depleted that fishing is severely restricted in attempts to allow their populations to recover. During the 1980s, pollution from acid rain emissions in Europe, caused among other things damage to fish stocks in Scandinavia, as well as damage to forests across Europe estimated at $30 billion a year.[2]

Commercial logging and conversion to agriculture have led to the destruction of at least 50% of the tropical forest that existed world wide in 1900; an area of tropical forest the size of Great Britain is still cleared each year. Tropical forests contain between 50% and 90% of the species that live on Earth, according to best estimates, and their destruction represents a massive loss of diversity. Moreover, their soils are of low fertility, and the agriculture that follows in the wake of forest destruction is often unproductive within a few years. The low productivity of cleared soils exposes them to erosion, which clogs rivers and damages fish stocks. Even where forestry plantations are established, these are based on just a few species of trees per hectare, rather than the 300 or more tree species found in each hectare of natural tropical forest.

In Temperate and Boreal Forests commercial logging and plantations are just as damaging.

The net result of this destruction is an economic loss at the national, and not just local, level. While individual timber extraction or agricultural operations may be profitable because of their specific conditions, these are obtained at the expense of other profitable uses of the environment[15] – for example, sustainable extraction of individual mature timber trees, forest products, or fisheries, which are destroyed by silt washed into rivers from eroding soils.[15] If silt reaches as far as coral reefs in coastal waters, it destroys those and their associated value in providing fisheries, coastal protection, and recreation.

But beyond even these serious losses, chronic and far less visible declines in biodiversity present a further threat. In the soils of Europe, a combination of factors including pollution, and intensive agriculture and applications of agrochemicals, has led to declines in soil fungi of as much as 80%.[14] Soil fungi, along with other soil organisms such as worms and bacteria, are key elements in the operation of detritus ecosystems which decompose dead organic matter, and recycle its nutrients back into living organisms.

Detritus ecosystems are essential for long-term maintenance of soil fertility. To a certain extent loss of soil fertility can be compensated by use of fertilisers, pesticides and mechanical tillage. But these require large amounts of energy, primarily from fossil fuels – already in intensive agriculture only twice as much energy is produced in cereal crops as is input overall in the form of fossil fuel energy – and destroy the detritus ecosystems that regenerate soil fertility, turning living soils into little more than chemical substrates. Furthermore, the intensive use of fertilisers and pesticides creates wider problems, such as damage to freshwater sources, and adverse affects on ecosystem and human health from the accumulation of toxic substances in food chains. And ultimately, the energy inputs needed compensate for damage to detritus ecosystem functions in intensive agriculture, cannot be sustained indefinitely.

Conclusions – the way ahead

If the world is to address seriously the problems of irreversible loss of biodiversity and damage to biogeochemical cycles, then business and industry, as much as any other sector of society, will have to undergo a fundamental shift in values, and work to reduce absolute consumption levels; to increase the efficiency with which a smaller quantity of resources are used in production; and to develop and implement cleaner production technologies.

In one form or another business and industry account for about 80% of world GDP, and if business and industry do not play their part in such action, the efforts of other sectors of society will have little impact. These measures are vital to reducing, and eventually eliminating unsustainable rates of use of renewable and non-renewable natural resource stocks, and unsustainable rates of emissions of pollutants and of generation of pollution and wastes.

The role of business and industry should be to minimise their direct and indirect effects on biodiversity and biogeochemical cycles, and to redirect their resources to the needs of sustainable development. This will require the development of new markets, and abolition of old unsustainable markets, as part of a move to cyclical industrial systems. Many companies may find that major changes in production processes and product design will be necessary to achieve this, and some products which are fundamentally unsustainable in use will need to be phased out.

The role of government is to set in place policies that give confidence to business and industry in making the investment necessary to achieve long-term national goals, as well as targets to which governments are committed in international agreements. Furthermore, the innovation that is essential to future competitiveness should be encouraged by tough but achievable targets, set nationally for standards of performance and environmental quality. Long-term wealth creation and industrial development will not be achieved without sustainable development.

The management tasks are clear. Many are essential right now for company survival and will bring direct 'bottom-line' benefits. These include activities like the environmental review or audit, recycling, energy management, setting an environmental policy, emission control, and waste management.[17] In the longer-term, those companies which look ahead and direct their R and D and capital investment programmes to meet the needs of sustainable development will be the successful companies in a future where environmental excellence and sound economic performance go hand-in-hand.

References

1 Schmidheing, Stephen, with the Business Council for Sustainable Development. *Changing Course*, MIT Press, Cambridge, Massachusetts (1992) *also* World Commission on Environment and Development, *Our Common Future*, Oxford University Press (1987).
2 Meadows, D. H., Meadows, D. L. and Randers, J. *Beyond the Limits: Global Collapse or a Sustainable Future?* Earthscan Publications, London (1992). MacNeill, J., Winsemius, P., Takushiji, T. *Beyond Interdependence: The Meshing of the World's Economy and the Earth's Ecology*, Oxford University Press (1991).
3 Parikh, J., Parikh, K., Gokarn, S., Painuly, J. P., Saha, B., Shukla, V. *Consumption Patterns: The Driving Force of Environmental Stress*, Indira Gandhi Institute of Development Research, Bombay, report prepared for UNCED Secretariat (1991).
4 Daly, Herman. 'Towards some Operational Principles of Sustainable Development', *Ecological Economics*, vol. 2, 1990, pp. 1–6.
5 *The State of the Environment*, OECD, Paris (1991).
6 Roberts, L., 'Counting on science at EPA', *Science*, vol. 249, (1990), pp. 616–618.
7 US Environmental Protection Agency (US EPA). *Unfinished Business: A Comparative Assessment of Environmental Problems* (1987).
8 Centre for Exploitation of Science and Technology. *Industry and the Environment: A Strategic Overview*, CEST, London (1991).
9 Houghton, J. T., Jenkins, G. J., Ephraums, J. J. *Climate Change: The IPCC Assessment*, Cambridge University Press (1990).
10 Leggett, Jeremy. 'Anxieties and Opportunities in Climate Change' in Prins, G. (ed.), *Threats without Enemies – facing Environmental Insecurity*, Earthscan, London (1993).
11 World Energy Council. *Energy for Tomorrow's World* (1993).
12 Advisory Council for Research on Nature and Environment. *The Ecocapacity as a Challenge to Technological Development*, publication RMNO, Netherlands (1992).
13 United Nations Environment Programme. *Environmental Data Report*, Third Edition, Basil Blackwell, Oxford (1991). Tolba, M. K. *Saving Our Planet: Challenges and Hopes*, Chapman and Hall, London (1992).
14 Tapper, R. 'Conserving Nature through Sustainable Development', *Environment Information Bulletin*, no. 21, (1993).
15 Ruitenbeek, H. J., *Economic Analysis of Tropical Forest Conservation Initiatives: Examples from West Africa*, WWF UK, (1990).
16 World Resources Institute, The World Conservation Union, and United Nations Environment Programme, 'Global Biodiversity Strategy' (1992).
17 Advisory Committee on Business and the Environment (1993), 'The Business Case for the Environment', London (1993).
18 Linda Taylor. 'Lessons from Japan – Separating Economic Growth from Energy Demand', Association for the Conservation of Energy, London (1990).
19 Colborn, Theo and Clement, Coralie (eds). *Chemically-Induced Alterations in Sexual and Functional Development: The Wildlife-Human Connection*, Princeton Scientific Publishing Co., Princeton, NJ (1992).

3

The philosophy and approach to modern environmental regulation

The experience of the European Community

STANLEY JOHNSON

Special Adviser on the Environment, Coopers and Lybrand

- The aim of the European Community's environmental policy is to improve the setting and quality of life.
- The EC's Fifth Environmental Action Programme, 'Towards Sustainability', runs from 1993 to 2000. It is the EC's response to Agenda 21, agreed at the UNCED Earth Summit in Rio de Janeiro in 1992.
- The Programme recognises that full economic benefits of the Single European Market are limited by environmental constraints. The whole internal market programme could be jeopardised if tolerance levels of the natural environment are reached.
- The Programme names five target sectors for special attention: industry, energy, transport, agriculture and tourism.
- It emphasises that public and private enterprise must be encouraged to take its environmental responsibilities seriously.
- To achieve the Programme's goals, measures to be introduced include environmental monitoring, scientific research, technology development, integrated sectoral and special planning, integration of environmental costs and risks in decision making, training of the workforce, and legislation.

Introduction

The European Community's environmental policy is over twenty years old. The twentieth anniversary of the adoption of the Community's first action plan for the environment fell on 22 November 1993. In some ways, the months and years following it are likely to present some of the most strenuous challenges for the development of environmental policy within the EEC that have yet had to be faced. The intense and prolonged debate over the ratification of the Maastricht Treaty has contributed to a degree of inertia, if not paralysis, in this as in other fields. Indeed, it is possible not only that further environmental initiatives will be stifled but that established Community policies, where the environment is concerned, will themselves be put into doubt.

What is the Community's environmental record to date?

All of the actions envisaged by the Community in the area of the environment have been defined and described in several 'action programmes', each of which represents the basic reference charter for Community Environment Policy. These programmes, which generally cover a period of four to five years, follow the example of other measures approved by the Council after being proposed by the Commission. The instrument used for the adoption of these programmes is usually a Council 'resolution', which shows the political will of the Member States to apply the measures contained in the programme, but which does not imply any legal obligation to do so. Since 1973, which marks the beginning of Community Environment Policy, five action programmes have been approved by the Council.

The 1973 first action programme (1973–1976)

The first programme adopted on 22 November 1973[1], covered the years 1973–76; it is the most important programme in that it defines for the first time the basic principles and objectives of Community Environment Policy and specifically described the actions to be carried out in the different sectors of the environment.

The objectives of community environment policy

The aim of Community Environment Policy as set out in the first Action Programme, is to improve the setting and quality of life, and the surroundings and living conditions. To this end it should:

- prevent, reduce and as far as possible eliminate pollution and nuisances;
- ensure sound management of and avoid any exploitation of resources or of nature which causes significant damage to the ecological balance;
- guide development in accordance with quality requirements, especially by improving working conditions and the settings of life;
- ensure that more account is taken of environmental aspects on town planning and land use;
- seek common solutions to environment problems with States outside the Community, particularly in international organisations.

The principles of Community Environment Policy

The general principles of Community Environment Policy as defined by the First Action Programme are as follows:

1. The best environment policy consists of preventing the creation of pollution and nuisance at source, rather than subsequently trying to counteract their effects. To this end, technical progress must be conceived and devised so as to take into account the concern for protection of the environment and for the improvement of the quality of life at the lowest cost to the Community. This environment policy can and must be compatible with economic and social development.
2. Effects on the environment should be taken into account at the earliest possible stage in all the technical planning and decision-making processes.
3. Any exploitation of natural resources or of nature which causes significant damage to the ecological balance must be avoided. The natural environment has only limited resources; it can only absorb pollution and neutralise its harmful effects to a limited extent. It represents an asset which can be used, not abused.
4. The standard of scientific and technological knowledge in the Community should be improved with a view to taking effective action to conserve and improve the environment and to combat pollution and nuisances. Research in this field should therefore be encouraged.
5. The cost of preventing and eliminating nuisances must in principle be borne by the polluter. However, there may be certain exceptions and special arrangements, in particular for transitional periods, provided they cause no significant distortion to international trade and investment.

6 In accordance with the Declaration on the Environment of the United Nations conference in Stockholm in 1972, care should be taken to ensure that activities carried out in one state do not cause any degradation of the environment in another state.

7 The Community and its Member States must take into account in their environment policy the interests of the developing countries, and must in particular examine any repercussions of the measures contemplated under that policy on the economic development of such countries and on trade with them.

8 The effectiveness of effort aimed at promoting a global environmental policy will be increased by a clearly defined long-term concept of a European environmental policy. In this respect the Community and the Member States must make their voices heard in the international organisations dealing with aspects of the environment and must make an original contribution in these organisations, with the authority which a common point of view confers on them.

9 The protection of the environment being a matter for all, public opinion should be made aware of its importance. The success of an environment policy pre-supposes that all categories of the population and all the social forces of the Community help to protect and improve the environment. This means that at all levels continuous and detailed educational activity should take place.

10 In each different category of pollution, it is necessary to establish the level of action (local, regional, national, Community, international) that befits the type of pollution, and the geographical zone to be protected should be sought.

11 On the basis of a common long-term concept, national programmes in these fields of the environment should be co-ordinated, and national policies should be harmonised within the Community. Such co-ordination and harmonisation should be achieved without hampering potential or actual progress at the national level. However, the latter should be carried out in such a way as not to jeopardise the satisfactory operation of the Common Market.

General description of the actions to be undertaken under the First Environmental Action Programme

The First Environmental Action Programme provided for the following three categories of action:

Action to reduce pollution and nuisances
Action to improve the quality of the environment
Action in international organisations

During the period of the first programme, the EC Council adopted legislation providing for air and water quality objectives, for emission standards in respect of certain hazardous processes and for product standards, particularly where different approaches to the requirements of environmental protection in the member states could lead to barriers to trade. The First directives on the disposal of waste and the prevention of noise pollution were adopted.

On 17 May 1977 the Council adopted in the form of a 'Resolution' the Second Environmental Action Programme of the European Communities for the years 1977–1981.[2] This programme basically represented a continuation and expansion of the actions undertaken within the framework of the First Programme of 1973. It reaffirmed in full the general principles and objectives of the 1973 programme. As far as the description and the timetable of the actions to be undertaken is concerned, it gave a certain priority to antipollution measures in the areas of water and air. It also provided for wider and more specific measures in the area of noise pollution. Finally, it reinforced the preventive nature of Community Environment Policy and gave particular attention to the rational protection and management of space, the surrounding environment and natural resources.

The Third Environmental Action Programme for the years 1982–1986 was adopted by a Resolution of the Council on 7 February 1983.[3] Under this Resolution for the first time a certain number of priority areas were established by the Council. These priorities were the following:

- integration of the environmental dimension into other Community policies;
- environmental impact assessment procedure;
- reduction of pollution and nuisance if possible at source, in the context of an approach to prevent the transfer of pollution from one part of the environment to another, in the following three areas: atmospheric pollution (especially by NOx, heavy metals and SO_2); freshwater and marine pollution; pollution of soil.
- environmental protection in the Mediterranean region;
- noise pollution and particularly noise pollution caused by means of transport;
- trans-frontier pollution;
- dangerous chemical substances and preparations;
- waste, and in particular toxic and dangerous waste, including trans-frontier transport of such waste;
- encouraging the development of clean technology, e.g., by improving the exchange of information between Member States;

- protection of areas of importance to the Community which are particularly sensitive environmentally;
- co-operation with developing countries on environmental matters.

The Fourth Programme developed and specified principles which already appear in the 1983 Third Action Programme, notably the need to integrate environmental protection into the other Community policies (employment; agriculture; transport; development, etc), or the need to intensify the global or integrated fight against pollution, in order to avoid the transfer of pollution from one area of the environment to another (air; water; soil). Finally, the Fourth Action Programme provided for a number of initiatives in new areas, in particular in the sectors of biotechnology and the management of natural resources; in the areas of, on the one hand, soil protection, and on the other, the protection of urban, coastal and mountain zones.

To what extent did the first four EC environmental action programmes succeed in meeting their objectives?

At the end of the eighties, looking back at the record, it was possible to argue fairly convincingly that environmental policy had been one of the success stories of the Community. As far as the passage of legislation was concerned, the record was remarkable. Though a formal legal basis for EC environmental policy was not enshrined in the Treaty of Rome until the amendments introduced by the Single European Act came into force on 1 July 1987, the raft of environmental measures on the EC statute book at that date was already substantial. EC environmental policy, driven by the Commission, had, in a very real sense, been the motive force behind the environmental initiatives of many of the member states. Though some Commission proposals were aimed at 'harmonising' different national approaches (i.e., action of some sort would have been taken at Member State level even without Brussels), in other areas the Commission certainly set the pace and, working together with the European Parliament, drove through legislation which would probably never have been enacted at Member State level without the impetus that the EC provided. At a rough guess, one could say that well over 50 per cent of recent environmental legislation had been Brussels-driven.

The record on implementation was patchy. In 1986 for example complaints and infringements detected by the Commission's own enquiries came to 192 and several of these were pursued as far as the European Court of Justice. In the Fourth Action Programme, the Commission was requested to give particular priority to the monitoring of environmental legislation.

Towards sustainability – the 1993 Fifth Environmental Action Programme

Towards the end of December 1992 the Council adopted the Fifth of its Environmental Action Programmes designed to protect and enhance the quality of the environment in the European Community. Due to run from 1993–2000 the Fifth Programme, entitled 'Towards Sustainability',[4] is a departure from the four previous programmes: the adoption of the Single European Act has for the first time given the Community a constitutional mandate to take environmental protection measures. The previous programmes were very much reactive, 'end of the pipe' strategies which responded to environmental problems after they had occurred. But the Fifth Action Programme is much more proactive, based on the thesis of 'sustainable development' as put forward by the 1987 Brundtland report,[5] whereby the root causes of environmental degradation are addressed before the problems become so pressing that they can no longer be ignored. Furthermore, the programme recognises that the full economic benefits of the single European market, in terms of continued growth and efficiency, are limited by environmental constraints, and that the whole internal market programme could be jeopardised if the tolerance levels of the natural environment are breached.

The main strategy behind the Action Programme is based on the realisation that environmental damage will never be halted, let alone prevented if behavioural patterns of producers and consumer, governments and citizens are not altered to take the environment more into account.

The Fifth Programme is to be seen as a Community response to Agenda 21, the action plan for integrating environment and development which was agreed at the United Nations Conference on Environment and Development (UNCED), held in Rio de Janeiro, in June 1992.

A new strategy for the environment and sustainable development

THE ACTORS

The main goal of the Fifth Action Programme is raising public awareness about the problems facing the environment, and changing people's attitudes and behaviour to become more environmentally-friendly. Three main groups of actors are identified:

- Public Authorities (central and local government) have a crucial role to play not only in enforcing legislation but also in making planning decisions, and informing and educating the public.

- As huge consumers of materials and producers of waste, Public and Private Enterprise must be both encouraged and obliged to take their responsibilities towards the environment seriously.
- The General Public, as voters, producers of pollution and consumers of goods and services, can exert a great influence on the future quality of the environment.

Selected target sectors

INDUSTRY

Given that industry is responsible for a significant proportion of environmental damage, the new strategy recognises that it is essential to work together with industry to achieve a solution, rather than simply relying on prescriptive measures. The European Community will encourage industry (including small medium-sized enterprises) to recognise its responsibility towards the environment by promoting:

- improved resource management;
- consumer confidence in environmental quality and choice;
- Community standards for production processes and products.

Achieving state-of-the art production methods which improve environmental protection should also help give Community industry a competitive edge in world markets.

ENERGY

The aim of the proposed energy strategy is to decrease demand and thus reduce emissions of CO_2 (which contribute significantly to the greenhouse effect and reduce emissions of other pollutants such as SO_2 and NOx which cause acid rain and its resultant problems). Improving energy efficiency is a key component of this strategy, as is producing more energy from renewable sources. Fiscal incentives (and disincentives) as well as research and development programmes should help to achieve this goal.

TRANSPORT

Transport is never environmentally neutral: the energy it uses gives off polluting emissions; transport infrastructure has a great impact both within urban areas and in the countryside; and noise is also a significant side effect. Unless measures are taken to curb demand for transport, which is expected to increase greatly in the single European market, the resulting environmental degradation will not be checked. The Commission has proposed a strategy in its green paper on the impact of transport on the environment, entitled 'Sustainable Mobility',[6] which involves influencing consumer behaviour to decrease demand, and transfer from the private car to public transport as well as options using fiscal measures such as road pricing, higher fuel prices, greater investment in public transport to make it more attractive, and information campaigns. Other measures proposed include encouraging more rail and waterways transport (for freight, too); stringent environment impact assessments on new infrastructure plans; and research and development to improve fuel efficiency and produce cleaner fuels.

AGRICULTURE

The agricultural sector has undergone drastic changes in the last few decades, many of which have been as a direct result of the Common Agricultural Policy, and many of which have had a negative impact on the environment, for example the demands for ever higher yields. This has led to soil and water pollution problems; loss of top soil; an increase in plant and animal diseases; drainage of wetlands; and clearance of forests, to name but a few of the problems. The strategy now put forward aims to minimise the impact of agriculture on the environment by encouraging farmers to see themselves as guardians of the countryside. Objectives include:

- reducing pollution from nitrates, phosphates, pesticides and livestock units by strict controls;
- financial incentives to encourage farmers to farm in an 'environmentally friendly' way, complemented by training measures; and
- new afforestation projects (mixed forests), action to prevent forest fires.

TOURISM

Tourism has great, and growing, significance in the economy of the European Community. With greater affluence, more and more people are taking holidays further afield, and taking more holidays each year. Certain areas of the European Community, most notably the Mediterranean coast and the Alpine zones, have been disproportionately affected by this increase in tourism and the attendant environmental impacts, such as destruction of natural habitats through hotel developments and transport infrastructure; greatly increased use of water, often in areas where water is already in short supply, and increased output of waste water and sewage. The strategy put forward includes:

- better management of mass tourism, with stricter controls for new developments and strict implementation of environmental standards in existing tourist areas;

- influencing tourist behaviours – staggering holiday periods, discouraging private car use, greater diversity of holiday options and general awareness-raising by promoting codes of conduct; and
- promotion of 'environmentally-friendly' forms of tourism, and training of tourism operators to be aware of the impact of tourism on the environment.

Apart from the target sectors tackled above, 'Towards Sustainability' also seeks to address various other issues of significance. Indeed, it stressed that designation of 'main' target areas does not mean that action in other areas is unnecessary. Environmental protection is essential in every aspect of European Community life: a non-exhaustive list of other themes and targets is also in the programme.

These include:

- **Climate change** – setting targets for reducing the emission of greenhouse gases such as carbon dioxide, chlorofluorocarbons, halons, nitrous oxide and methane.
- **Acidification and air quality** – setting targets for reducing the emission of sulphurdioxides, nitrogenoxides and volatile hydrocarbons which cause the phenomenon of acid rain with all its attendant potential for damaging forests, crops and human health. Based on the principle of 'critical loads' it is possible to indicate the levels of the individual pollutants which different types of ecosystem can bear – the aim is to decrease emissions to within the critical load amount. Emissions of other hazardous pollutants such as dioxin and heavy metals are also to be reduced significantly.
- **Protection of nature and biodiversity** – protecting flora, fauna and their habitats is essential to preserve the ecological balance and to maintain the irreplaceable genetic bank upon which scientific progress depends. A Natura 2000 network is to be created to protect habitats of particular ecological significance in the European Community, while other species and habitats will be maintained, and where necessary, restored. The trade in wild species of both flora and fauna is to be strictly controlled.
- **Management of water resources** – prevention of pollution of all water sources and ensuring that water demand and water supply are brought into equilibrium by more rational use and management of water resources.
- **Urban environment** – the strategy for improving the quality of the urban environment for the 80% of the Community's population which lives in urban areas is based on the Commission's green paper on the urban environment.[7] This green paper suggests such measures as

improvements in town and country land-use planning; better management of industrial growth, energy consumption, and waste production; rationalisation of urban transport requirements; and proper recognition of the importance of preserving the urban heritage. As achievement of the goals set out for the five target areas will also benefit the urban environment, no specific goals are set for this area – except with regard to achieving noise abatement, which is seen as a primary objective, given the negative impact this can have on public health.

- **Coastal zones** – minimisation of the pressures to which coastal zones are particularly prone, namely the demands of increased development and tourism, by means of a framework strategy for integrated management of coastal zones.
- **Waste management** – reduction in the escalating production of waste by means of clean technologies, eco-labelling, and behavioural changes of both producers and consumers. Encouragement of re-use and recycling, by means of easy access to recycling facilities, separate collection of different materials and by increasing demand for recycled materials. Safe disposal of dangerous substances is also to be a priority.

Management of risk and accidents

An important part of the Fifth Environmental Action Programme is the prevention of environmental risks, and, if accidents nevertheless occur, ensuring a prompt and efficient response to minimise the after-effects. The strategy identifies four main areas of concern:

- **Industrial accidents and hazards.**
- **Chemicals** – the number of different chemicals produced has soared, and about 100,000 are currently in use in manufacturing industry. Most of these chemical agents can be hazardous if incorrectly applied or released in large quantities and their effects not known and/or long-lasting. The Commission is in the process of dealing with existing chemicals, assessing their relative risk potential and targeting 50 particular chemicals for comprehensive risk reduction programmes.
- **Biotechnology** – it is still not always known what effect certain genetically modified organisms could have upon the environment in the long term and it is therefore necessary to apply a standardised set of environmental risk assessment requirements and safety measures as regards these biotechnology products.
- **Product labelling** – to identify dangerous substances quickly in the event of an emergency. Part of this strategy also involves a commitment to reduce by 50% the number of vertebrate animals used for experimentation purposes.

NUCLEAR SAFETY AND RADIATION PROTECTION

After the Chernobyl nuclear reactor accident and given the growing concern about the poor safety levels of many nuclear power stations in Central and Eastern Europe, the European Community feels it must show leadership in issues to do with nuclear safety. Nuclear energy is a significant source of energy in the European Community and this importance is unlikely to diminish in the short term at least. The Commission is therefore calling for an upgrading of safety measures (with particular respect to the countries of Central and Eastern Europe), greater verification of monitoring procedures, and strategic management of radioactive waste. Public information campaigns are also seen as vital, giving information on radiation protection measures in the event of an accident and providing courses on protection.

CIVIL PROTECTION AND ENVIRONMENTAL EMERGENCIES

Apart from nuclear accidents, other environmental disasters such as forest fires and oil spills at sea can have catastrophic consequences. The European Community has to ensure that it is in a position to respond quickly and effectively to such emergencies. The Commission is therefore proposing the establishment of European task forces to deal with the different types of accidents, providing better training to personnel who would have to deal with the after-effects and generally facilitating better communication mechanisms between the key actors involved in emergency situations.

BROADENING THE RANGE OF INSTRUMENTS

As already stated the Commission recognises that environmental protection cannot rely on legislative measures alone, although these will continue to be a major weapon in the Community's green arsenal. Encouraging changes in public attitudes and behaviour will also play an important role in environmental protection, as will forming new relationships with industry to work together to ensure economic development which is truly sustainable. But a wide range of 'instruments' will also be necessary if genuine sustainability is really to be achieved. In the fifth Environmental Action Programme the Commission identifies seven instruments other than legislation:

1 **Higher quality environmental data** in greater quantities, gathered and interpreted in a standardised manner by designated bodies in each Member State so that true pan-European comparisons are possible. The recent setting up of the long-awaited European Environment Agency will be essential for this process.

2 **Scientific research and technological development** are crucial for identifying cause and effect in environmental problems and for finding solutions to those problems; for developing the so-called clean manufacturing technologies.

3 **Integrated sectoral and spatial planning**, achieving the right mix of industry, employment, habitation and leisure facilities will make a great contribution to achieving the potential for sustainable development in a locality, region, country and in the European Community as a whole. Integrating environment impact assessment into the planning process at all levels and into all aspects of Community policy formation is an important part of this.

4 **Integrating the true environmental costs and risks** into all economic activity will help make decision-makers aware of the implications of their policy decision, although obviously the price of many environmental assets can never be determined. However, merely the attempt to set a cost can show just how priceless, literally, some environmental goods are. The use of economic and fiscal incentives, (such as eco-taxes and levies, state subsidies and tax concessions, as well as tradeable permits as regards pollution emission limits) can also provide the necessary incentive to encourage producers and consumers to choose the more 'environmentally friendly' option. Environmental audits can also be used as performance indicators which will reassure workers, shareholders and the general public that industry is making its contribution by the concept of environmental liability/ shared responsibility, although this latter should be seen very much as a last resort, for punishing those polluters who have failed to ensure proper prevention measures are in place.

5 **Informing the general public**. The strategy for achieving sustainable development can only be really successful if the public can be persuaded that there is no alternative to the action proposed. Therefore the public must be informed about the issues and the means for protecting the environment, and, crucially, they must be involved in the process.

6 **Informing the workforce.** Not only the general public needs to be informed, but the workforce in both the private and public sectors must be *trained to think in environmental terms*, and more specially trained workers in certain fields (e.g., biologists) will be needed.

7 **Funds.** Last but certainly not least, some funding will be required to help ease the transition to sustainable development. The strategy as a whole should pay for itself, in terms of costs saved, damage avoided and so on. Nevertheless, the Community may wish to fund particular pilot projects and promote models of production to demonstrate what is possible. The LIFE financial instrument will be the means to this end, while the new cohesion fund agreed at the Maastricht summit in

December 1991 will help the less-developed regions of the Community achieve their development goals while ensuring environmental protection. Projects financed by other Community funds, such as the structural funds and the European investment Bank will have to fulfil environmental criteria before being given the go-ahead.

SUBSIDIARY AND SHARED RESPONSIBILITY – IMPLEMENTATION AND ENFORCEMENT

The Commission is well aware of the importance of subsidiarity – that is the carrying out of an action at the most appropriate level of government, be that European, national or local. But in the context of the environment the concept of shared responsibility is seen as having more relevance. All sectors and levels of society should work together to protect the environment: their actions should be complementary and as all are working towards the same goal, no conflicts of areas of responsibility should develop.

The wider international arena

In Part Two of 'Towards Sustainability' the Commission looks at the European Community's in the wider international arena:

- it is a major contributor to global issues such as climate change and deforestation and must therefore be instrumental in addressing these issues;
- it has responsibilities towards the developing countries and must show that it is not just preaching at those countries but is willing to change lifestyles and cut consumption levels. Partnership with the developing countries to achieve sustainable development at a global level is essential;
- the newly democratised countries of Central and Eastern Europe are understandably looking to the Community for assistance in clearing up their many environmental disasters;
- the 1992 United Nations Conference on Environment and Development in Rio de Janeiro and the follow-up work are providing the Community with the opportunity of showing the rest of the world just how committed Europe is to achieving sustainable development.

Looking to the future

The EC Fifth Environment Action Programme was formally adopted by the Council in May 1993 (OJ C138, vol 36, 17 May 1993). As before,

Environment Ministers agreed on the accompanying Resolution. In that Resolution it is possible to detect the fall-out from the Maastricht debate. Whereas in the past, as noted above, EC environmental policy had been to a considerable extent Brussels-driven, the EC Council in adopting the Fifth Environment Programme referred specifically to the agreement on subsidiarity reached at the EC Edinburgh summit in December 1992 and called on the Commission to 'ensure that all proposals it makes relating to the environment fully reflect that principle'. The Council warns that Environment Ministers themselves will scrutinise all Commission proposals for consistency with it, Admittedly, the Resolution goes on to pledge that application of the subsidiarity principle 'will not lead to a step backwards in Community policy or hinder its effective development in the future'.

At the time of writing, the future of EC environmental policy looks uncertain. Under the new post-Maastricht regime, it is not clear that major new initiatives (such as those which led in the early nineties to the adoption of the EC Habitats Directive) will again see the light of day. The likelihood is that they will be stifled at birth on first discussion in the Council, or even more probably, smothered in the womb by the Commission. Other major proposals which are at the moment on the table of the Council, such as the proposed EC carbon/energy tax, may be sidetracked for political (including 'subsidiarity') as well as economic reasons. We may even find that a nervous, punch-drunk Commission, begins increasingly to shirk its primary duty under the Treaty: namely to ensure the enforcement on an equal basis among all the Member States of EC environmental measures. The cop-out may join the opt-out as a tool of policy. The Fifth Action Programme would remain very much a list of pious hopes and good intentions

The alternative scenario is, of course, a more hopeful one. We could imagine that with the Maastricht Treaty finally ratified by all Member States and a new Commission in place in Brussels, EC environmental policy gains a second and third wind and enters its third and fourth decades with restored energies and vigour.

References

1 OJ No C 112 of 20/12/73.
2 OJ No C 139 of 13/6/77.
3 OJ No C 46 of 17/2/83.
4 COM (92) 23 Final.
5 1987 Report of the World Commission on Environment and Development.
6 COM (92) 46.
7 COM (90) 218.

US regulatory approaches

The implications for Europe

BRADFORD S. GENTRY

Partner in charge,
Land Use and Environmental Law Group for Europe,
Morrison and Foerster, London

- Current priorities for environmental protection in the US are the clean-up of contaminated land and cost recovery from those who are liable; also manufacturers' compliance with environmental regulations. Environmental problems caused by the sale, distribution and recycling or disposal of products is of particular concern in some states.
- These priorities reflect public perception of the most important environmental risks, its lack of trust in the ability of governments to address environmental issues adequately; and belief that 'polluters' should pay for pollution control.
- Approximately US$1.40 billion is spent each year in the US on pollution control. This represents 2% of the GNP.
- Greater attention is being paid to ways of reducing the environmental impact of products, but huge costs are involved.
- US businesses are required to comply with an extensive set of environmental self-monitoring and reporting requirements.
- In terms of company planning, understanding the US experience in preparing for and responding to such situations can be invaluable as Europe moves towards more formal enforcement actions and expanded liability for environmental damage.
- European experience with environmental labelling and other initiatives is a harbinger of things to come in the US. Management programmes designed to address these issues in Europe may provide a basis for planning for, and responding to, initiatives as they develop in the US.

Introduction

This chapter outlines the differences that currently exist between the regulatory approaches to environmental issues in Europe and the US. It begins with a discussion of the different priorities assigned to environmental issues in Europe and the US. It then considers the specific techniques which are used in the US to control or respond to environmental problems, and examines the governmental structure under which the US laws are enacted and implemented. In particular, it highlights the problems created by this structure.

Finally, the major implications of these factors for efforts by European businesses to manage the environmental risks facing their product sales or manufacturing operations in the US and elsewhere, are presented.

DIFFERENCES IN PRIORITIES BETWEEN THE US AND EUROPE

At first glance, there are many similarities between the European and US approaches to environmental regulation. The same basic environmental issues are faced: pollution of air, water and land, among others. Many of the same general regulatory tools are applied, including authorizations or consents, emission or discharge limits, enforcement where breaches of these requirements occur, and liability for damage caused by pollution.

This surface similarity, however, hides major differences in the priority accorded to different environmental problems and different techniques for controlling pollution, as well as the very different societal and historical context in which these efforts occur.

In the US, the current priorities for environmental protection are:

1 The clean-up of contaminated land and the recovery of clean-up costs from those legally liable for the costs. This is by far the greatest concern to the public, to governments and businesses. Businesses are particularly concerned about the possibility of unknowingly acquiring liabilities running into millions of dollars with their US operations.

2 Compliance of ongoing manufacturing operations with environmental laws and regulations is the second priority, given the aggressiveness of US enforcement efforts (see below) and the ever-increasing costs of meeting new operating requirements (such as those under the recent Clean Air Act Amendments).

3 Attempts to address the environmental problems caused by the sale, distribution and recycling or disposal of products, so familiar in Europe, is a distant third on the US federal scene (although individual states are very active).

Not surprisingly, the environmental management programmes adopted by US businesses reflect these priorities. As such, they are overwhelmingly geared to the issues of contaminated land and operating compliance.

These priorities are, in large part, a reflection of the following attitudes of the voting public:

- perceptions of the most important environmental risks, even where those perceptions are inconsistent with the risk weightings by the government's environmental professionals;
- lack of trust in the ability of the government to adequately address environmental issues; and
- belief that other parties, notably 'polluters', can, do, and should pay for pollution control efforts (underscoring aversions to new environmental taxes and, hence, greater attraction to tradeable emission rights).

Clearly there are signs, however, that the US's historical environmental priorities are shifting. Greater attention is being paid to preventing environmental problems from arising, and to ways of reducing the environmental impacts of products. In addition, and as I will discuss later, as a result of the huge costs associated with the US approach to environmental protection, more and more questions are being raised about whether there might be more efficient ways of applying society's resources in order to address the most important environmental risks.

Techniques used in the US to control or respond to environmental problems

1 LITIGATION

The first thing anyone outside of the US thinks about when they consider the American legal system is the huge amount of litigation and the shockingly high awards made by juries.

In many ways, however, these classic personal injury horror stories are only the tip of the iceberg in describing the importance of litigation to the functioning of the American environmental regulatory system.

In this context, I am using the word 'litigation' to encompass a wide range of formal proceedings, hence formal relationships, between people involved in many aspects of US pollution control efforts. For example, litigation plays a major part in the:

i Development of US environmental legislation, through formal hearings in Congress, formal administrative proceedings to develop new regulations, and a multitude of lawsuits by industry, environmental groups and other interested persons arguing that this or that piece of legislation or regulation should be ruled unconstitutional or otherwise illegal.

ii Enforcement of environmental requirements, where, should a breach be identified, the first step at virtually any level is the filing of one or more formal actions to remedy the violation and impose penalties.

iii Recovery of damages for a wide range of personal injuries, damage to private property, damage to the environment, or contributions to clean-up costs from legally responsible parties.

Why this emphasis on litigation? It is certainly tempting to say that it is because there are so many lawyers. But, while this certainly is a contributing factor, there are a number of more fundamental reasons why litigation is especially prevalent in the environmental arena:

i Since the government is always under pressure to prove that it is being tough on 'polluters', the filing of formal enforcement actions provides a quick and easy method for 'bean counting' and numerical proof to Congress, the media and the public that action is being taken. (See press release from the Department of Justice in the Appendix to this chapter.)

ii The threat of lawsuits is a major influence in the government's decision-making process, whether it be on new regulations, other administrative decisions or even new legislation.

iii Given the huge costs involved in many US environmental problems, there is a strong economic incentive to spread the cost as widely as possible by suing as many other parties as possible.

The results of this reliance on litigation are of tremendous concern to European businesses. In addition to the fact of litigation and the associated problems in particular cases (such as obtaining insurance coverage), it also leads to:

- extremely formal interactions between the government and companies, companies and environmental groups, and all of the parties potentially at odds over environmental matters because all sides always have at the back of their minds the idea that litigation might be just around the corner, therefore what they say today may be used against them in court tomorrow;
- limited flexibility in relations among the interested parties and limited opportunities for developing sensible regulatory and enforcement policies; and
- an intense focus on the process of environmental regulation, but not necessarily on the substance of the environmental ends to be or actually being achieved.

This gap between process and substance can be extremely difficult for non-Americans or even US engineers to accept, but it is a fact of life which must be recognized and accommodated when considering how best to manage US environmental issues.

2 INFORMATION

In the US, there is a strong presumption that any information held by the government should be freely available to the public. During litigation, there are also broad rights to compel the other side to provide you with information in their possession which may be helpful to your case.

In addition, an extensive set of environmental self-monitoring and reporting requirements are imposed upon US businesses. Such requirements include:

- monitoring of air emissions and waste water discharges for compliance with permit limits and the regular reporting of monitoring results to the government;
- reporting of any accidental spills of hazardous materials, as well as any information which comes to one's attention concerning the presence of historical contamination; and
- reporting on regular emissions of several hundred 'toxic' chemicals to employees, neighbours and governmental agencies under the Toxic Release Inventory or 'TRI'.

As a result, huge amounts of information on environmental matters are generated by the government and industry, virtually all of which is freely available to the media, competitors and the public.

This can place great pressures on companies which are used to operating in a system where less information is readily available to the public. It leads to a situation where a company must assume that virtually any document it prepares may, eventually, find its way into the public domain and, as such, plan accordingly. Such planning can include questions as to whether one actually needs to create the documents, whether they should be developed under the protection of attorney-client privilege, and an imperative for follow-up on any problems identified.

3 PUBLIC INVOLVEMENT

The US system of environmental regulation is also characterized by extensive involvement of the public, including individuals, local pressure groups and national/international environmental organizations.

The public is given broad rights of access and involvement in the:

i Development of environmental laws, through formal and informal access to legislative proceedings, regulatory proceedings and court challenges thereto.

ii Enforcement, as virtually all federal environmental laws expressly authorize citizens to bring actions directly against parties which are in breach of the laws (so-called 'citizen suits').

iii Actions to prohibit the continuation of a polluting activity, requirement that restorative work be done, or the obtaining of compensation for moneys expended.

This broad involvement of the public builds on the scope of the information which is freely available, as well as on their access to the courts. For example, the self-monitoring reports which must be submitted to the government by industrial dischargers:

- are freely available to any member of the public who wants them; and,
- under the citizens' suit provisions of the federal Clean Water Act, can serve as the basis for an automatic finding of liability and the award of the legal fees for the cost of bringing the action.

As a result, it may be said that the US has over 200 million environmental enforcement officials. Environmental groups are a powerful influence on the development of environmental legislation at all levels of government, as well as on local issues of facility siting and enforcement.

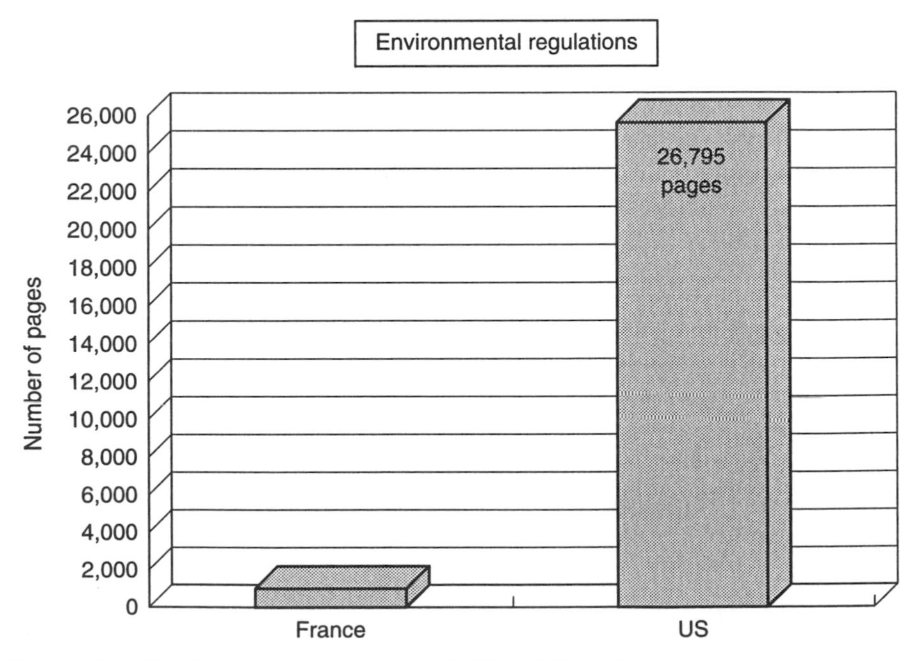

Figure 4.1 Chart comparing scope of USA and French environmental regulations

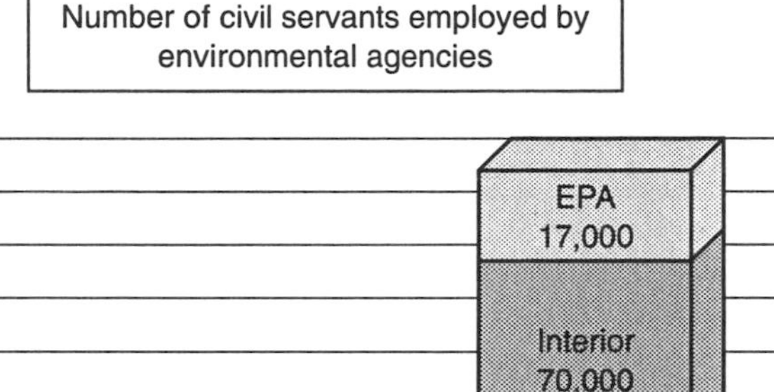

Figure 4.2 Chart comparing number of USA and French civil servants employed by environmental agencies

The government structure and its impact on the way in which US Environmental Laws are enacted

In addition to a predisposition towards litigation, the free accessibility of information and the potential for 'citizen suits', there is yet another complication. It is that there are opportunities for many different governmental agencies in the US to be involved in the same environmental issue.

It is true that the US has a strong federal system. Under the US Constitution, the federal government has direct enforcement authority for its environmental and other programmes. Just to make the point – although possibly an unfair comparison – while the EC's environmental directorate generate has approximately 300 employees, the US Environmental Protection Agency (EPA) has approximately 17,000. (Figs. 4.1, 2 and 3 are charts which compare the scope of the US and French environmental programmes.)

However, this is not the complete picture. Those federal EPA employees are mirrored by environmental officials and requirements at the state and local levels. As a result, multiple and overlapping jurisdictions and responsibilities are still a major issue in US environmental programmes. Even where the federal government has legislated in a particular area, most federal environmental laws leave the states free to adopt stricter standards. Where the federal government has not acted, or has not completely filled the area, states have always played an important role as 'laboratories' for new legislative provisions and enforcement tools.

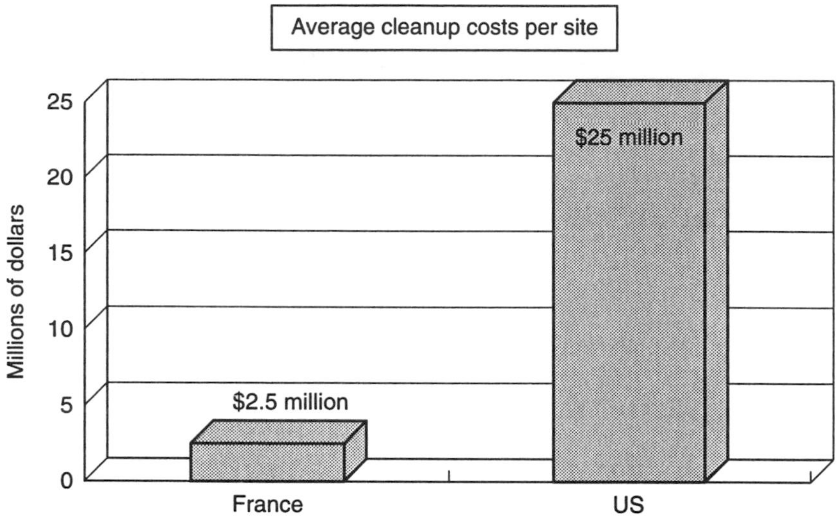

Figure 4.3 Chart comparing USA and French average clean-up costs per site

This has been particularly true in the environmental area during the Republican presidencies of the 1980s. Finally, there are many opportunities for independent enforcement actions to be brought by different public authorities under these different laws.

These overlapping jurisdictions create clear problems for US manufacturing facilities during permitting, enforcement and clean-up activities.

They are also a growing problem for the distribution of products throughout the US. This is because individual states, rather than the federal government, are taking the lead in developing controls over packaging, used product disposal, recycling, environmental advertising and related matters.

While many Europeans may think that the situation on product regulation in the EC is difficult, at least the EC institutions recognize the importance of harmonization in a single market.

The allocation of environmental clean-up costs in the US

Historically, the cost of addressing environmental problems in the US has not been a major concern to either the public or the government. For example, a *New York Times* opinion poll conducted in June 1989 found the 80% of those polled agreed that, when it came to protecting the environment, 'requirements and standards cannot be too high, and continuing environmental impovements must be made regardless of cost'. (*New York Times*, 2 July 1989, pp. 1, 18.)

This has been true both of the substance of US environmental laws, and of the process for making and enforcing them. With respect to the substance, Congress has responded to public concern about environmental risks by adopting tighter and tighter environmental standards, and requiring EPA to implement more and more detailed legislation on tighter and tighter timeframes. The general view has been that 'polluters' should and do bear the costs of these more stringent requirements.

The same applies to the process for making and enforcing the laws. The government has a clear incentive to prove to the public that it is taking action against 'polluters'. Citizen groups have been utilizing their broad procedural rights aggressively to promote their environmental goals. Again, the belief appears to be that only 'polluters' (read *industry* not *individuals*) bear the costs.

Not surprisingly, the result has been that huge amounts of money are spent in the US to address environmental problems. For example:

- Approximately $140 billion is spent each year in the US for pollution control, a figure which represents 2% of the gross national product and which continues to rise.
- In 1992, the US federal government recovered over $2 billion in civil and criminal penalties, environmental damages and clean-up costs, in addition to the amounts recovered by state and local enforcement agencies.
- Up to $700 billion is estimated to be spent cleaning up all of the different categories of contaminated sites in the US.
- Of the money which has been spent on contaminated sites to date, approximately 70% has gone to litigation and other transaction costs and not to clean-up.

These are huge amounts of money for US society. While most US citizens agree that environmental protection is and should remain an extremely high priority, many are also asking themselves whether the US has the mix right. Should priorities be changed, with less money going to the clean-up of contaminated sites and more to address indoor air pollution, for example? Can the money which is spent be spent more wisely by reducing transaction costs or using the market to allocate the responsibility for reducing pollution (for example, through a system of tradeable emission rights)?

How this debate will turn out, given the entrenched public perceptions of important environmental risks and positions taken by many in the environmental community, remains to be seen. For Europeans with major operations in the US, it does open a door for bringing into the States examples of the different priorities and techniques used in Europe to respond to the same basic environmental issues.

Implications for European business

Obviously, given all of these factors and costs, environmental issues pose a major risk to European companies operating in the US market.

At the risk of over-generalizing, but from the point of view of company planning, it seems fair to characterize the US approach as:

- confrontational;
- process oriented;
- compliance and liability focused;
- fragmented, both in terms of overlapping governmental responsibilities, and for requirements affecting products; and
- extremely costly.

In order to begin to manage these issues, European businesses need to understand these differences, particularly the traps for the unwary which exist. For example, many foreign investors were caught completely unaware by the liability which faces current, 'innocent' owners of sites which were contaminated in the past.

Understanding the differences then allows the company to plan its response, which can vary from staying out of all or portions of the US market, to adopting state-of-the-art, US-based environmental management systems.

In doing so, however, European companies are increasingly being presented with an opportunity to learn from the different approaches being taken by Europe and the US, and to build upon those experiences. For example, as Europe moves increasingly toward more formal enforcement actions and expanded liability for environmental damage, US experience in anticipating, preparing for and responding to such situations can be invaluable.

At the same time, the greater European experience with environmental labelling, used-product take-back schemes and related product initiatives is increasingly looked to as a harbinger of things to come in the US. As a result, management programmes designed to address these issues in Europe may well provide you with a basis for anticipating, planning for, and responding to such initiatives as they continue to develop in the States.

Appendix: Press release from the US Department of Justice

FOR IMMEDIATE RELEASE ENR
THURSDAY, OCTOBER 29, 1992 (202) 514–2007
TDD (202) 514–1888

DEPARTMENT OF JUSTICE ANNOUNCES
RECORD $2 BILLION YEAR FOR ENVIRONMENTAL
ENFORCEMENT

WASHINGTON, D.C. The Department of Justice announced today that in fiscal 1992 it achieved a series of record successes in enforcing the nation's environmental laws, including the recovery of more than *two billion dollars* in monetary payments.

The Department's two billion dollar recovery, a new annual record, comes from criminal penalties, including fines and restitution; civil penalties; Superfund cost recoveries and court-ordered hazardous waste cleanups; and natural resource damages. Combined with the three preceding 'billion-dollar years,' the 1992 results brought the Department's environmental enforcement record to over five billion dollars won during the past four years.

Several other records were set in both criminal and civil enforcement.

With respect to criminal enforcement, Acting Assistant Attorney General Vicki A O'Meara of the Department's Environment and Natural Resources Division said: 'Over half of all the indictments and convictions in the entire history of the program – and 94 per cent of the fines and penalties, and 69 per cent of the actual prison time to be served for environmental crimes – have come during the last four years. By any measure, the Department's commitment to tough enforcement of our environmental laws has been demonstrated by our record. These extraordinary results are a tribute to the hard work by our staff attorneys and United States Attorneys, as well as the Environmental Protection Agency (EPA) and the Federal Bureau of Investigation (FBI), which are largely responsible for investigating and referring these cases to us.'

Of civil enforcement efforts, O'Meara said, 'The development of strategic enforcement efforts has helped us achieve these results. Among the areas we focused on last year were recalcitrant environmental violators, illegal transportation of hazardous wastes to and from Mexico, enforcement under the Clean Air Act regarding the chemical benzene, disposal of primary metals and industrial chemicals, and industrial waste pretreatment plants.'

Some of the enforcement accomplishments for fiscal 1992 include:

- A record 191 criminal indictments.
- A record $163,064,344 in criminal penalties.

- A record $65.6 million recovered in civil penalties for environmental violations.
- A record $923 million recovered for natural resource damages.
- The largest environmental criminal penalty ever imposed, $125 million, and the largest single civil monetary settlement in history, $900 million, both arising out of the Exxon Valdez oil spill. Exxon will reimburse the US and the State of Alaska for all of their cleanup and damage assessment costs, and will restore, replace, or acquire the equivalent of the natural resources affected by the spill.

CIVIL ENFORCEMENT

In civil enforcement, over $1.953 billion was recovered in civil penalties, defendant cleanups, EPA cleanup cost recoveries, and natural resource damages.

New records achieved in the civil enforcement of the nation's environmental laws include civil penalties assessed and natural resource damages recovered. In fiscal 1992, nearly $66 million in civil penalties was imposed for violations of the Clean Air Act, the Clean Water Act, the Resource Conservation and Recovery Act, the Comprehensive Environmental Response, Compensation, and Liability Act (CERCLA, also known as the Superfund Act), and the Safe Drinking Water Act. The Department was also successful in recovering over $923 million arising out of damage to natural resources.

In addition, the civil enforcement accomplishments included the filing of 297 new cases, including 154 lawsuits filed under the Superfund Act, and the continued litigation of ongoing cases under all environmental statutes. The 154 Superfund suits filed in the past year matches the 154 filed in fiscal 1991, which set a record for the filing of Superfund cases, and represents one Superfund lawsuit filed against defendants every two and one-half days of the year

Below is a summary of fiscal 1992 civil enforcement statistics.

Enforcement Lawsuits Filed

Superfund	154
Clean Air Act	64
Clean Water Act	52
Resource Conservation and Recovery Act	13
Safe Drinking Water Act	7
Toxic Substances Control Act	2
Oil Pollution Act	2
Marine Protection Resources and Sanctuary Act	3
TOTAL	*297*

Enforcement Litigation Results

Superfund	
EPA Cleanup Cost Recoveries	$202,967,554
Court-Ordered Cleanups by Private Parties	761,904,116
Civil Penalties	9,701,808
Natural Resource Damage Recoveries	923,390,974
Civil Penalties	
Resource Conservation and Recovery Act	6,079,375
Clean Air Act	29,764,389
Clean Water Act	19,937,728
Safe Drinking Water Act	145,632
TOTAL	*$1,953,891,576*

CRIMINAL ENFORCEMENT

The Justice Department continued its strong emphasis on criminal prosecution of illegal pollution, as reflected in the 191 indictments returned, a new record, and 104 guilty pleas or convictions, the second highest annual number over obtained, during fiscal 1992. This brings the total criminal enforcement during the past four fiscal years to 551 indictments and 392 guilty pleas or convictions.

Further, the Department's fiscal 1992 criminal prosecution efforts yielded a record $163 million in penalties, as well as 34 years of prison terms actually to be served. Even without counting the extraordinary criminal penalty in the Exxon Valdez oil spill, the Division's prosecution efforts yielded a record $38 million in criminal penalties. Last year brought the total criminal enforcement results record over the past four years to $224 million in penalties and 142 years of actual jail time to be served. This represents 94 per cent of all criminal penalties ever imposed under the nation's environmental laws, and 69 per cent of the actual jail time ever imposed in the history of US criminal environmental enforcement.

Below is a summary of the Department's enforcement results for environmental crimes.

Fiscal Year	*Indictments*	*Pleas / Convictions*
83	40	40
84	43	32
85	40	37
86	94	67
87	127	86
88	124	63
89	107	101
90	134	85
91	125	96
92	191	104

Penalties Imposed		*Prison Terms*			*Actual Confinement*		
FY 83	$ 341,100	11 yrs			5 yrs		
FY 84	384,290	5 yrs	3 mos		1 yr	7 mos	
FY 85	565,850	5 yrs	5 mos		2 yrs	11 mos	
FY 86	1,917,602	124 yrs	2 mos	2 days	31 yrs	4 mos	12 days
FY 87	3,046,060	32 yrs	4 mos	7 days	14 yrs	9 mos	22 days
FY 88	7,091,876	39 yrs	3 mos	1 day	8 yrs	3 mos	7 days
FY 89	12,750,330	53 yrs	1 mo		37 yrs	2 mos	
FY 90	*29,977,508	71 yrs	11 mos	3 days	48 yrs	1 mo	1 day
FY 91	18,508,732	24 yrs	8 mos		22 yrs	8 mos	
FY 92	+$163,064,344	37 yrs	6 mos	1 day	34 yrs	1 mo	1 day

* This total includes a $22 million forfeiture that was obtained in a RICO/mail fraud case against 3 individuals and 6 related waste disposal and real estate development companies. A major portion of this forfeiture is expected to be designated for hazardous waste cleanup upon liquidation of assets. Included in the jail terms are two 12-year/7-month sentences against two individuals in the same RICO/mail fraud case.

+ This total includes $125 million of a $250 million criminal assessment against Exxon Corp. and Exxon Shipping Co. for the *Valdez* oil spill. The other $125 million was remitted for pledges by Exxon to make expenditures far exceeding this amount on environmental safety projects, a contribution to a response fund for large-scale oil spills, and a commitment of 25% of their total research expenditure on environmental and safety research.

5

The growth in environmentally-responsible investment

TESSA TENNANT

Head of Green and Ethical Investments, NPI

- Environmentally-responsible investment is the integration of environmental criteria into the investment decision-making process.
- Some US$650 billion of investment worldwide is subject to screening for social and/or environmental responsibility, and an increasing range of green investment products is now available. Finance houses are paying greater attention to this added dimension.
- Firms showing good environmental performance are often attractive propositions on ordinary investment analysis. Managers in industry alive to the environment and making a considered and practical response to it are likely to be forward-thinking in other areas.
- Of 300 American shareholder resolutions submitted in 1993, 22% covered environmental topics.
- Financiers are the pump primers of the global economy. They can influence industrial development so that it is compatible with the sustainable development agenda.
- Institutional investors are starting to focus on the value of environmental data in their assessment of the prospect of companies. The US Securities and Exchange Commission requires environmental disclosure by companies, and groups in the UK are recommending that the London Stock Exchange should consider adopting similar standards as a requirement of its listing practices.
- Companies are beginning to discover that it is insufficient to demonstrate that current performance is 'on the right track'. They must also demonstrate to investors that they have started strategic planning to ensure their place in a green world.

Environmentally-responsible investment defined

Environmentally-responsible investment is the integration of environmental criteria into the investment decision-making process. Through any combination of: shareholder activism, direct investment in local initiatives and investment in companies listed on major stock markets, individuals and institutions aim to benefit from, and support the transition to, sustainable development.

Investors usually seek to achieve returns which compare favourably with the market. A few choose to invest in specialist funds with lower than average returns on the basis that they are supporting projects with very high positive social and environmental impact.

The origins and present status of environmentally-responsible investment

> Like it or not, the days when financial decisions could be made in a complete moral and social vacuum are numbered.
>
> *Financial Times* 14 April 1990

The view that it is possible to do well in business by doing good is not new, but it found new strength in the 1960s with the emergence of the social investment market in the USA. Many socially conscious investors were shocked to discover they were investing in the chemical companies producing Agent-Orange used in the Vietnam War. The first funds to be created aimed to avoid companies involved in the war effort and in South Africa. South Africa has remained the single most important divestment issue, although a wider ethical and social agenda for rating companies has also developed. Essentially the early funds worked on the basis of an avoidance approach, screening out unacceptable activities rather than directing investment more positively. Although environmental issues were also considered, it was not until the late eighties, and the massive surge worldwide in green consciousness, that environmentally-responsible investment came of age.

The growth in social investment in the US is now driven 'not just by softhearted baby boomers or guilty 1960s radicals, but by people in their fifties, by the elderly, and by institutional investors which control – and own, through you – the majority of the investment resources in this country. All of these investors are coming to the conclusion that there may be different and better ways to do business'.[1] The social investment market elsewhere is about five years behind the US and in the UK environmental concerns are being driven more by individual rather than institutional investors. One survey indicated that the majority of investors adopting a green investment policy were female and under 45 or over 60.[2] Another, in 1989, by Scottish Equitable, revealed that 96%

of people interviewed indicated areas where they would not want to invest for 'ethical' reasons. Environmental concerns topped the list. However the survey also revealed that only 28% of those interviewed were aware that ethical investment opportunities existed, and only 10% currently held 'ethical' investments.

Table 5.1 Listed Green and Ethical funds

		£ million
Europe:	UK	400.00
	France	78.21
	Luxemburg	76.69
	Sweden	53.93
	Netherlands	37.85
	Denmark	30.97
	Germany	27.70
	Austria	5.35
	Norway	3.70
	TOTAL:	714.40 (1)
Rest of World:	USA	11,000 US$ milllon(2)
	Canada	213.7 Cd$ million (3)
	Australia	44.5 Aus$ million (4)

Sources: (1) EIRIS latest estimate, all European figures taken from A Survey of European Funds, The Merlin Research Unit, January 1993 (2) US Social Investment Forum, 1991; (3) The Canadian Social Investment Organisation (4) Money Matters Inc., NSW Australia.

The figures in Table 5.1 show the amount of money under management in listed green and ethical funds such as unit trusts, mutual funds and investment trusts. Table 5.1 does not include green and ethical deposit accounts held by co-operative and 'alternative' banks in the US and Europe, nor does it include the substantial sums held in institutional funds with green and ethical policies.

There are no reliable figures for the size of the entire environmental investment industry. The difficulty arises because investment managers are not obliged to declare whether their portfolios have specifications beyond the requirement to maximise financial return. However, it has been estimated that more than £51 billion of investment in the UK is subject to some form of screening for social responsibility (3). In the US, Peter Kinder of social investment consultancy Kinder Lydenberg and Domini guestimates that some $500 billion is invested with an environmental and/or social remit.

As the figures illustrate, the lion's share of 'responsible money' is not in the listed funds but in the large institutional funds held mainly as pension funds and trusts by charities, churches, universities, municipalities and some companies. While most of these funds invest in publicly traded securities chosen as a result of negative screens and an avoidance strategy, a few enterprising funds are developing a positive and dynamic approach as active shareholders supporting corporate change and directing capital towards 'high social' impact projects, such as Community Loan funds in the USA, and The Ecology Building Society in the UK.

Mirroring this gradual change in institutional investment policy is the increasing range of green investment products which are now available to the consumer: savings schemes, life, health and household insurance policies, mortgages, bank accounts and pension plans. All at no greater risk and at little or no extra cost to other products on the market. If only for competitive reasons, established finance houses are having to pay greater attention to this value-added dimension to investment products.

Driving forces

The emergence and growth of environmentally-responsible investment is still seen as insignificant by many people in business and finance. They see it as the domain for the woollyheaded idealist and as an investment activity which will always remain marginal. There are a number of reasons why this assessment is incorrect.

1 GREEN BUSINESS IS GOOD BUSINESS

As demonstrated elsewhere in this book, firms showing good environmental performance often turn out to be attractive propositions on ordinary investment analysis. Managers in industry alive to an issue such as the environment and making a considered and practical response to it are likely to be equally forward-thinking in other areas.

Responding to environmental concerns may, for example, open up new markets, lead to savings in resource costs and enhance a company's image. Moreover, a firm failing to respond may lose market share or face costly penalties. Indeed, with the possibility of stricter liability regimes being introduced, banks are already being deterred from lending to certain industries for fear of liability for clean-up costs. This may happen if they realise land as security in a loan default situation. Uncertainty over future conditions is also impeding the development of the market for environmental liability insurance.

Thus, as investment managers are finding, the double due diligence of environmental and financial evaluation is likely to lead to better stock selection.

2 DEVELOPMENTS IN INTERNATIONAL POLICY AND REGULATION

> Business is a large vessel; it will require great common effort and planning to overcome the inertia of the present destructive course, and to create a new momentum toward sustainable development.

Changing Course, The Business Council for Sustainable Development, MIT, 1992

The role of finance in sustainable development was recognised at The Earth Summit, The United Nations Conference on Environment and Development held in June 1992. Several contributions were made by the financial community, including the *Rio Resolution* from the international social investment community – representing funds under management of more than $650 billion[4] – and the *Bankers' Charter*. The Charter was signed by 40 banking institutions from around the world, including National Westminster Bank, Deutsche Bank, Royal Bank of Canada, Hongkong & Shanghai Banking Corporation, and Westpac Banking Corporation.

The letter from the international social investment community to Maurice Strong, Secretary General to The Earth Summit, stated that

> 'Contrary to acting as an impediment to growth, we believe that protecting the Earth is the only way to assure that a healthy economy – and people's livelihoods – can be sustained. Financiers are the pump primers of the global economy; they can withdraw funds or give their full support to any enterprise. They can therefore uniquely and powerfully influence the course of industrial development so that it is compatible with the sustainable development agenda'.

The letter also pointed out that

> 'Social investors are at the forefront of developing methodologies for the assessment of corporate performance from an environmental and social perspective. This analysis is the bedrock from which companies can evaluate and adjust their activities in relation to environmental and cultural priorities. Such analysis deserves wider recognition as a key mechanism by which sustainability can be achieved.'

At The Earth Summit, the international community reached a set of five accords designed to shape future modes of global environmental management. The accords were: Rio Declaration on Environment and Development, Agenda 21, United Nations Framework Convention on Climate Change, Convention on Biological Diversity, and a Statement of Principles on the Management, Conservation and Sustainable Development of All Types of Forests. Implicit in the accords of 1992 are new premises and procedures for a global order based on enhanced environmental responsibility at all levels. This emphasis on responsibility will influence the development of environmental legislation around the world and it can be expected that companies will be required to account for their actions and to internalise costs as never before.

Already the requirement to internalise environmental costs into business accounting is being pursued by the European Community. Its recent Eco-Management and Audit Regulation is an example of the way policy is developing. However, industry and government can only do so much without the support of the financial community. Only recently has this been widely understood. Consequently, policy-makers are turning their minds to ways of ensuring that financiers participate in the process. The Dutch government's current examination of the tax status of green funds and their consideration of an eco-labelling scheme for financial products is therefore significant. Recent recommendations from the UK's Advisory Committee on Business and the Environment's Finance Sector working group are also an indication of what is likely to come. The group noted that 'institutional investors have not fully focused on the value of environmental data in their assessment of the prospects of companies'. The report recommends that 'environmental disclosure should be readily accessible'. The report endorses the statement by the UK's '*Hundred Group of Finance Directors*' which encourages companies to report on their environmental performance.

3 ENVIRONMENTALLY-RESPONSIBLE INVESTMENT IS MARKET DRIVEN

The first funds to pay any attention to environmental concerns did not arise from regulations. Indeed the first funds had to fight existing laws and interpretations of the law to win the right for existence. In the UK, it took Charles Jacob and his colleagues seven years to persuade The Department of Trade and Industry to accept the principle for ethical investment. Friends Provident was then able to launch the first fund in 1984.[5] In Germany, environmental technology funds are permitted but funds which place emphasis on environmental responsibility and ethics are still forbidden.

The first environmental funds were a response to a market demand and their growth has continued to be market driven. For the first time investors had a choice; people with concerns for the environment and human welfare found these new funds were more attractive. Public interest in these funds is likely to grow in line with the green consumer trends described in Robert Worcester's Chapter 1. These funds have grown rapidly in recent years to reach the levels shown in Table 5.1. In the UK the growth has been 120% in 3 years.[6]

The financial services sector is an intensely competitive industry and environmentally-responsible investment represents a value-added dimension to finance. The Co-operative Bank in the UK reversed their losses to profits in 1992/3 by publicising their ethical stance. Their retail deposits increased by 13% on a base of 2.3m customers with half their new customers mentioning the Bank's ethical stance as a reason for joining. Significantly, customers which the Bank has attracted tend to be wage-earning, responsible and honest, and are typical of the types which are coveted by any retail bank.[7]

Hit by the scandals of the eighties – BCCI, Maxwell, Poly Peck and Drexel Burnham to name but a few – the industry experienced loss of consumer confidence. Thus, competition and the need to re-establish core values of integrity are likely to act as powerful incentives for further developments in the field of green and ethical investment. Already financial advisers who previously disregarded environmentally-responsible investments are becoming more confident in including this option in the package which they present to clients.

4 THE PERFORMANCE OF ENVIRONMENTALLY-RESPONSIBLE INVESTMENT

Performance depends as much on the abilities of the fund manager to time and interpret the market as it does on the overall theme of a fund. Even so, there are already signs that environmentally-*responsible* investments compare favourably with regular investments. This is resulting in improved methods for tracking performance. In the USA, for the period 1992/3, the Domini 400 Social Index (DSI) showed a gain of 3.06% compared with Standard & Poor's 500 Index (S&P) which gained 2.68%.[8]

These results should not be confused with the performance of some of the environmental technology funds which were launched in the late eighties where whole sectors, such as waste management for example, were regarded as environmentally-responsible regardless of the quality of environmental performance of the companies in the sector. Some of these 'quick-buck' funds have performed very poorly.

Environmental evaluation techniques

The criteria used by different funds and investment managers vary widely and interpretation is often determined as much by those who provide the research as by the financiers. Few investment houses have their own environmental and social research teams and most research is carried out by independent organisations such as EIRIS, Ethical Consumer and New Consumer in Europe, and CEP, ICCR and IRRC in the USA. Some investment managers such as FIFEGA in Europe, and Franklin and KLD in the USA, also provide research. The criteria being used are becoming more sophisticated in line with developments in analysis, corporate accounting and the definition of environmental objectives by environmental organisations and government.

The systems versus the issues approach

A differentiating feature of environmental funds is whether they take an issues or a systems approach, or a combination of the two. For the issues approach, a checklist of criteria are specified which the fund may select or avoid, such as tropical rainforest products, acid rain, animal welfare, recycling, carbon dioxide generation, ozone depletion or nuclear power. The UK research organisation, EIRIS has pioneered the quantification of corporate involvement in issues which are commonly raised by ethical investors. Consequently this approach has dominated many of the UK-based green and ethical funds. The issues approach has been favoured because it makes environmental objectives easily manageable – a company is either using CFC's or it is not. However, it can lead to some absurd anomolies such as avoidance of a company which may still be using an ozone-depleting substance in a closed-system, because there is no alternative and at the same time may have some very innovative practices in the areas of waste recycling and resource conservation.

The systems approach aims to prevent this situation arising by considering a company in its entirety. Anything a company says or does may come in for scrutiny but will be taken in the context of the whole company and the industry in which it is operating.

Systems analysis requires heavy investment in research to take account of a wide range of issues which may be relevant to a particular industrial activity. It may also include comparative analysis of companies operating in the same field. Without this depth of research the systems approach can lose its value because it is less dependent on absolute standards. Examples of companies which have pioneered the systems approach are Franklin Research and the Calvert Group.

Strategy versus day-to-day operations

There is another differentiating feature between environmental investment managers. Some managers simply want to know that the company is in compliance and gradually improving its performance through waste minimisation programmes. Others are interested in matters concerning day-to-day operations but they also want to understand how the company fits in a sustainable economy. For these managers, it is insufficient for a company to demonstrate that current performance is on the right track. They must also demonstrate that they have at least started the process of strategic planning to ensure their place in a green world.

Use of a wide range of sources

The prevailing academic view of finance is based on the Efficient Capital Market theory which posits that markets operate as perfectly as it is possible because they incorporate all currently 'knowable' information about a particular security. It does not take much research to see the degree to which this is untrue. Company analysts receive two kinds of information: the first from the companies themselves, the second from other sources – governmental, media, interest groups, etc. It takes little imagination to understand the enormity of the task of monitoring all the information which is generated from the second category. Consequently, company analysts may rely on the on-line data services provided by companies such as Reuters, Extel and Bloombergs. A brief examination of the media directories which make up these on-line services reveals the paucity of the coverage of specialist periodicals covering issues of concern to environmentally-responsible investors. Until such time as the market is properly informed about environmental, development and social issues it cannot be said that it is working perfectly.

Thus, a distinguishing feature of environmentally-responsible fund managers is that they are taking on board the views of environmental and community groups as well as the findings of governments, business and academic institutions. The corporate view-point is less dominant in the information which they receive.

Rating systems

Investment managers, like anyone else, want an easy life. As mentioned previously, many buy their environmental research from external agencies and prefer corporate environmental accounts which give simple bottom-line answers. Demand has therefore grown for rating systems. However, there are problems. For example, there is no common agreement on how to provide a comparative value system which takes in a wide set of environmental performance indicators, such as land and resource use, environmental management systems, waste emissions, and the environmental performance of a product range. Consequently the rating systems which do exist are either qualitative or define environmental performance in narrow terms and only by what is measurable, such as waste emissions.

Table 5.2 Environmental rating systems

Organisation	*Description*
Council on Economic Priorities USA	Information profiles. Summary comments: top, medium, bottom ranking. No numerical values.
ECO-rating Switzerland	Percentage involvement in environmental technologies.
EIRIS UK	Only certain aspects of environmental performance. Percentage interest in certain activities, use of substances and compliance records.
Franklin Research & Development Corporation USA	Information profiles. Summary qualitative ratings (see Table 5.3).
Investor Responsibility Research Centre USA	Detailed environmental profiles of S&P 500 companies. IRRC Emissions Efficiency Index and IRRC Compliance Index provide quantitative ratings for these two criteria.

The Investor Responsibility Research Centre in the USA has developed the closest real values rating systems with their Emissions Efficiency Index and Compliance Index. These indices are calculated from all the data which is available from US public registers. Although the measurements are only valuing limited aspects of environmental performance, they do produce a comparative scoring system and benchmarks for performance of different industrial sectors. Table 5.2 summarises some of the rating systems in use.

Table 5.3 Franklin Research and Development Corporation: company ratings for AT&T Telecommunications Company, June 1993

	Franklin's Insight Ranking	
	2	
South Africa 2		Human Rights 2
Employee Relations 1		Energy 2
Environment 2		Product 2
Citizenship 2		Weapons 3
	(1 = excellent, 3 = average, 5 = bad)	

Sector research

The existence of green funds does not mean that industry has suddenly become environmentally-responsible. Even The Body Shop, a favourite green stock in the UK, has said there is no such thing as a company with no impact on the environment. Furthermore, it is impossible for a company to become green overnight. Environmentally-responsible investors are therefore interested in best practice against a set of environmental criteria. This leads to comparisons between companies and their competitors. For example, National Provident Institution looks at the environmental impacts of different industrial sectors. A detailed questionnaire is sent to companies. The replies are used alongside other research and conclusions are drawn about the companies in the sector which are preferred.

Franklin Research (Table 5.3) looks at companies in a similar manner and also publishes research on 'new' sectors such as recycling and mass transit, thereby helping to define the sectors and the market opportunities. As previously mentioned, the IRRC indices also provide guidance on the performance of companies relative to their sector.

Ethics and environment

Ethical investments have been available longer than environmental investments. The former tend to screen out companies which are deemed to be unethical, such as those trading in South Africa, involved with the arms trade or in tobacco, while the latter use their screening to select companies which are trying to help the environment and are moving towards sustainable development. The distinction is becoming blurred because of a growing realisation that environmental equity is unlikely to exist without the first principles of human equity.

Ethical norms are fundamental to any green investment analysis. They find expression in numerous ways such as the integrity of the vision and leadership of management, openness and accountability to employees and society, and corporate policy towards overseas trading relationships. Without high standards in areas such as these the chances of an enduring environmental management programme are minimal.

Corporate environmental reporting

> Reports *per se* are not what is required: public accountability and targeted action are.[9]

The key point for the business manager is to understand the importance of developing environmental accounting techniques for his or her business.

As previously mentioned, law-makers are increasingly considering ways to ensure that business internalises environmental costs. Information transparency is the name of the game and experience shows that disclosure is one of the more effective measures for making industry cleaner. The impact of the US Toxic Release Inventory on American business is a good example which is being emulated elsewhere.

The EC Eco-Management and Audit Regulation is an indication of the bottom-up, plant-by-plant requirement for environmental accounting. Although it is currently a voluntary measure, few doubt that it will become a statutory requirement in due course. Complementing this measure are the pressures for a top-down, total corporate account. The UK's ACBE Finance Sector Working Group considered the extent to which companies disclose information about their environmental performance and its integration with the investment management process. They concluded that the level of disclosure by companies is still low, with no standard for the quality of environmental reporting and varying quality of disclosure between companies.

While the production of regular environmental performance reports is not yet mandatory in the UK, it may well become a statutory requirement in the future. The ACBE Finance Group observed that while 'practice in this area is developing fast, the Group is very clear that it wishes to see all companies publishing environmental reports. Action is now required to achieve this.' The Group goes on to recommend that 'some standards will be required and verification may be appropriate; legislation may also be necessary'. Following from the example of the Securities and Exchange Commission (SEC) in the US, the Group also recommended that the London Stock Exchange should consider adopting standards of environmental disclosure as one of the requirements of its listing particulars.

In the US, companies have had to report more fully for the SEC and now the Coalition for environmentally-responsible Economies, better known as CERES has raised expectations dramatically. As the shareholder actions indicate, CERES is a flagship for investors to rally around and is an important catalyst for corporate America in responding to environmental affairs. However, the general view of US corporate environmental reports is that, with a few exceptions like Polaroid, they still tend to cherry-pick and do not report in a systematic way. In February, an appeals court upheld a 1992 ruling that the company, International Paper, had 'materially misrepresented its environmental record' in its response to a shareholder proposal that it adopt the CERES principles.[10] The ruling is seen as a milestone which could set the stage for more lawsuits if companies do not increase disclosure in their accounts. Beyond SEC requirements, systematic environmental reporting is not yet mandatory in the US or Canada. In response to growing criticisms, the SEC has stepped up its review of the adequacy of environmental disclosure requirements, using Environmental Protection Agency (EPA) information on companies and EPA staff to train its staff for environmental liability disclosure review.

In Europe, there is no benchmark like CERES although there are numerous codes of conduct. The European view is that environmental reporting is a new and rapidly evolving science, requiring input not just from companies but also feedback and assistance from the accounting and financial communities, and the public at large. It appears that legal requirements will be considered if voluntary reporting agreements do not work.

At present there is less reporting from Japanese companies possibly because there has been a less active environmental campaigning community and calls for accountability have not been so great. However, this situation is changing although Japanese companies are very wary about the competitive implications of environmental reporting.[11]

Types of investment

Environmentally-responsible investors are little different from other investors in the types of investment they choose. Investments range from gilts, bonds and shares in companies listed on the major markets, to private placements, venture funds, property and in some cases high social impact funds.

Some types of investment have been criticised. For example, money market funds divert money away from more community-based banking systems and lend it back to large corporate creditors in the form of low-interest commercial paper. Thus community access to finance is reduced. Government bonds are also questioned – there are, after all, few governments which have fully embraced the quest for sustainable development. However, some bonds such as those issued by American municipalities and European agencies are issued to raise finance for specific projects. These may fit well with sustainable development goals such as improving rail stock or pollution control measures.

Investing in companies listed on major stock markets

This approach to managing portfolios accounts for most of the money under environmentally-responsible investment management. The traditional approach to stock selection was to avoid companies in obviously unacceptable areas such as arms and the nuclear industry. More recently a more proactive approach has emerged where companies are selected for positive reasons.

Companies may be selected because they are active in an industrial sector which fits with the sustainable development theme. For example, these sectors include the gas (in the interim) and water industries, railways and telecommunications, or emerging sectors such as pollution control, recycling and renewable energy. Companies may also be selected because they are seen to be 'green chips': those companies in other established sectors which contribute to human welfare and are frontrunners in environmental protection.

Many companies are still avoided, particularly by the specialist funds, because their activities are unacceptable. However, due to their large size, this avoidance approach is inappropriate for the large institutional funds. Consequently, the more enlightened have adopted active shareholder policies.

Shareholder policy

In March 1990, the international accountants and business consultants, Touche Ross conducted a survey (Table 5.4) which looked at management attitudes to environmental issues. It found that only 9% of companies in the UK had felt pressure from shareholders to change the environmental impact of their products or processes, compared to 70% in Denmark.

Table 5.4 Management attitudes to environmental issues

Question: *Have you had to, or do you intend to change products or processes in response to pressure from shareholders?*

	UK	*Germany*	*Denmark*	*Netherlands*	*Belgium*
%	9	50	70	20	20

The Touche Ross study indicates the way shareholders in different European countries view their responsibilities. An active shareholder policy, or as it is known in the US, 'relationship investing' can have a marked effect on management response to a range of issues including executive accountability, environmental performance and community and labour relations.

In continental Europe the relationship between financiers has traditionally been a close and very closed-doors one. On the plus side, this has resulted in financiers taking more interest in supporting companies for the long term. On the minus side, the lack of transparency makes it difficult for outside interests – politicians and the community – to know what is going on. The UK market is more like the American one, although American investors have been quicker to take a more active interest in companies. Although short-termism is seen to be endemic, the system is more open and accountable to other interests.

American investors have been at the forefront of shareholder action initiatives. In 1993, 300 shareholder resolutions were submitted and 22% covered environmental topics: 60% of the environmental resolutions focused on corporate environmental reporting requirements as defined by CERES. The rest covered issues such as toxic and radioactive waste, toxic-release reporting worldwide, mining on native American lands, sustainable energy policy and health, environment and safety standards at work.[12]

Over the years, The Interfaith Centre for Corporate Responsibility has been an active campaigner for South Africa and most recently, the CERES principles. Institutions such as the Council on Economic Priorities and the Investor Responsibility Research Centre in the US, and Pensions Investment Research Consultants (PIRC) in the UK, provide a wide range of services on environmental and social affairs to the shareholding community. Following the launch of the CERES principles in 1990, PIRC introduced the Environmental Investor's Code for pension funds in the UK. The code is now supported by funds under management exceeding £10 billion and has been instrumental in changing corporate activities, such as the peat extraction policies of Fisons plc.

In the US, the New York state pension fund and California state pension fund have been particularly active shareholders. New York state for example, was responsible for ensuring that Exxon implemented board level responsibility for environmental affairs in the wake of the *Valdez* disaster.

Venture funding

Venture funders have a notorious reputation for turning down projects and expecting very high returns from the projects which are supported. When fingers have been burnt several times with businesses which have failed, this behaviour is perhaps not suprising. Few environmental venture funds exist and they tend to back new technologies, rather than new companies, with a more environmentally sensitive approach to making established products and services. Table 5.5 lists some of the funds:

Table 5.5 Some venture funds

Name	*Country*	*Contact*
Alex Brown European Environmental Fund	UK	071–828 8001
Calvert Social Venture Ptnrs	USA	1–301–718 4272
Hambrecht & Quist	USA	1–415–576 3310
First Analysis Corporation	USA	1–312–372 3111

High impact

This is a new generation of investment funds which have emerged as a response to the 'capital gap', a term used in the USA to describe the inability of existing financial institutions to provide capital to certain communities and enterprises which are deemed to be uneconomic or un-creditworthy. Examples of these institutions which include, 'alternative' banks, co-operatives, credit-unions and loan funds are shown in Table 5.6. The experience of these institutions is demonstrating that conventional analysis is often wrong and many successful enterprises have grown from this more supportive seed-funding.

Critics argue that these initiatives are simply providing cheap money and spawn enterprises which are cushioned from the market place, unable to face market realities. However these projects tend to generate fewer externalities than mainstream business and, as in the case of urban regeneration, often convert the costs of contemporary business practice, such as mass redundancies or factory relocation, into productive economy again. Throughout his election campaign, President Clinton referred to the success of the South Shore Bank in Chicago and in his State of the Union inaugural speech he expressed his wish to leave a network of at least 100 community development banks as a legacy of his administration.

Table 5.6 Examples of high social impact funding

	Country	*Contact*
Triodos Bank	Holland	31–3404–16544
Mercury Provident	UK	0342–823739
ACCION International	USA	1–617–492 4930
South Shore Bank	USA	1–312–288 1000
Grameen Bank	Bangladesh	880–2–803 559
Calvert High Social Impact	USA	1–301–951 4800
Shared-Interest	UK	091–261 5943
EDCF	Holland	31–33–633 122

For a more complete list, contact INAISE or the Social Investment Forum in the UK or the USA.

In many ways this grass-roots funding, which has embraced environmental and social objectives from the beginning, is seen as a cutting edge for sustainable development in developing and developed economies alike. It is centred on empowerment, not charity, with the objective of 'trade not aid'.

Implications of environmentally-responsible investment for business

There is little doubt that environmentally-responsible investments will grow in size and influence. The effects will vary depending on the size of business.

For *small companies* with clearly defined markets and activities which fit with the goals of sustainable development there are funding opportunities. In some instances there has been fair criticism that these opportunities are no greater than from the market as a whole.

For *medium-sized companies* with activities which fit the ethos of green funds, there are great opportunities to raise the company profile and attract loyal, long-term investors. This is especially the case if the company is listed on a stock market or coming to the market. However, environmentally-responsible investors are not easily duped and as much as they are loyal, they can be highly – and publicly – critical if actions do not keep up with words, or there are false claims.

For *large companies* there are often difficulties fitting with the sustainable development agenda. The environmentally-responsible investor's assessment of corporate environmental performance can act as a barometer and sounding-board for strengths and weaknesses relative to competitors. Large companies should recognise the significance of being included in a green portfolio, it is not a matter for complacency but an opportunity to stay on course for future success.

It is probably fair to say that just as the 1970s will be remembered as the decade when industry made its first faltering steps along the green road, so the 1990s will be seen as time when the financial community comes to terms with its role in protecting the natural world. Increasingly, investors will be asking awkward questions about environmental performance. Business should be honest, open and prepared.

References

1 Kinder P. D, Lydenberg S. D, Domini A. L. *The Social Investment Almanac: A Comprehensive Guide to Socially Responsible Investing*, Henry Holt, New York (1992) pp. 8–23.
2 Survey of UK financial advisers promoting social investment, conducted by The Merlin Research Unit, January 1993.
3 Pensions Investment Research Consultants, London.
4 US Social Investment Forum.
5 Conversation with Charles Jacobs, 26 July 1993.
6 The Association of Unit Trusts and Investment Funds, *Daily Telegraph*, 30 June 1993.
7 *Financial Times*, 25 March 1993.
8 Press Release, KLD & Co, Inc., 14 June 1993.
9 Deloitte Touche Tohmatsu International. IISD and SustainAbility. *Coming Clean: Corporate Environmental Accounting*, DTTI, London (1993).
10 IRRC Investor's Environmental Report, vol. 2, no. 1.
11 Deloitte Touche Tohmatsu International. IISD and SustainAbility. *Coming Clean: Corporate Environmental Accounting*, DTTI, London (1993).
12 Communication with Interfaith Centre for Corporate Responsibility, 29 July 1993.

Recommended Reading

Advisory Committee on Business and the Environment, Report of the Financial Sector Working Group (London: DTi & DoE) 1993.

Bruyn, S. T. *The field of social investment*, Cambridge US: Cambridge University Press (1987).

Campanale M. *et al. Green and Ethical Funds in Continental Europe*, Merlin Research Unit, London (1993).

CERES 1991: *Environmental Performance Report Form*, CERES, Boston (1992).

Deloitte Touche Tomatsu International, IISD and SustainAbility. *Coming Clean: Corporate Environmental Accounting*, DTTI, London (1993).

EA051192 Environmental Accounting, Fédération des Experts Comptables Européens, Brussels (1992).

Gray R. H. *The Greening of Accountancy: The Profession after Pearce*, ACCA, London (1990).

Hundred Group of Finance Directors. *Statement of Good Practice: Environmental Reporting in Annual Reports*, HGFD, London (1992).

Kinder, P. D., Lydenberg, S. D. and Domini, A. L. *The Social Investment Almanac: A Comprehensive Guide to Socially Responsible Investing*, Henry Holt, New York (1992).

Owen, D. L. *Green Reporting: The challenge of the nineties*, Chapman & Hall, London (1992).

Schmidheiny, Stephan, with the Business Council for Sustainable Development. *Changing Course: A Global Perspective on Development and the Environment*, MIT Press, Cambridge, Massachusetts (1992).

Social Investment Forum. *The Rio Resolution* SIF, London (1992).

Part II

■

DEVELOPING BUSINESS OPPORTUNITIES

'There will be 900 budding capitalists out there somewhere thinking how can they plug the hole in the ozone layer to fix the rainforest in the Amazon . . . entrepreneurs with an idea and who want to make a buck will make it happen.'

Bernard Levin, *Business Review Weekly*, 21 July 1989.

'On a long downhill stretch, you can create more energy than the car can use, thereby gaining a little on the miles to – empty readout.'

Georg Kacher, automotive journalist, on the new Mercedes-Benz Vision-A-concept car, *Car* magazine, November 1993.

Introduction to Part II

■

This section sets out the ways in which companies in six different sectors have each treated the environment, and seen the pressures for changes outlined in Part I as a major business opportunity. Together the six chapters demonstrate the importance of being proactive in fitting opportunities to existing corporate priorities, and of building on core competences when making up-front investments.

The chapters on energy conservation and on waste management, emphasise the role of alliances between service providers and customers, as a way of reducing business uncertainty in the face of tightening environmental legislation and related drivers. Motivation of staff through both training and ongoing management programmes are important in making real progress, and are touched on in the chapters on industrial products and on the International Hotels Environment Initiative.

Through this, these companies have adopted flexible approaches that enable them to test out new opportunities, and to develop and refine management systems and new technology to achieve real environmental and business benefits. Phasing the introduction of new opportunities in relation to current business activities is an important element of success that is covered in these chapters, and which is also dealt with in detail in David Ballard's Chapter 19 in Part IV.

One way in which companies have handled this has been to introduce environmental products and services as part of an integrated package of options marketed to clients. This approach is illustrated in the chapters on energy conservation, on consultancy services and on banking, The other feature of this part is the wide range of different business benefits that companies identify as flowing from their handling of environmentally-linked business opportunities. These include increased customer loyalty, contract security, increased employee motivation and improved recruitment, identification of new opportunities for marketing and technology development, as well as direct cost savings.

Executive summaries

■

6 Developing business opportunities in energy conservation

An energy marketing company as a pace-setter in energy conservation can seem to be a paradox for conventional marketing people. That paradox does not exist for British Gas. Since the oil crises of the seventies the apparent problems of energy conservation have become classic marketing opportunities for the company.

Energy-conservation-based marketing and below-the-line activities have produced rewarding business opportunities, including new 'products', that range from the British Gas School of Fuel Management to computer-based simulation models for energy management. In the domestic sector particularly, some business opportunities have proved to be ephemeral, and important lessons have been learned. Although energy conservation offers valuable PR opportunities and can be used to establish business relationships in unlikely sectors, on the whole the UK market for energy conservation goods and services is stagnant. There are some prospects.

7 Business opportunities in waste management

Public perception of the need to protect the environment continues to intensify. Major changes in attitudes and resourcing are necessary to enable society to cope with increasing problems and, in particular, waste management issues. This evolution is creating global pressure, and thus opportunities, for industry. EC legislative policy is highlighted against industry growth prospects and, in Chapter 7, a description is given of the positioning and portfolio of services required for any company to be successful in future years. The chapter concludes with an outline review of some of the key waste-management options that exist today, and a profile of the annual UK waste arisings.

8 Developing environmental opportunities in industrial products: a look at the life cycle model

Chapter 8, explaining environmental opportunities in industrial products, concentrates on a 'Life Cycle Model' that was developed by 3M's

Corporate Product Responsibility Group. The model gives an overview of items to consider when doing a life cycle analysis. These include all phases from product concept to final disposition.

The chapter also talks about 3M's very successful 'Pollution Prevention Pays Program', which began in 1975. The key elements of this program are identified as well as the cost savings that have been generated from it.

The intent of the chapter is to show how a firm can make environmentally intelligent products for the industrial market-place by going through a life cycle analysis and evaluating every component of that analysis.

9 Developing environmental opportunities in a service business: accounting and consulting services

The driving forces for improving environmental performance can be just as important, strategically and operationally, to service businesses as well as to industrial companies. Waste treatment and disposal services, although virtually industrial processes, are directly regulated under most regimes of environmental law. Advisory or contracted-out services are more likely to be affected by customer requirements, often influenced by environmental law and a drive towards quality management.

KPMG Peat Marwick is a service business providing a wide range of accounting and consulting related services on a major international scale. In the UK in particular, but also in several other major countries, the firm believes that opportunities lie in offering environmental management consulting services and specialist advice alongside services. It is also recognised however that in areas such as financial audit, corporate finance, receiverships and management consulting, the environment must be considered as a matter of course alongside other elements of business performance and risk.

KPMG in the UK, has invested in establishing a specialist team – The National Environment Unit (NEU). Chapter 9 explores the thinking behind the original decision taken in 1989, and examines some of the NEU's work with accountants and in integrating the environment with the day-to-day functions of management.

For *accountants*, whatever their line of work in the firm's different business units, the financial implications of a number of issues need to be considered, notably:

- current plant may be inadequate to comply with legislation;
- some raw materials will be discontinued;
- markets for certain products will grow or shrink/disappear;

- prior or current operations may have led to contamination of the site;
- fines and penalties for breaches in legislation may be punitive.

10 Environmentally-friendly management in hotels

Environmental concerns are now firmly established on the global political agenda of the nineties. Global warming, depletion of the ozone layer and the destruction of the rain forests are no longer considered to be cries from prophets of doom. They are fast becoming proven scientific realities and, as such, they can no longer be ignored.

Hotels are ideally placed to play a crucial role in educating the world at large on the major issues. They can influence their suppliers to produce and supply products with the least impact on the environment. They can train and manage their staff to use resources wisely and sparingly; to reduce waste, recycle materials and re-use rather than throw away. They can influence the thinking of their customers and show their shareholders a better return on their capital as a result.

Chapter 10 sets out to provide advice for hoteliers on setting up an environmental management system which addresses the issues that influence the hospitality business. The main areas of concern centre around waste management and energy and water conservation, and the chapter discusses current, proven practices, new technology, functioning, maintenance and design.

11 Environmental management in a leading bank

Environmental protection is becoming increasingly important in all sectors of business and industry. For a company in the financial services sector this means tackling environmental problems both in-house – for example by implementing measures to reduce the use of resources – and in its business with corporate customers. For a bank and its customers environmental protection has become a business factor presenting opportunities and risks. Capital investment in environmental protection offers the bank an interesting new market to tap. To support companies with environmental problems, special products have been developed, such as information services. On the other hand, there are potential risks for a bank in environmental protection which should not be overlooked.

The protection of our environment has become one of the most important questions facing our society, and this obliges us all to act in a responsible fashion. The current situation is analysed in Chapter 11 from the perspective of a bank in Germany.

6

Developing business opportunities in energy conservation

ROBERT J. JONES

Consultant, Energy Policy Studies

- Energy conservation has become a classic marketing opportunity for British Gas since the oil crises of the 1970s.
- Ensuring that customers are conscious of energy efficiency leads fuel buyers to choose natural gas and demonstrates publicly the important role that British Gas has in UK's energy strategy.
- It is a mistake to over-play your hand in this area of marketing support, especially if the sales function is successfully winning support through provision of information.
- For most customers, energy conservation business opportunities need to be exploited by positioning products and services in terms other than cost savings. Energy sold on a long-term basis needs to be sold responsibly to help customers minimise running costs and to maximise return on investment.
- Integration of field sales forces with the company's Technical Consultancy Service enables customers' needs to be identified and met, and is also a source of new ideas for products and applications.
- Investment in energy efficiency and for domestic gas appliances is essential to maintain the dominance of natural gas in this sector.
- In EC countries, the multiplier effect of government investment in energy conservation can induce total expenditure in energy efficiency goods and services of up to five times the government investment. In the UK, a government budget of £300 million could therefore generate a market for energy-saving goods and services of some £1.5 billion per year.

1 Introduction

A PARADOX FOR AN ENERGY SUPPLIER

Staff working in a major fuel-supplier's energy conservation department soon come to expect questions such as, 'Why does an industry whose business is selling energy spend so much effort on encouraging its customers to use less?'

A story that circulates among the 'Energy Conservation Mafia' is of the highly-placed government official saying to his Secretary of State, 'The problem is, Minister, that British Gas *owns* energy conservation'.

Even back in 1986, the *Financial Times* declared, 'The activities of British Gas (in energy conservation) dwarf those of the Government.' The Company remains as proud of that independent accolade as it does of the civil servant's rueful suggestion of 'ownership'. The Company's programmes for encouraging energy conservation have a long and successful history, cover a wide range of business and social activity and have contributed handsomely to its marketing success. It is, perhaps, a text-book marketing case study, demonstrating how the most successful business leaders see challenges in terms of opportunities rather than threats.

MARKETING SUPPORT OPPORTUNITIES

In the Seventies, during the Oil Crises, British Gas took a positive decision to support what soon became the Government's 'Save It' campaign but decided to do so in support of marketing objectives. The creation of the British Gas School of Fuel Management in 1975 was the first manifestation of this policy and set the trend for imaginative responses to the energy conservation challenge that can be followed to more recent initiatives. One such recent example is the high-profile partnership with government departments that produced the British Gas Awards for Defence Energy Efficiency. These now yield great advantages in a sensitive and very attractive energy market. Another is the British Gas initiative that has become the government's Energy Saving Trust.

Natural gas is a clean, competitively-priced and highly controllable energy source but it has always been sold in strong competition with other sources, so that British Gas has developed an aggressive and successful marketing tradition. Some sectors of its markets present fewer challenges than others but, overall, market share has to be earned and loads gained have to be vigorously defended.

Friends and allies have to be cultivated in a climate where hostility arises not only in the market-place but in an increasingly complex and interventionist socio-political scene. Ensuring that customers and the general public are 'energy literate', helping them to be energy-efficiency conscious, not only leads fuel buyers and specifiers to choose natural gas, but also enables the public and politicians to appreciate the Company's marketing stance and its important role in energy strategy. For these reasons it makes good commercial sense for a premium fuel-supplier to be strongly involved in fuel conservation and energy efficiency, seeking out and developing business opportunities, both in new products and in above- and below-the-line support for the core business. Furthermore, energy supply is not a short-term business. Investments have formidable lead-times, are high risk and large scale. The customer, too, makes a long-term commitment in making a fuel choice; either as an individual householder investing in a domestic heating system that may have a life of some twenty years, or as a major manufacturer committing millions of pounds to a gas-fired production line, or to a large-scale combined heat and power (CHP) system.

Energy sold on a long-term basis needs to be sold responsibly, having regard for the customer's need to minimise running costs and thus maximise return on investment. An energy supplier cannot remain in business on the expectation of a quick return from a gas-guzzling installation. The load gained will be exposed to predatory competition and will do no good to the Company's reputation in the market-place.

Of necessity, most of the Company's current energy conservation marketing activities cannot be examined in detail because they are part of its competitive marketing. Instead, this chapter uses case studies that have either ceased to be part of the current strategy or are so strongly established that their purpose is self-evident or, perhaps, their value unassailable by competitors. However, the business opportunities that they illustrate remain relevant to current marketing needs.

2 Opportunities in the Industrial and Commercial sectors

OPPORTUNITIES IN TRAINING AND EDUCATION

This section is chiefly a case study of the British Gas School of Fuel Management and its associated products. The Company's investment in the School recognises a market for energy efficiency training; it also recognises that a high-profile presence in energy training may be used for many below-the-line marketing projects in support of selling into the industrial and commercial markets. It can also support the creation of new markets and can contribute to PR and reputation-building in the eyes of major-user fuel-specifiers, and in the eyes of those operating in the political area that impinges on the Company's business.

During the Seventies the Oil Crises prompted the government and public to recognise that energy efficiency and fuel conservation were issues that were vital to national security and economic stability. In those days the euphemistic 'Policy Gap' (political-speak for supply shortage) of the 1978 Green Paper was a grim reality. Several years before, the Three-Day-Week had reminded us that society could grind to an uncomfortable and socially dangerous halt if deprived of energy – with or without a 'Policy Gap'. British Gas recognised that business people at the highest level were not equipped to cope with such problems. Even those who were aware of their own 'energy illiteracy' also knew that there was a chronic shortage of knowledge and skill among their staffs to improve energy performance, even after the boardroom had recognised that improvement was needed. To meet these needs, the School was set up in 1975, linked to the Company's Midlands Research Station (MRS). Being adjacent and institutionally linked to MRS, to some extent it was similar to the research-teaching relationships of a university.

BRITISH GAS SCHOOL OF FUEL MANAGEMENT

The early courses were devised in close co-operation with government and enjoyed considerable support from officials and ministers as a contribution by British Gas to the national 'Save It' campaign. As far as British Gas marketing interests were concerned, existing and potential industrial and commercial gas-users were being exposed to the very latest technical advances in gas utilisation, just at a moment when R & D was opening up vast new potential markets and when substantial supplies of natural gas were becoming available.

In those days the Company was still a nationalised industry, so there were differences in marketing strategy and selling programmes compared to today's conditions. Competition was real and new loads had to be fought for in a lively market-place but there were restraints that do not apply today. For example, business people and politicians were sensitive to a major nationalised industry's activities that were marginal to its core business or which appeared to be being sold 'on the margin'; especially if these activities were in competition with those of independent operators who existed in a less protected environment. For this reason the full potential of the School of Fuel Management had to be curtailed whenever its initiatives were seen to be disadvantaging private competitors. This could be discouraging for managers, whose imaginations and efforts had developed energy efficiency 'products' only to be reined-back to allow others a share of the rewards. Once the School had courses well organised, energy consultants moved into the market with comparable services. As this happened British Gas tended to move out and to devote its energies to filling different training needs. However, even this mode of operating could be exploited to strengthen the Company's standing with government and to open avenues into the minds of fuel-specifiers, thus easing the job of the selling force. The history of the government's own 'Energy Managers Courses' illustrates the Company's exploitation of such an opportunity.

THE NATIONAL ENERGY MANAGEMENT COURSES

In 1978 a government advisory body concluded that there was a need for government-backed training in energy management, not only to do a training job but also to make a strong signal to industry and commerce that the subject was important. The services of the School of Fuel Management were offered to the Department of Energy and its facilities were used for the first courses. Staff of the School managed the courses, administered bookings, accommodation, payment of speakers and so forth. No mention of British Gas was made on promotional, administrative or teaching literature, but everyone who mattered knew where the credit lay.

Newcomers in this area of marketing support often make the mistake of over-playing their hand in this sort of sponsorship situation. It is important to know when to keep a low profile. If the Company is really doing a good job and making a valuable contribution, it can be counter-productive to shout too loudly about it, especially if the sales function is quietly getting over a lot of product information to influential participants and winning friends. Where the Energy Management Course was concerned, staff at all levels had to be briefed and supervised firmly to keep this low-profile, which some could see as alien to the aggressive marketing stance then being encouraged. Even so, the very use of School premises was clearly too 'British Gas' for government's purposes and a programme was introduced to enable their own officials to take over running the courses. British Gas used its business contacts to bring together the Department of Energy and the British Institute of Management (now the Institute of Management) and brokered a deal by which the Institute took over the courses, which then became the BIM/D.En. 'Energy Managers Workshops'.

Much credit had been gained from being prime movers, from being seen to give generous support to a government initiative, and from having an opportunity to study the newly emerging profession of Energy Manager. 'Motherhood and Apple Pie' activities, such as supporting government or public initiatives, are entirely legitimate marketing activities and are cost-effective, so long as one keeps the 'feel good' element in perspective. Business should respond positively to social responsibility – and British Gas staff have always been strong in this area – but there are also shareholders to consider. The Company is not investing resources only for the public good but also to reap marketing advantage.

In British Gas there has been a long tradition of not being shy of seizing business opportunity in supporting 'feel good' energy conservation activities. It is true that the public and government have benefited greatly alongside the company's gains. An often quoted Company joke sums up the attitude: 'Our energy conservation activities are for the public good as well as ours. It is "quid pro quo" but, for the sake of the shareholders, let's make it "*two* quid pro quo!". In this area we have also taught a lot of charitable or non-profit-distributing bodies to recognise their worth to potential sponsors, and to try to maximise returns in seeking sponsorship. For our part, we expect to pay commensurately for the commercial benefits that they offer us.

SUPPORTING THE NATIONAL ENERGY MANAGEMENT MOVEMENT

From experiences and contacts made by the School of Fuel Management and the National Energy Managers Courses, it became evident that, in those days, an important new element was emerging in the industrial and commercial energy fuel-choice process. This was the Energy Management Movement, which offered further business opportunities arising from energy conservation.

Apart from benefit to be gained from supporting the Energy Management Movement itself, experience suggested that this skilled and technically competent group of customers would be receptive to a computer-based energy business game. This could be used as a marketing support initiative and might also be developed as an energy-efficiency product in its own right.

The UK Department of Energy had started its Energy Management Groups in 1978. The idea was to encourage like-minded people, with energy management interests and problems, to meet on a regular but voluntary basis to exchange ideas. This was much the same as many other business interest groups, such as accountants, transport managers, personnel managers, and so forth, where companies generally encourage their own professionals to attend, to assist organisation and to hold office. The Department of Energy recognised the potential benefit of such activity to energy managers and encouraged the formation of Energy Management Groups with modest budgetary support and with administrative help from the government's Regional Energy Efficiency Officers (REEOs).

There were soon around 80 groups, divided into ten regions and with a membership of over 10,000. They met at members' business premises and, of course, a session at the British Gas School of Fuel Management was soon a prized event, and local British Gas Industrial and Commercial Sales Managers were offering local facilities for lectures and demonstrations.

Each Group was invited to nominate a member to the National Energy Management Advisory Committee (NEMAC), which met regularly, at government expense, with Department of Energy ministers and officials who represented the views of experienced energy managers. The movement had quickly become an important aspect of the government's energy-efficiency programme and an influential source of expert opinion for government policy-makers. Needless to say, energy pricing became a major agenda item, and the movement was clearly a force to be reckoned with for the energy marketeer. A long and respected tradition of support for energy conservation gave British Gas a considerable advantage in responding to this challenge, with constructive, but business-generating, support.

THE BRITISH GAS ENERGY MANAGERS' COMPETITIONS

As it is past history it is probably no secret that in around 1982 the Energy Management Groups were experiencing something of a crisis of identity, stagnant membership and falling numbers at meetings. Their new political masters had been largely responsible for this, by responding negatively to expert criticism, but Department of Energy officials were still anxious to encourage the movement and to reap the benefits for the energy efficiency of UK plc. In consultation with the NEMAC Chairman and Department of Energy officials, British Gas produced a competition in energy management, as a revitalisation exercise for the Energy Management Groups. A computer-based mathematical model had been produced at the Midlands Research Station, which could be used in a management-game mode to produce a stimulating and instructive competition for individuals or for group players. It became known as 'IMAGE' (Investment and Management Game in Energy) and was offered to NEMAC, through the School of Fuel Management, as an exclusive activity for the Energy Management Groups, with substantial prizes. By turning the model into a competition exclusive to Energy Management Groups, British Gas gave the movement a substantial boost in membership and prestige. It also gave local sales staff a legitimate reason to become active with the Groups without being seen to be too aggressively commercial. A package was assembled and offered to NEMAC as, 'The British Gas Energy Managers' Competition 1983'. The offer was accepted and eighty-two teams entered.

(The game was essentially like a conventional management game. The computer model simulated the energy use of a large industrial site. Teams could take decisions on energy saving management and investment and the computer made calculations that represented a year of operation. Over the five-month period of the competition, teams took monthly investment decisions, each month representing a year of real time.)

Each team had a starting budget of £10,000 and, in successive years, could invest 50% of the value of energy savings made in the previous year, up to a value of £40,000. At the end of the competition, the winning team, Wessex, had saved a staggering total of over £530,000. The second team, Gwent, was only £400 behind. It was so close that both teams each received the top prize of a home computer package for each team member. British Gas was also a winner! Throughout the period of the competition it was arranged that the newspaper, *Energy Management* would carry a progress report. Ministerial endorsement appeared regularly, praising British Gas's initiative, and local sales managers reported high levels of interest and contact. By popular request of government, NEMAC and British Gas's own sales staff, the competition was repeated in 1984/85 on a Mark II IMAGE program. It had certainly been a well exploited business opportunity.

The simulation program itself also revealed a further business opportunity. The Mark I program was re-styled as an energy investment teaching model, marketed under the name of 'IMAGE' and made available through the British Gas School of Fuel Management. It was also sold overseas. The School itself made a teaching package of the program, which is still in its product range. A low-priced postal version was devised for use by colleges. Overall IMAGE, in its day, was a very low-cost investment that was exploited to make a substantial impact at many levels. It won the company many friends, made the politicians of a new administration quickly aware of British Gas competence in this area, showed them the worth of British Gas support and gave local sales people valuable leads among fuel decision-makers.

CONVENTIONAL ENERGY CONSERVATION PRODUCTS IN THE INDUSTRIAL AND COMMERCIAL SECTORS

There have been other, more conventional, business opportunities for British Gas in energy conservation in the industrial and commercial energy market, arising from the development and exploitation of new energy-using products. Most of these have been straightforward developments arising from the activities of the Midlands Research Station (MRS) and the Technical Consultancy Service (TCS) of the Industrial and Commercial Marketing Department. The Company has generally pursued a policy of joint-ventures with gas equipment manufacturers, producing award-winning new products in burners, controls and process systems. Products are not only developed profitably in their own right but a high-profile approach to publicising them and the awards they win enhances the overall gas sales effort. It demonstrates that the Company's marketing strategy is based on enabling customers to use the fuel at maximum efficiency and thus at lowest operating cost.

Even these conventional product developments have led to a less conventional marketing opportunity in the form of the *British Gas Directory of Energy Efficiency Equipment*. Some ten years ago the Company produced a comprehensive directory of UK-based manufacturers of energy efficiency equipment for industry and commerce. It was an immediate success. The UK government bought copies for all overseas trade missions and UK demand was such that a new edition was produced within two years, and it is now regularly updated. Besides being a product in its own right, it earns the Company prestige and helps its customers, on whose commercial survival the Company's own prosperity depends.

3 Opportunities in the domestic sector

AN OPPORTUNITY OR AN ILLUSION?

In the domestic sector the Company has an equally long history of support for energy efficiency but the market opportunities are entirely different from those offered in the industrial and commercial sector. Individual gas loads are far smaller, where some 17 million users jointly consume little more gas than only 600,000 in the industrial and commercial market. The marketeer is confronted with mass markets and relatively low-value products. Although there have been many business opportunities for British Gas that have yielded valuable below-the-line benefit, the Company's experiences have highlighted the pitfalls that await business people who think that mass markets exist merely because energy conservation and environmental protection are continuously appearing as 'flavours of the month'. Among well-informed, thinking people the case for purchasing energy-saving devices for the home is compelling.

Unfortunately, the market-place reveals that the market either does not perceive that it needs to invest in this way or, if it does, it does not have the discretionary buying-power to do so.

Commonly available market research shows that large percentages of dwellings still need double-glazing, weather-proofing, cavity-wall fill, storm lobbies and even substantial top-up to existing roof-space and hot-water storage insulation. These measures all offer business opportunities. Tens of thousands of dwellings are still inadequately heated, or their heating and hot-water systems could benefit from energy-saving alternatives or modernisation. For example, replacing an ageing conventional gas-boiler with a Heatsaver condensing model could reduce a gas bill by more than 20 per cent.

Domestic users consume over 40 per cent of the UK's fuel. Reducing that consumption would make a worthwhile contribution to climate and pollution problems, besides giving actual cost savings to each householder. In spite of this, the markets for domestic energy-saving products remain stagnant and have a history of some twenty years of project failure.

British Gas experience in these potential markets has demonstrated that profitable business opportunities only exist for certain small-scale or niche operators and that the mass markets are not there, though there are useful opportunities for below-the-lines activities in support of domestic gas marketing and load retention. There are two main reasons why the apparent potentials are not realised. They could be expressed as, 'Those who can afford it cannot be bothered; those who need it cannot afford it.'

THE FUEL POOR

Social research (chiefly Boardman, 1991[1]) reveals that some 7.1 million households in the UK live in what has become known as 'fuel poverty'. Fuel poverty arises from a combination of having insufficient income to heat a house that also needs too much energy to heat it. Those with least wealth live in the houses that cost most to heat. The problem is compounded because they do not have the money to make investments to improve the energy efficiency of their homes. Over a third of the UK's householders have to try to live in this situation. For the energy conservation marketeer this is a modern phenomenon that has to be appreciated. No matter how attractive a home energy-saving product or service may be, a third of the market cannot afford to buy it. Some small-scale operators make a living on the small percentage of these households that receives government assistance for basic measures (less than 1 million annually), but it will take a massive public-works programme, or a phenomenal improvement in UK social conditions, to create attractive, mass market business opportunities in this sector.

THE FUEL RICH

For the two-thirds of householders that could afford to invest in energy conservation products or services, the picture is equally dismal for business opportunities. Because domestic energy prices are low in the UK and, at least for gas, have been falling in real terms, the fuel bill represents a small and declining element of the household budget. This means that even the most cost-effective energy conservation measure has a saving potential of barely £100 a year; far less than is spent on petrol for the family car and of far less financial significance than the food and veterinary bills for the family dog! As if to exacerbate the situation, both government and energy-efficiency pundits tend to emphasise the 'payback' or financial benefit, which is the weakest aspect of the selling story. The experience of the market-place should warn the would-be energy conservation marketeer that this is the wrong emphasis. For example, one of the least cost-effective energy saving measures is retro-fit double-glazing, yet this has enjoyed a high level of sales success. Clearly, householders do not buy retro-fit double-glazing to save money.

Energy conservation business opportunities need to be exploited by positioning products and services in terms other than money-saving for the majority of customers. However, there is an interesting paradox: even if the money saved is not sufficiently significant to stimulate a market, where governments intervene to subsidise costs the market responds by buying. An example of this has been the subsidy of gas-fired condensing boilers by other EC governments. Not only does the subsidy in the Netherlands prompt purchasing, but some 60 per cent of those prompted do not even bother to collect the subsidy, suggesting that it is not so much the cash value as the social value of an official 'signal' that has most influence on this sector.

BRITISH GAS SUPPORT FOR DOMESTIC ENERGY CONSERVATION

Faced with these difficulties in the domestic market, British Gas has adopted a two-pronged aspect approach to supporting successive government initiatives; first, by adopting a 'technical fix' solution and second, by giving advice and education. Both approaches have been exploited as business opportunities to sell new gas and to retain established customers.

The 'technical fix' solution simply means investing in R & D and supporting gas appliance manufacturers and gas system installers so that the customer has access to the most energy-efficient appliances and systems. This is sound business common sense, as already mentioned with regard to the much larger industrial and commercial customers.

Where advice and education for energy efficiency are concerned, British Gas has had a continuous programme of successful initiatives, ever since it first chose to give strong support to government's 'Save It' campaigns of the Seventies. As with the industrial and commercial market, the 'energy literate' householder is a far easier proposition for the gas salesperson and is more receptive to the sales pitch, 'clean, controllable and economic'. Large sums have been spent on educational support projects, for primary schools, through secondary, higher and further education, to investment in continuous professional development (CPD) for fuel-specifying professionals. Leaflets, booklets and videos have been produced for established domestic customers, which have also been successful in attracting new users, using energy conservation as a socially beneficial, and entirely legitimate, vehicle to secure the domestic energy market.

THE MARKET FOR ENERGY-EFFICIENCY PACKAGES

Besides giving energy-efficiency advice, British Gas is experienced in attempting to stimulate the business of selling energy-saving products and services.

It has come to the conclusion that, in the present political climate, there are few business opportunities in this area. Unsuccessful energy-efficiency package pilot schemes have only served to reinforce the hard messages of the real market-place. The market for household energy-saving packages is largely ephemeral; as is the market for the most recent energy-conservation service, the Home Energy Audit or Energy Labelling.

British Gas initiatives to create a market for domestic energy-efficiency products and services offer a useful case study in this sector.

Since the mid-Seventies all regions of British Gas had provided energy-efficiency advice to their domestic customers through leaflets, posters, media-advertising, showroom displays or a combination of all or some of these measures. Some were attempting to sell energy-saving materials and services, either provided directly or by sub-contractors. Generally this was done, more or less successfully, to support gas-appliance sales.

Where packages were intended to be profit-making in their own right, it was usually a local enthusiast who was responsible for entry into these markets. There was no uniformity of approach and few real sales. The business was kept going, one felt, chiefly by the need to earn PR points at a time when businesses that made their livings from selling energy sensed themselves to be in an ambiguous position, as the world began to beat its breast for past energy profligacy. Against this background, the Company decided to get some control over and uniformity into the situation and to put marketing science behind an activity that was patently dependent on a lot of local whim or enthusiasm.

'DON'T WASTE YOUR ENERGY' AND OTHER PACKAGE INITIATIVES

In the summer of 1979 a pilot scheme was launched in one region, with both operational and research objectives:

- to demonstrate that the Company actively supports the responsible and wise use of gas in the domestic market;
- to ensure that energy conservation measures, if taken, are taken expertly and do not inhibit gas and gas-appliance marketing;
- to establish the market potential for package sales of insulation and controls;
- to provide showrooms with additional customer traffic.

A package of energy-saving goods and services was prepared and promoted under the theme, 'Don't Waste Your Energy'. Three major showrooms were equipped with Energy Advice Units, with a technical adviser in attendance, eye-catching displays and racks of energy-conservation materials and devices. Strong local media advertising was implemented and week-end, out-of-town energy conservation cash-and-carry fairs were organised in service depots that were not operational on Saturdays. Joint ventures were established with local insulation and double-glazing contractors.

In another region a similar pilot was arranged, to run concurrently with a different design style and based on a free Home Energy Audit, rather than the showroom Energy Advice Unit.

Neither exercise was successful in terms of products sold. Even the free Home Energy Audit attracted few customers. (Now that there is the

Government's Standard Assessment Procedure (SAP) and more professional products available (National Home Energy Rating Scheme and 'Starpoint'), the market may be more receptive. At least 'free' would now be seen to be worth around £60.) After a year, both pilot projects had been closed down. The Company concluded that energy-saving measures and services, however accessible and attractively packaged, were not a merchandisable product. Some years later, for purely political reasons, and after advising the instigators not to do it, the Company assisted the government to offer the Home Energy Audit on a fee-paying basis through a company known as 'HEAT Ltd'. In spite of extensive market support it failed as a commercial venture. About half a dozen audits were sold to a potential market of over 25,000.

'Thermsaver'

One spin-off of the two 1979 packaging exercises was the evolution of the brand-name 'Thermsaver', to give a generic name to the various products and services on offer. In spite of the undeniable failure in sales terms, the political climate continued to require a highly visible presence by British Gas in the energy-efficiency product market. A brand name, linked by the word 'Therm' to natural gas, gave a useful identity and differentiated the activities from those of competitor fuels.

Both public and staff quickly grasped the name and concept. The establishment of the brand-name made it easy to talk and think about energy-efficiency activities and produced a much stronger overall PR impression, although few products were ever sold, in support of the Company's stance of promoting what was then termed, 'the responsible and wise use of gas'.

Because there was so little real sales success, the brand-name was useful in maintaining sales staff interest in what was by then a PR activity.

In 1991/92, with the Company now long out of its nationalised industry situation, it was the Regulator rather than the government that encouraged British Gas into yet another flirtation with the domestic energy-efficiency package as a profitable business opportunity. Two British Gas operational districts were chosen, a business partner was involved, a reputable marketing consultancy handled the promotion but, yet again, the ephemeral nature of this market was revealed. Both British Gas and its business partner needed to make the exercise a success, but the resulting sales were derisory and those of us who are closely involved in this sector remain convinced that there is no great market opportunity. The 'low-energy society' is unlikely to arise if it relies solely on market forces to motivate the domestic customer to buy energy-efficiency products: 'Those who need it cannot afford it. Those that can afford it do not want it.' However, given public support and subsidy, the situation could change and it would be wrong to dismiss entirely the energy-conservation business opportunities.

4 General marketing prospects

A DECLINING MARKET

Like all other traders, those engaged in the energy-conservation products and services sectors have suffered from the UK recession. The Association for the Conservation of Energy has reported that demand for its members' products fell by 28 per cent between 1989 and 1992. But the Association also claims that recession has only been part of the problem. Relatively low energy prices, and a decline in government energy-efficiency activity, have been the main reason for the contraction, at a time when prospects for similar markets are brighter on mainland Europe. At the time of writing, there seems little prospect of improvement in the UK but it is possible to detect some hope. There is growing pressure for the UK government to take positive measures to meet its CO_2 reduction targets. The prospect of higher energy prices, through such measures as the imposition of VAT, or even a Carbon Tax, could awaken public interest in investing to save energy. Both the electricity and the gas industry are introducing energy labelling of appliances, in advance of EC requirements, which should generate public realisation that there are differences between the energy running costs of apparently similar appliances. This could stimulate the up-grading and replacement market and, itself, raise public awareness of the attractions of investing to save energy costs. There are now energy auditing services available for householders, which themselves represent a new market, and which should stimulate sales of products and services to meet the needs that are established by such audits – more commonly termed, 'Home Energy Labelling'. However, these are merely prospects of 'green shoots'. The reality is that energy conservation markets remain in decline. Although most households could cut their energy costs by 20 per cent without much effort or inconvenience, with quite attractive pay-back periods; and although similarly attractive prospects exist for industrial, commercial and public service energy users, it will need substantial price-hikes or significant government action to get household energy conservation markets moving in the UK.

PROSPECTS FOR CHANGE

Other EU countries have shown that multiplier effects from government investment in energy conservation can be as much as one to five; that is to say, every £1 of investment could induce a total expenditure of £5 in energy-efficiency goods and services. The Danish government found that for every £1 of public money invested in household energy saving, £4 of private investment was made; even though the energy standards of their buildings are already far higher than those of the UK.

Although there seems little prospect of a UK government spending to stimulate the energy-efficiency goods and services market, the marketeer would do well to watch the development of the newly formed Energy Saving Trust. This body has been set up chiefly to administer the spending of money raised by the electricity and gas suppliers in the UK, specifically to subsidise energy saving business. At present it has a modest budget of only a few million pounds but this is likely to rise to hundreds of millions by 1997. Through the multiplier effect, a budget of £300 million would generate a market for energy-saving goods and services of some £1.5 billion per year. The Trust's significance in the energy conservation markets would be huge and those engaged in, or contemplating entering, them would be well advised to give priority attention to the development of the Trust.

Conclusions

For the marketeer, the energy-conservation market is complex, paradoxical and apparently moribund. For energy suppliers it can also appear to be threatening, even when it is depressed and not generating significant demands for energy-efficiency products. At present its main attractiveness is for below-the-line initiatives, especially exploiting growing 'feel good' needs from global warming and environmental considerations.

As far as energy conservation goods and services are concerned, the market is likely to remain depressed in the absence of government action. The 7.1 million UK households in 'fuel poverty' are excluded from participating in any market for energy-saving measures. Others have few inducements to buy. Most industrial and commercial fuel users are unwilling to invest in the hope of savings that have pay-back times that might exceed their business survival expectations. Only the best, and niche operators, are making good livings in this climate. However, UK companies should not forget that mainland EU countries are far more active, and that markets for energy-efficiency goods and services enjoy better prospects across the Channel. In Eastern Europe, most countries are in dire trouble in energy consumption, especially in the industrial and commercial sectors. They could present excellent market prospects but there are severe capital limitations which inhibit progress. However, some companies are demonstrating that success is possible. The British Gas division, Global Gas, is among those who have successfully taken energy-conservation products into these markets.

In the UK the market situation would certainly improve if government adopted a more interventionist policy. The newly-formed Energy

Saving Trust is a suggestion that things are moving in that direction. If the general public itself became aware of what harm atmospheric and local pollution is doing to it through profligate energy use, the market prospects in energy conservation would be exceptionally good.

References

1 Boardman, Brenda. *Social Energies Policy,* Policy Studies Institute, November 1991; also other publications by Dr Boardman, Environmental Change Unit, University of Oxford.

7

Business opportunities in waste management

WILLIAM SEDDON-BROWN

Director, European Government Affairs,
Waste Management International, Brussels

- Almost half of the key environmental issues perceived by Government environmental protection agencies in the US and EC are directly related to waste management.

- Of the 10,000 environmental technology and service firms today serving European markets, 18% are in direct waste management, with a further 46% in water and waste-water treatment; major consolidation of up to half the companies involved can be expected over the next decade.

- The key for the fastest growing environmental services companies is in applying worldwide expertise to the needs of local markets, maintaining an advanced technology base, and handling the whole waste-management cycle so as to bring economies of scale.

- The international economy will become increasingly dependent on the environment over the next decade, and 'green industry' will play a key role in job creation.

- Public authorities are looking towards integrated private enterprises to cope with increasingly difficult waste-management responsibilities, in the face of tough financial constraints and more stringent environmental standards.

- Progress on waste reduction, recycling or resource recovery is likely to be slow until landfill disposal prices are increased from their current level. This will best be accomplished by rapid application of tighter landfill design and operating standards, which will both tighten environmental protection and raise landfill disposal costs.

- The EC 'ladder' concept for waste management puts increasing stress on disposal, incineration, useful application, re-use – and prevention, which is the overall goal of EC waste policy.

- Rigorous enforcement of strict standards will create a demand for increased investment in proper waste disposal equipment and services; it will also provide opportunities for development of an active environmental pollution control industry.

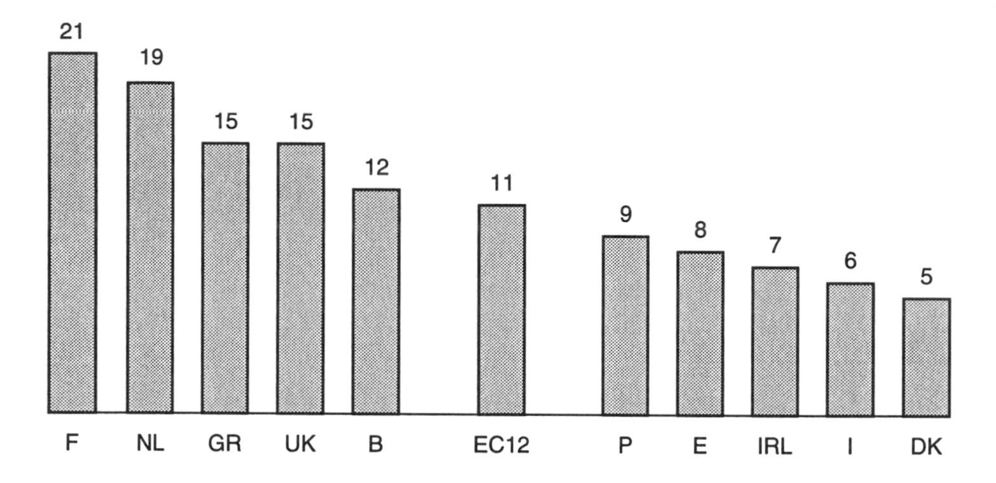

Figure 7.1 Protecting the environment – an immediate and urgent problem. Percentage of national shifts betweeen 1988 and 1992

A society of limits and chance

From New York to California, from East and West Germany to the crowded cities of Japan, society is becoming less and less able to cope with growing environmental problems and, in particular, waste management. Although these 'people needs' represent immense opportunities, major changes in attitudes and resourcing are required. In what amounts to a disguised form of privatisation, the private sector must mobilise to serve human and financial needs. The state can no longer cope alone.

At a time of increasing political Balkanisation, raising threats to trade and giving greater emphasis to national interests, business is moving from the consolidation period of the 1980s, to the new strategic alliances of the 1990s. The challenge for the private sector must be 'how to respond globally?'

From global pressures to broader opportunities

In its annual report, the World Watch Institute of Washington[1] states 'The international economy is set to become increasingly dependent on the environment over the next decade, and the "Green Industry" will be playing a key role in job creation.' This is not a short-term effect but a shift in societal direction which will underwrite the need for, and the viability of, a growing waste-management sector. Public perception of the need to protect the environment continues to grow (Fig. 7.1) as a European Community study shows.[2] In all member states at least seven out of ten citizens share the opinion that protecting the environment is an immediate and urgent problem.

Opportunities are not just within national boundaries but also across borders. It is noticeable that non-governmental activities are also globalising, as are the facilities and communications of governments. In the US Environmental Protection Agency, of the Science Advisory Board's list of key environmental issues, almost half are directly related to waste management, and these issues are similar to the listings in European countries, such as the UK (see Table 7.1).

Table 7.1 Key US environmental issues

No.	Issue	No.	Issue
1	Hazardous waste operation	16	Sewage treatment
2	Abandoned hazardous waste sites	17*	Vehicular emissions
3	Industrial water pollution	18	Pesticides in food
4*	Occupational environment	19*	Greenhouse
5	Oil spills	20*	Drinking water
6*	Ozone layer	21*	Wetlands
7	Nuclear accidents	22	Acid rain
8	Industrial chemical accidents	23	Urban runoff
9	Rad. wastes	24	Solid waste sites
10*	Factory air emissions	25	Biotechnology
11	Leaking underground storage tanks	26*	Indoor air pollution
12	Coastal contamination	27	X-ray
13	Solid waste	28*	Radon
14*	Pesticide workers	29	Microwave ovens
15	Agricultural runoff		

*Items on USEPA, Science Adv. Board List

A further area of future pressure will be from the stockholders of individual companies. In the USA today, institutional investors now hold three trillion dollars in major American corporate assets. New York City funds own stock in approximately 2,300 corporations, and the $45 billion New York City pension funds have about 3 per cent of all assets invested in European equities.[3]

Eugene J. Wingerter, Chief Executive Officer of the US National Solid Waste Management Association at the ISWA (International Solid Wastes Association) 1992 congress, expressed the core elements for an effective waste-management program as understanding:

1 the demographics of waste generation and the populations served;

2 the technologies for effective environmental solutions;

3 the political realities of the communities served;

4 how to mobilise active public participation.

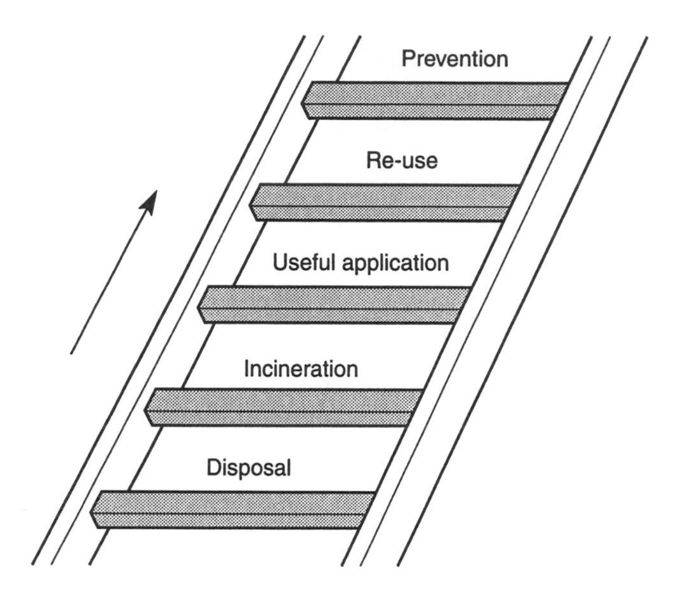

Figure 7.2 The EC 'Ladder Concept' for waste disposal options
(Source: European Commission)

In the European Community, these global requirements are equally valid. Environmental policy requires an integrated approach more and more. The European Commission has based its strategy on the so-called 'Ladder Concept' (Fig. 7.2) where policy and measures will strive to push an ever-increasing amount of waste, up the ladder, towards the ultimate goal of complete waste prevention.

How will these trends and policies translate to waste management requirements and opportunities?

A major industry segment and growth prospects

Before proceeding to some of the specific waste management options which are the basis for many of the business opportunities, let us examine the degree to which increasing environmental awareness, and more stringent standards, have created a new investment requirement.[4]

In 1988, environmental control expenditures by government and industry were estimated at 120 billion Ecu. Of these, the European Community represented 35.2%. Of total Western European expenditure in the same year, 16.9% was in the UK, 31.2% in Germany, 13.4% in France, 8.8% in Italy and 6.5% in the Netherlands.

Growth in this huge market could exceed 7% per year to the end of the decade, and waste management and recycling should grow from 24% of the total to more than 34% in the year 2000 (Fig. 7.3).

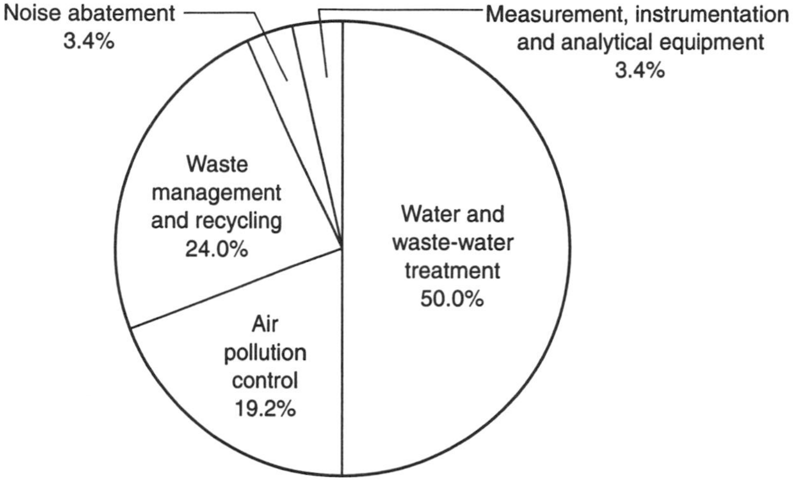

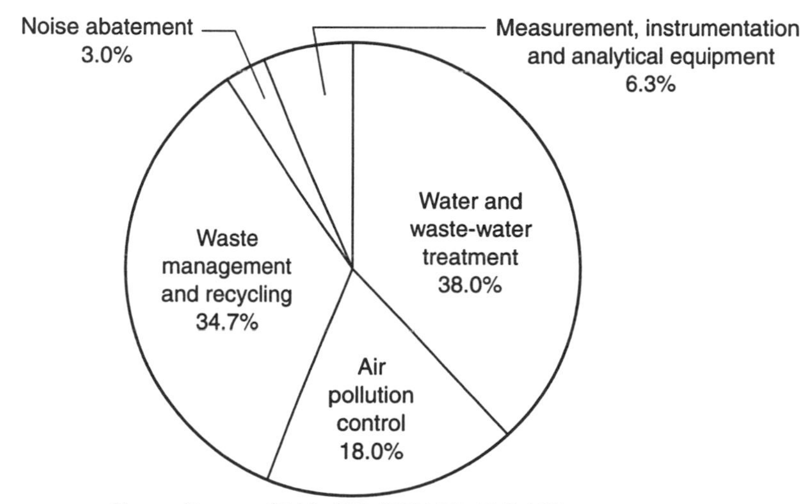

Note: $1.27 = Ecu 1

Figure 7.3 Growth of environmental control expenditures in Western Europe: 1987 and 2000

(Source: Arthur D. Little, Inc. and HKU)

Of the 10,000 environmental technology and service firms today serving the European markets, 18% are in direct waste management, with 46% in water and waste-water treatment. Major consolidation of up to half the companies involved can be expected over the next decade. As Joseph Holsten, Vice President of Waste Management International, one of the world's fastest growing environmental services company, says: 'The key is applying worldwide expertise to the needs of local markets'. He adds, 'Our approach is to build a full set of services within each market. The customer likes it because one supplier can handle the whole waste-management cycle. We like it, because it gives us economies of scale. Services that may not be economical on their own, can become profitable when they are added to an existing infrastructure.'

To this basic corporate consolidation must also be added the trend for public authorities to look towards an integrated private enterprise to cope with the waste-management responsibilities of an increasing difficulty, in the face of tough financial constraints and more stringent environmental standards, and the consequent need for greater technical expertise. However, no comment on the extraordinary nature of the growth of this sector, would be complete without a word of warning. The technology requirements, the liabilities and the sensitivity of permit applications, and the links with authorities and the general public, frequently cause problems.

Positioning for waste management opportunities: 1987 and 2000

Three aspects of environmental management directly concern a company looking for opportunities:

1 visible and demonstrable commitment;

2 compliance and stewardship;

3 priorities on the waste ladder and shared responsibility.

Visible commitment must come from the top. A company well placed to enter the environmental services market should have an open management process with a clear environmental mission statement, endorsed by the Chief Executive. The company's commitment should go 'one further' than just legal or regulatory compliance. Table 7.2 shows a typical plant site checklist.

Table 7.2 Key areas of compliance concern: a typical site checklist

- Right to know
- Special waste
- Tanks (above and underground)
- Surface water management
- Spill prevention control and countermeasure plan
- Air quality management
- Permits and authorisations
- Hazardous waste generator standards
- Recyclable materials

The International Chamber of Commerce's new publication 'From Ideas to Action' says: 'Waste is a double loss: we must replace a valuable raw material lost in the production process, and we run out of places where we can dispose of our garbage. Therefore there is a common interest by producers, consumers and society at large to reduce waste at the source.[5] This shared responsibility is in line with the EC 'ladder' concept and should also be related to the principle of proximity and self-sufficiency. The growing support for the minimisation of waste is underlined by the series of case studies in the ICC report where major companies demonstrate their public performance. There is an opportunity here for the environmental service industry, e.g., through the Waste Audit. Here a service company works with a generator to identify possibilities to reduce waste, but almost always emerges with a contract to manage the ongoing waste management process.

With the EC twelve member states plus Austria, Switzerland and Scandinavia pumping out 2 billion tonnes of waste per year, or 65 tonnes per second, every day; there is ample scope for waste minimisation and shared responsibility.

Different options on the ladder

In 1976, the UK Government first published a *Waste Management Paper*. The latest 1992 'Review of Options'[6] reminds us that 'Waste management in the UK is divided between three sectors: local government, which is responsible for the collection and disposal of household waste; the waste-disposal sector of private industry, which deals primarily with industrial and commercial waste but which increasingly is becoming involved in the collection and disposal of household waste; and producers of waste in large quantities, which are dealt with 'in house' by individual sectors of industry themselves.'

In looking at the different steps on the waste ladder, from waste reduction through stages such as re-use, recycling, energy recovery and incineration on to landfill, economies play a vital role. A Coopers and Lybrand report for the UK Government notes that little progress in waste reduction, recycling or resource recovery will be made until these options become economically viable. This viability is not likely to be achieved until landfill disposal prices are increased significantly from current levels.[7] This will best be accomplished by rapid application of tighter landfill designs and operating standards such as a ban on co-disposal of liquid and solid wastes. These changes will increase landfill pricing and improve the environmental performance of remaining landfills.

Unlike arbitrary disposal taxes, the raising of landfill design and operating standards and their strict enforcement, has the dual effect of tightening levels of environmental protection, as well as raising landfill disposal costs. The rigorous enforcement of strict standards will create a

Figure 7.4 Recycling household waste

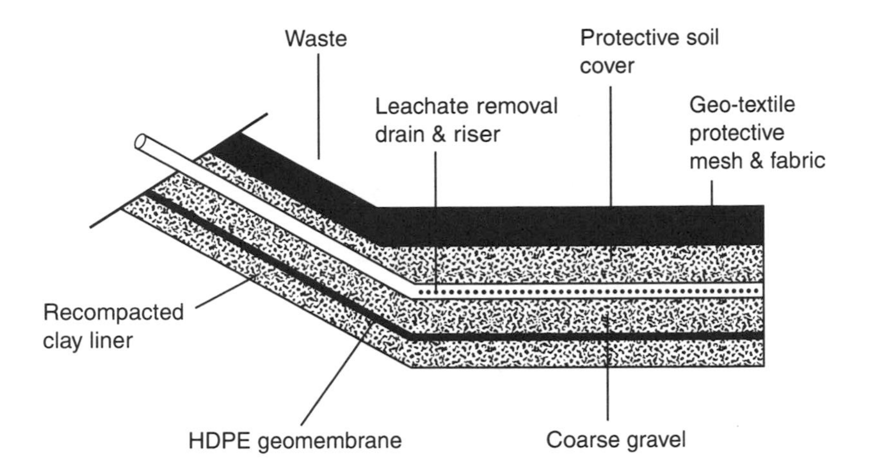

(a) Typical leachate collection system detail

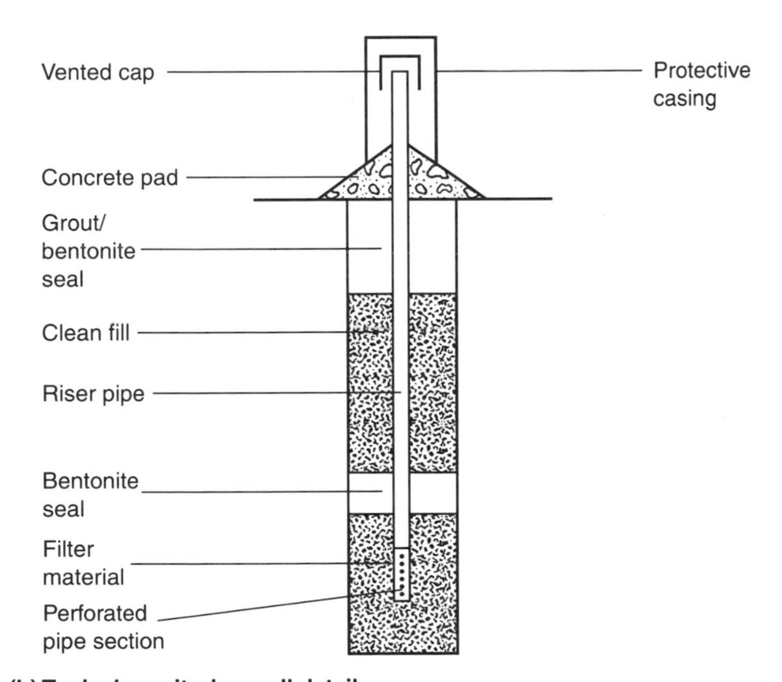

(b) Typical monitoring well detail

Figure 7.5 Preventing contamination of ground and groundwater

demand for significant increased investment in proper waste-disposal equipment and services which in turn will result in opportunities for the development of an active environmental pollution control industry.

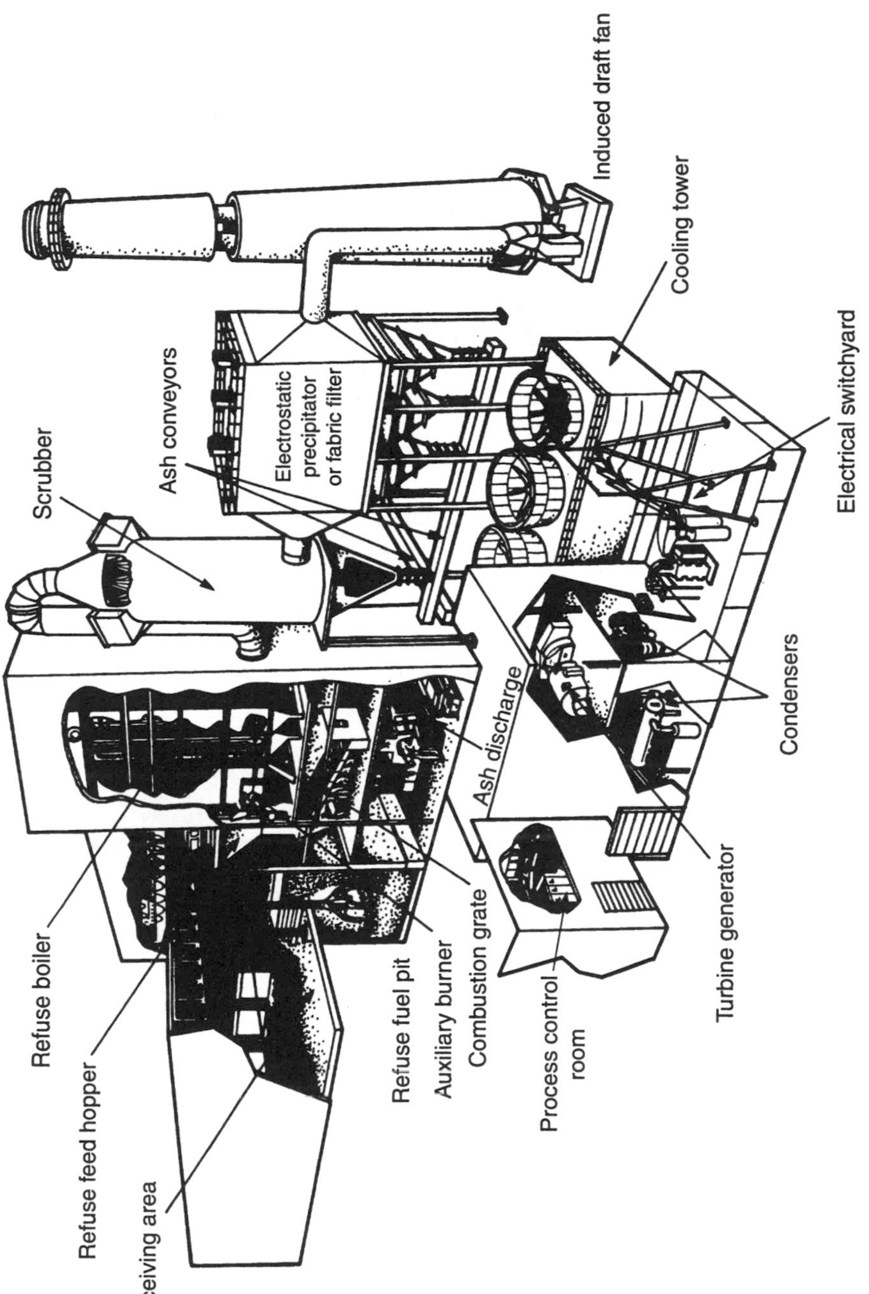

Figure 7.6 A trash-to-energy system

(Source: Wheelabrator Environmental Systems Inc.)

A range of options

Unquestionably, one of the necessary characteristics of a company wishing to take advantage of future opportunities will be an integrated capability: 'One-stop shopping with consistency of service, regardless of the location.' High environmental standards require the most up-to-date technology and integration of service:

1 Recycling of household waste (Fig. 7.4).
2 Modern landfill systems must also ensure that no pollution passes from the waste into surrounding/underlying ground and groundwater (Fig. 7.5). The design of an appropriate liner system is determined by the type of waste which the facility will accept, also the hydrogeology of the site, and the site's location.
3 Well engineered systems for leachate collection, groundwater monitoring, and methane-gas recovery are also vital to high environmental standards.

Waste to energy facilities are also an area of growing importance (Fig. 7.6). Modern control systems and technology are an increasingly effective alternative source of energy, with the capability to maintain a high degree of environmental pollution control.

Many other areas of waste-management expertise will be open to the integrated, high-technology service company of tomorrow. For example, a contract filtration service may be designed to meet client-targeted recycling, re-use, or disposal goals (Fig. 7.7).

Major opportunities will be open to those companies with an advanced technology base, and a wide spread of integrated services. These companies must offer the highest standards with respect to legislative and regulatory compliance, and a comfort factor to clients with respect to liability. The variety and size of the waste sector (see Table 7.3) will ensure that great potential is open to service companies with the appropriate range of technologies.

Table 7.3 Annual UK waste arisings

Waste Source	*Percentage of total arisings*	*Arisings – millions of tonnes per annum*
Controlled wastes		
Household	4	20
Commercial	3	15
Industrial	14	69
Demolition and construction	6	32
Sewage sludge*+	–	1
Other wastes		
Agriculture	48	250
Mining and quarring	21	108
Dredging Spoils*	4	21
Total	100	516

* Dry weight
+ Sewage sludge is only a controlled waste when landfilled or incinerated but not otherwise.

Source: *Waste Management Paper No. 1*, 'A Review of Options', Department of the Environment, HMSO, 1992

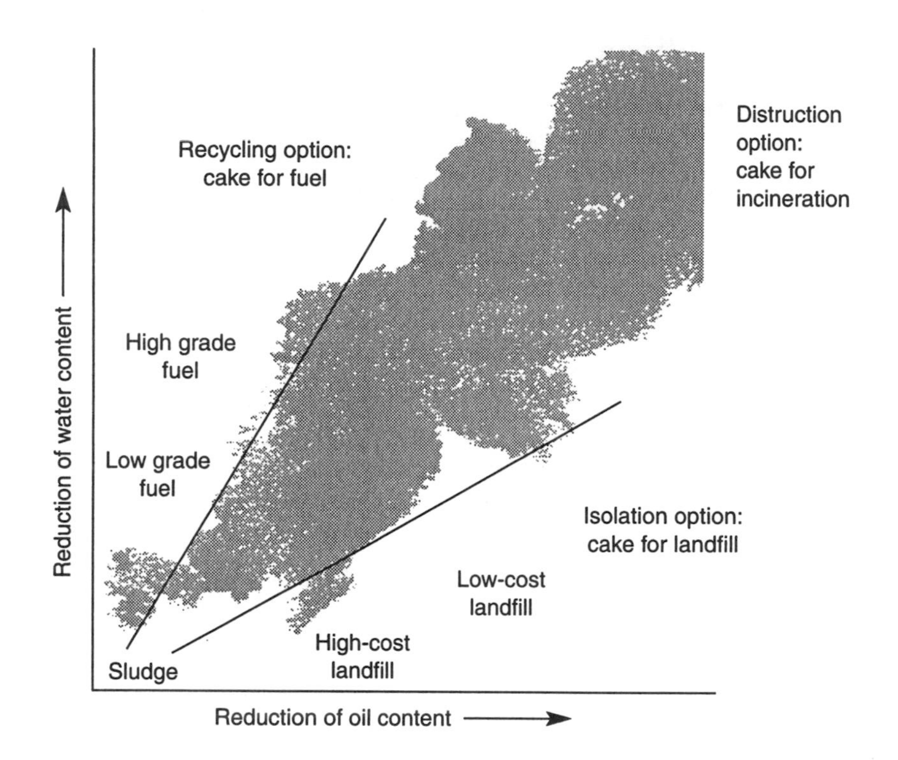

Figure 7.7 Recycling, re-use or disposal?

References

1 World Watch Institute of Washington annual report 9 January 1993. (See also article in *Europe Environment* No 402, 19 January 1993.)
2 Eurobarometer study of March 1992 (EC Commission), 13,000 people in 12 member states. Socio-demographically representative with individuals aged over 15 years.
3 Remarks by New York City Controller, Elizabeth Holtzman, before UK presidency and EC Commission conference, 'European Business and the Environment', November 1992.
4 *Spectrum,* Environmental Management Industry, issue 23 September 1992. Source Arthur D. Little Inc. (Authors Bernhard H. Metzger, and Jonathan Shopley). Articles on European Environmental Control market.
5 Willums, Jan-Olaf and Golüke, Ulrich. *From Ideas to Action: Business and Sustainable Development*, ICC report on the Greening of Enterprise. ICC Publishing and Ad Notam Glydendal, Oslo (1992).
6 Department of the environment (UK). 'A Review of Options', *Waste Management Paper No. 1*, guidance on the options available for waste treatment and disposal, HMSO (1992).
7 Letter from E. Falkman (Chief Executive, Waste Management International plc) to the Rt Hon. Michael Howard, QC, MP, Secretary of State for the Environment, Department of the Environment, London.

Recommended reading

Bird, I. B. *European Community Law after 1992. A Practical Guide for Lawyers Outside the Common Market*, Ch. 8 'European Community Environmental Policy and Law', Kluwer, London (1993).

Daley, P. 'Setting Environmental Priorities: Are We Meeting the Challenge?' The Royal Academy of Engineering and the Royal Society Lecture (1993).

Department of the Environment (UK). 'A Review of Options', *Waste Management Paper No 1*, guidance on the options available for waste treatment and disposal, HMSO, London (1992).

Elkington, J. and Burke T. *The Green Capitalists: How Industry Can Make Money and Protect the Environment*, Gollancz, London (1989 reprint).

Environmental Resources Ltd. *Economic Instruments and Recovery of Resources from Waste*, a study for the Department of the Environment (UK), HMSO (1992).

Falkman, E. G. 'The Environment and Business', *Finance International*, (November 1992).

Meadows, Donella H., Meadows, Dennis L. and Randers, Jorgen. *Beyond the Limits: Global Collapse or a Sustainable Future*? Earthscan Publications, London (1992).

Multi-author, including reports on UNCED Rio Conference and Globe 92 in Vancouver. *Environment Strategy Europe*, Campden Publishing (1992).

Renaux, G. *EC Waste Policy*, study written and published by Club de Bruxelles (1993).

Willums, Jan-Olaf and Golüke, Ulrich. *From Ideas to Action: Business and Sustainable Development*, ICC report on the Greening of Enterprise. ICC Publishing and Ad Notam Glydendal, Oslo (1992).

8

Developing environmental opportunities in industrial products

A look at a life cycle model

ALLEN H. ASPENGREN

Manager, Environment, Health and Safety,
3M Europe

- Environmentally sound products and processes can mean higher quality, lower costs, better marketability, less regulatory impact, fewer liabilities, better employee morale, and a better company image.

- The race is on to develop products that can be manufactured, used and disposed of with the least possible effect on the environment.

- 3M's environmental philosophy is to solve pollution and conservation problems at their source whenever possible. Pollution-prevention at source is carried out through product reformulation, process modification, equipment design, and resource recovery.

- The 3M Life Cycle Model puts this into practice, and looks at each phase of product development – product concept, design, process, distribution, use and disposal – and helps reduce the impact of regulatory compliance on 3M and its customers.

- Examples of the use of the Life Cycle Model include:
  - ensuring adequate chemical management throughout a product's life cycle, where performance requires the use of a toxic chemical;
  - designing and manufacturing products so as to manage risks to employees, users and the environment;
  - examining all aspects of customers' use and handling of products to reduce waste, by-products and emissions;
  - conducting extensive risk assessments covering all pertinent types of risks once a product design is finalised.

- 3M establishes systems to anticipate and respond to significant changes throughout a product's commercial lifetime, for example, from changes in markets, customer applications, process, volume, or manufacturing sites.

Introduction

The demand for environmentally improved products – from both governments and customers – will continue to grow rapidly around the world. The race is on to develop products that can be manufactured, used and disposed of with the least possible effect on the environment – 'sustainable products', as they are called now. Indeed, the age of the truly recyclable car and monocomponent computers and copiers is virtually upon us.

As companies consider opportunities for making and marketing sustainable products, two issues are crucial. First, to be credible, top management must vigorously and *visibly* support continuous environmental improvement throughout the company. Second, this commitment should be expressed in a clear, consistent environmental philosophy which is reflected in the company's policies and investment priorities. Usually, employees, customers and regulators can tell whether a company's environmental commitment is genuine or merely an opportunistic attempt at exploiting a social trend.

At 3M, this philosophy is best stated in the *environmental policy* we adopted in 1975. In the policy we made a commitment:

- To solve our own pollution and conservation problems.
- To prevent pollution *at the source* whenever possible.
- To develop products that have a minimal effect on the environment.
- To conserve natural resources through reclamation and other methods.
- To assure that our facilities and products meet and sustain the regulations of all country, state and local governments.
- And to assist government agencies and other organisations in their environmental responsibilities.

This philosophy, as expressed by the policy, has stood the test of time. Nearly 20 years after it was first put down on paper, it still has the unwavering support of 3M management. It has also spawned a network of policies and practices which affect virtually every operation at 3M.

The life cycle model

Clearly, this philosophy is at work in the model for product development – called the *Life Cycle Model*[1] – developed by 3M's Corporate Responsibility Group. As a model, it is not meant to be all-encompassing, but to provide a checklist of environmental issues to be considered when developing a product at 3M:

Table 8.1 Life cycle model

Product Life Cycle Phases	*Key Elements*
Product Concept	Define customer performance requirements Anticipate public and governmental product safety and environmental requirements
Product Design	Select components: – Minimal toxicity – Renewable resources where possible – Re-usable/recyclable components – Reduced energy usage – Safety Packaging: – Minimal packaging – Renewable/recyclable/disposable – Non-toxic Product Risk Assessment
Process	Safety/health/environment assessment Waste and emission reduction Efficient use of raw materials and energy
Distribution	Safe transportation/handling
Use	Safe handling recommendations Potential misuse identified and addressed Waste and emission reduced Energy use minimized
Dispositon	Re-usable/recyclable/returnable/degradable Disposal recommendations

This model provides a clear set of product-development guidelines:

1 Ensure that 3M products can be manufactured, distributed, used, and disposed of properly.

2 Select safer and more environmentally responsible options among materials, components and processes.

3 Make efficient use of energy and natural resources.

4 Emphasize our environmental hierarchy: source reduction, re-use, recycling and safe disposal.

5 Reduce the impact of regulatory compliance on 3M and our customers.

These are dynamic concepts that must be applied repeatedly as the product changes during development.

Let's look at each of these phases – product concept, design, process, distribution, use and disposal.

PRODUCT CONCEPT

In this stage we gather input from marketing, technical and manufacturing functions, from regulatory groups, and especially from our customers. Customer input is very important because it helps determine the design and product performance expectations. Here is an example from our Home and Commercial Care Division. Our surveys showed that consumers were dissatisfied with ordinary steel wool pads, which rust and splinter over time. With this is mind, we set about developing pads which will not splinter or rust. We developed the ScotchBrite Never-rust Soap Pads. In this case, we were able to make the new soap pads entirely of recycled materials – ground-up plastic soda pop bottles.

As we consider customer preferences, we also need to look at the potential safety, environmental and regulatory issues associated with these preferences.

DESIGN

This stage involves designing both the product and the manufacturing process. Here are some of the issues that should be considered during this phase:

- Select materials and components to minimize toxicity in the manufacturing process and final product.
- Consider alternative chemistries that reduce risks – of harm to health or the environment, of regulatory liability, and of adverse public reaction. Where possible, choose chemicals that are less toxic, that do not appear on various toxic chemical lists, that are not thought to deplete ozone, and that are easy to handle.
- Where performance requires use of a toxic chemical, ensure adequate chemical management *throughout the product's life cycle*. Some chemicals warranting special attention include: volatile organic compounds, ozone-depleting chemicals, heavy metals, persistent and bio-accumulative chemicals known as carcinogens, and new chemicals of concern to regulatory bodies.
- Consider the use of recyclable, renewable or degradable materials. For example, 3M has developed a line of Post-it brand notes from recycled paper.
- Consider the environmental hierarchy in selecting components to facilitate re-use or recycling; use the least number of materials in both the product and package. Is there an opportunity to conserve natural resources by incorporating recycled materials?
- Design and manufacture the product in such a way as to manage risks to employees, processors, users and the environment. What are the

product's characteristics? Where is the potential for misuse? For example, 3M is making a strong effort to reduce use of hydrocarbon solvents in its manufacturing processes.

- Examine all aspects of the customer's use and handling of the product to reduce waste, by-products and emissions. Can customer's energy use be reduced?
- Ensure that employees involved in the product and process development fully understand any hazards associated with the materials they are using and take appropriate precautions for safe handling.
- Conduct an extensive risk assessment when the design is finalized. Consider all pertinent risk parameters (human health, environmental compatibility, physical safety, regulatory compliance, public perception and strict liability) at *each stage* of the product's life cycle.
- Establish systems to anticipate and respond to significant changes throughout the product's commercial lifetime. This might be a change in markets (i.e., from industrial to consumer, from health care to industrial or consumer, etc.), in customer applications, in process, in volume, or in manufacturing sites.

PROCESS

This third stage includes all potential air, water and waste emissions from the manufacturing process. At 3M, once these are assessed, we concentrate on finding ways to eliminate or reduce the pollutant at its sources, rather than merely finding ways to control it. Eliminating pollutants at the source is the backbone of *3M's Pollution Prevention Pays (3P) program*, which has produced outstanding results since it began in 1975.

Table 8.2 The 3P record 1975–1993

Pollution prevention	*US*	*International*	*Total*
Air pollutants	158,000 tons	19,000 tons	177,000 tons
Water pollutants	16,800 tons	1,400 tons	18,200 tons
Sludge/solid waste	433,000 tons	26,000 tons	459,000 tons
Wastewater	2 billion gals.	0.70 billion gals.	2.7 billion gals.
Approved 3M projects	1,173	2,990	4,163
Savings	$577 million	$133 million	$710 million

The figures above represent savings only from the first year of each project. Projected over a period of several years, the pollution prevention becomes even more significant. The benefits from this program as noted above have resulted in a better environment, conserved resources (both natural and financial) and improved technologies.

The 3P concept is carried out in four main ways:

- *Product reformulation* – This involves the product design phase where non-polluting products are developed by using different raw materials or feedstocks.
- *Process modification* – Looking at the process and finding ways to control by-product formation or incorporate non-polluting or less polluting raw materials.
- *Equipment design* – Modifying equipment to minimize pollution.
- *Resource recovery* – Recycling materials is the key to this element.

Some examples[2] are as follows:

1 Product reformulation

- Many 3M products, in Europe and elsewhere, that originally had to be made using solvent-based coating solutions, have been reformulated to use water-based coating solutions. The scrap coating materials and clean-up wastes from making these products are no longer classified as hazardous.
- Many 3M products, such as tapes, are coated with pressure-sensitive adhesives. In some cases, a hot-melt adhesive is substituted for a solvent-based adhesive. As with water-based coating solutions, the scrap coating materials and clean-up wastes for applying the adhesive to these products are no longer classified as hazardous.

2 Process modification

- Copper sheeting must be thoroughly cleaned before it can be used for making electronic products. Formerly, the sheet was sprayed with ammonium persulfate, phosphoric acid and sulfuric acid. This created a hazardous waste that required special handling and disposal. That procedure has been replaced by a specially designed new machine with rotating brushes that scrub the copper with pumice. The fine abrasive pumice material leaves a sludge that is not hazardous and can be disposed of in a municipal landfill. This process change eliminates the generation of 40,000 pounds of hazardous waste per year and saves $15,000 per year in raw material and disposal costs. The capital cost was $59,000.

- In 3M's plant in Germany, cleaning kettles used to require 800 hours and meant filling them with 110 tons of solvent annually. The cleaning process involved filling the kettles with the solvent, stirring the solution with a mixer, draining the kettles and even having workers entering the kettles to do some hand scraping. Cleaning one kettle took one employee three hours. A high pressure, rotating spray head is now being used to clean the kettles. Use of this head has greatly reduced the amount of cleaning solution required, the amount of time required for cleaning the kettles to 10–20 minutes and has eliminated the need for workers to enter the kettles. All the solvent is saved for re-use by passing it through a sedimentation tank to remove solids. First-year savings were $61,500. The capital cost was $69,000.
- Adhesive at another 3M plant is made in batches and then transferred into a large storage tank before use. If one batch did not meet the required quality standard, it would have spoiled the entire contents of the storage tank. The rejected material contained solvent and would have to be disposed of as a hazardous waste. A technique was developed for rapidly running a quality control test on freshly made batches of adhesives so that they could be tested and either accepted or rejected before being placed into the storage tank. This has reduced the amount of rejected material that must be disposed of by approximately 110 tons per year at an annual cost saving of about $207,000.

- An adhesive used in manufacturing abrasives that remained in the coater feed tank at the end of the week was sent out for disposal as hazardous waste because it could not be kept over the weekend. A method has been developed for cooling the adhesive, storing it as a solid then reheating it for use on Monday, thereby eliminating the need to scrap it out for disposal. This procedure has prevented the generation of 6 tons per year of hazardous waste and saves the company approximately $10,000 annually.

3 Redesign of equipment

- A Teflon rope-packing seal on a product dryer was found to be contributing to contamination problems with a particular product. The seal was replaced with a mechanical seal which reduced contamination of the product, thereby reducing the amount that had to be sent to waste for disposal. This equipment change led to annual savings of $322,000 through increased yields and reduced maintenance and disposal costs.
- When sampling a particular liquid phenolic-resin product, using a tap on a process flow line, some of the product had to be wasted before and after the sample was actually collected. A funnel was installed

under the sample tap and piping was connected back into the process so that when samples were being taken, no product would be lost. This prevented the generation of about 9 tons of chemical waste per year and saved approximately $22,860 annually in increased yield and decreased disposal costs. The capital cost was about $1,000.

- A product-coating solution cart was redesigned to eliminate places where solution could become trapped. This change has made it possible to use less solvent when cleaning the cart between coating runs, thereby reducing the amount of contaminated solvent that must be disposed. Better cart cleaning has also greatly reduced the manufacture of rejected product. Use of the redesigned cart has eliminated the generation of about 600 pounds of hazardous waste per year and saves approximately $58,000 annually through increased product yield and reduced disposal costs. The capital cost was about $1,200.

4 Reclaim/recycle

- One 3M plant uses a thin-film evaporator for recovering solvents from scrap adhesive. This reduces, by over 80%, the 1,300 tons per year of rejected adhesive (hazardous waste) produced by the plant. At the same time, approximately $480,000 worth of solvent is reclaimed for re-use.

- A liquid chemical product requires final filtration. Filter elements must be changed from time to time and, when this is done, the filter housing must be drained. The chemical product drained from the filter housing used to be rejected and sent out for disposal, but now piping has been installed to capture it and return it back into the upstream equipment. About 35 tons of product are recycled back into the process per year and no longer need to be disposed of. This produces an annual saving of $37,000 per year and only cost $200 for piping.
- Several 3M waste-solvent streams have been found suitable for use as boiler fuel. At one plant it is planned to use these waste solvents in the boiler on site. This will greatly reduce the amount of waste having to be shipped off-site to a disposal facility and will also reduce fuel costs. In this case, 30,000 tons of solvent will be burned per year, producing a net savings in disposal and fuel costs of approximately $630,000 annually. Operating costs will be about $30,000 per year. The total capital cost for boiler modifications, piping, etc., was $330,000.

DISTRIBUTION

Here it is vital to identify ways to handle and transport the product safely – including types of vehicle used to transport it, routes the vehicle will take, proper packaging and labelling, and proper hazard communications for both transporters and customers. Will it be transported through neighborhoods? Will it be used by trained workers in one application and ordinary consumers in another?

USE

The customer needs clear instructions in using and disposing of the product safely. Generally, Material Safety Data Sheets should be included with the products. Beyond that, manufacturers need to think about possible *mis*use of their products. For example, aerosol products using solvents as carriers should carry clear warnings against breathing fumes over a protracted period.

DISPOSAL

The time to think about recycling or disposal of a product and its packaging is in the design stage. Designing a product or package so that it can be re-used benefits both the environment and your customers. For example, 3M ships large numbers of blank videocassettes to companies which duplicate movies for renting. We designed a re-usable package for bulk shipments, so instead of having to dispose of the packages, customers simply return them to us for re-use. This eliminates much of the waste we had created for them.

Real-life processes

That is the Product Life Cycle Model developed at 3M. It is only a model. In real life, the process may not be this orderly.

In going through this life cycle model one can see how a truly intelligent product can be made. As this is just a model it is important to remember, that while going through the various steps, and stopping and restarting the whole process in order to develop a better product that is environmentally safe and sustainable, may sometimes be necessary.

Moreover, this whole process should be linked to a total quality environmental program. Environmentally-sound products and processes can mean higher quality, lower costs, better marketability, less regulatory impact, fewer liabilities, better employee morale, and a better image for the company.

Employees need to see that developing environmentally-sustainable products is important – to the company, to the customer, to our environment and to themselves.

References

1 3M Product responsibility guidelines, *Life Cycle Model.*
2 Hunter, John S. and Benforado, David, M. (3M Company). *Life Cycle Approach to Effective Waste Minimization*, unpublished presentation at the 80th Annual Meeting of APCA, 21 June 1987.

9

Developing environmental opportunities in a service business

Accounting and consulting services

MARTIN HOULDIN

National Environmental Unit,
KPMG Management Consulting, London

- For services businesses, environment can mean two things:
  - no option but to adopt new standards and techniques, and to manage environmental pressures (or fail to compete or minimise liabilities);
  - new opportunities through development of new markets or new ways to differentiate products and services.
- Meeting ever higher standards cost-effectively depends very much on managements' ability to develop 'cleaner' processes and to reduce or eliminate problems at source.
- Environmental management is now moving into purchasing, supplier relationships, information systems, and finance, and is influencing changes to processes and management practices faster than many organisations expect.
- In areas such as financial audit, corporate finance, receivership, and management consulting, environment must be considered alongside other elements of business performance and risk, as a matter of course.
- There is growth potential in development of specialist services, but there could be even greater market potential gained from integration of environment with some other service areas.
- Environmental management has more to do with integrating environmental considerations with normal management functions and in all business areas, than with development of separate management systems.
- Using, and developing, appropriate environmental performance measures is a key aspect of meeting regulatory requirements, and of internal reporting, performance management, public relations, benchmarking and decision-making.
- Environmental management and performance measurement require the availability of the right information – companies need to understand and manage their information needs, and to put in place processes and systems for this.

The environment and the service sector

Service businesses are now so integral with industry and commerce that any issue of importance, such as new legislation or the development of new technologies, is likely to present opportunities and risks to at least some service providers. The environment is no different in this respect. Indeed since the late 1980s we have seen a dramatic shift in environmental management thinking, with the effect that managers now recognise the much wider business implications of new legislation and a growing general interest in the environmental performance of business. Over a relatively short period a large number of businesses have come to see environmental protection as a business and management issue rather than solely one of science and technology. Improving environmental performance and, more to the point, using this to commercial advantage, involves activity in a range of business functions, such as product development, sales and marketing, public relations/external affairs, financial management (especially relations with investors and the financial community), among others. The environment has therefore become firmly rooted in business and how it is managed.

This broader, more holistic perspective brings a whole range of service providers closer to the environmental challenge. These businesses are provided with an opportunity to contribute to the environmental performance of their customers/clients. However they must also be able to respond to new service needs and changes in the specification and contractual performance of existing services, to remain competitive. In some cases services providers could be just as much at risk as enjoying the opportunities, as a result of their clients' environmental position – lending services are one example. Table 9.1 illustrates how some of the services which a typical industrial organisation uses might be affected by new environmental legislation and developments in environmental management.

Perhaps the more obviously affected businesses are in the legal profession, also banking and waste. We have, in just a few short years, seen major changes in the standard and value of waste treatment and disposal services. As a result, landfill costs are expected to double by the end of the century. The waste industry itself has undergone rapid rationalisation; consolidation has resulted in there now being only a handful of major players.

Similarly, it would be hard to find a city-based or large provincial law firm that had not focused on new and emerging environmental legislation, either by creating an environmental team and/or by providing specialised briefings to clients. One might argue that they have little option but to take account of any new legislation. There is strong evidence however that in a number of cases the EPA has encouraged them to branch out beyond traditional areas of service; some are even carrying out environmental reviews and site assessments for clients.

In addition, the banking sector has responded in different ways; some are making major changes to routine procedures and practices but a significant number of large banks seem unconvinced that the environment represents either an opportunity or a threat. There is, however, a growing number of venture capital funds devoted specifically to investment in companies/businesses which are perceived to be operating in growth areas. There seems little doubt, by consensus, that the environmental technology and services sector is set to grow (worldwide) from around $200bn in 1990 to an estimated $300bn by the year 2000 (OECD figures). One suspects that banks will not miss this kind of opportunity – some are already ahead of the game. On the risk side, banks are also gradually introducing the environment into credit risk analysis and, much more so, into investment appraisal and due diligence work for acquisitions and Management buy-outs, and other corporate deals.

So, from these examples, it is clear that the service sector is not only affected but recognises the commercial opportunity. Furthermore, we can imagine that some of these service businesses will strongly influence manufacturing and commercial businesses, either through the availability of service (new treatment technologies) or through their own decision-making (investment and lending against environmental risk).

In conclusion of this overview of the interaction between the environment and services businesses, it is clear that the driving forces for environmental management (as illustrated in Fig. 9.1) affect service businesses as well as industrial and commercial businesses, either directly or indirectly. Indeed, these forces and the businesses are so intertwined that service businesses are just as much players, and can also be driving forces in their own right. What does the environment mean for service businesses? Two things, and often a mixture:

1 No option but to adopt new standards and techniques or fail to compete, and to manage the potentially negative effects of environmental pressures.

2 New opportunities through the development of new markets or as a result of new ways to differentiate product and service.

The remainder of this chapter examines one particular type of service, namely that of accountancy and management consulting. The case is provided by the experience of KPMG Peat Marwick and the development of its National Environment Unit (NEU) in the UK. We will see how parts of the business have needed to introduce new practices, how others are competing more strongly as a result of being able to offer environmental advice as an integral part of service, and how the environment has led to the opportunity to develop a new source of income.

Table 9.1 Implications of environmental legislation and management for service businesses

Services	*Opportunity*	*Threat*
Legal	• Service differentiation • New sources of income	• Failure to give adequate advice • Unable to provide full service
Banking (Lending)	• Growth in environmental business sector	• Credit risk
Waste treatment/ disposal	• Growth in demand for higher value services (e.g., incineration)	• Potential legal liability • Failure to compete on standards and service quality
Personnel/training	• Growth in demand for recruitment of environmental skills • Demand to develop environmental skills	• Failure to advise clients (e.g., re training needs)
Engineering (and related)	• Demand for design/ installation of cleaner process • Demand for new energy and water management services	• Unable to provide full service/ advice on latest developments
Cleaning (industrial and offices)	• Service differentiation	• Failure to meet new client standards • Potential non-compliance on waste disposal
Corporate Finance	• Service differentiation • New source of income	• Failure to meet reporting standards/requirements • Failure to advise clients on material issues
Corporate Recovery/ Receivership	• Service differentiation	• Potential direct liability • Failure to advise clients on major issues
Management Consulting	• Service differentiation • New source of income • Market profile (environmental management issues)	• Failure to advise clients (e.g., business strategy)
Market Research	• Service differentiation • New source of income	• Failure to recognise significance as business/political/public issues
Contract Processing (e.g., power plant operation)	• Service differentiation (e.g., understand regulatory requirements/advise customers)	• Failure to meet regulations/ client standards (e.g., potential EPA prosecution)
Catering (industrial)	• Service differentiation (e.g., contribute to client environmental performance	• Failure to respond to new clients requirements

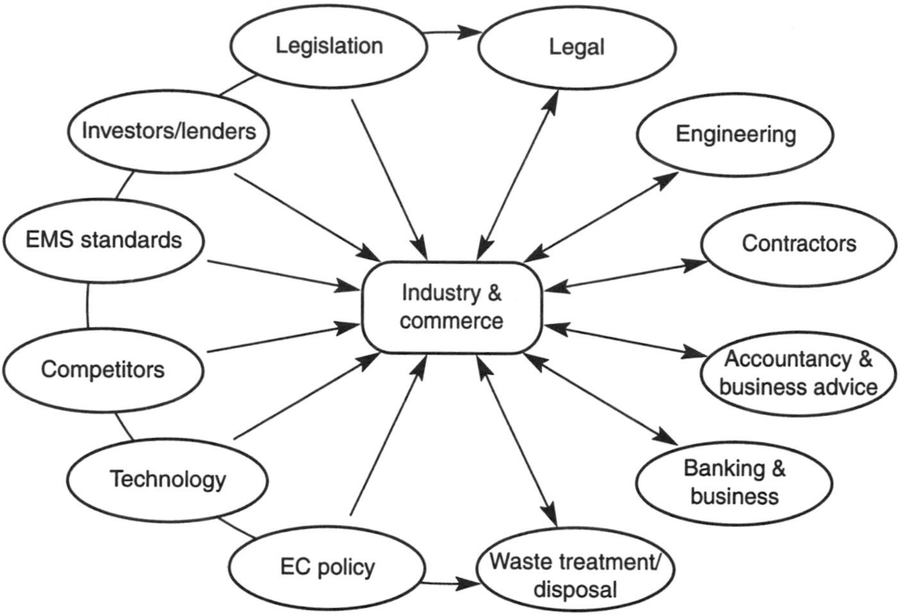

Figure 9.1 Driving forces for environmental management/change

About KPMG and the environment

Those people who know of KPMG's National Environment Unit, either through direct contact or references in environmental publications and directories, could be forgiven for thinking that the environment and KPMG is essentially about creating a successful and growing environment management consulting unit. Just as important to the firm, however, is the other side of the National Environment Unit, integrating environmental management issues throughout the services which the firm provides. Also important is its role of managing KPMG's own environmental policy and performance-improvement programmes.

KPMG is a firm which operates worldwide in 124 countries and with over 73,000 staff, generally within the accountancy and consulting fields. Services are mostly in the areas of:

- financial audit;
- tax consultancy;
- company receiverships;
- corporate finance including special accounting investigations, and due diligence for mergers/acquisitions and MBOs; and

- management consulting in information and systems management, business strategy, financial management and human resources management.

In 1989 senior management recognised the growing importance of environmental legislation and the profile that some environmental issues were getting, and that these would have important implications for all types of businesses and organisations – our clients. A small environmental team was formed, combining business, management and environmental skills. The research and analysis carried out in those early days led to three key conclusions which overall indicated strongly that environmental issues would be of major significance to many of our clients, and therefore to the way in which KPMG delivered its services:

- the solutions to environmental problems would be as much managerial as technical. In the past there had been a tendency to develop predominantly end-of-pipe solutions to waste and emissions problems. We recognised that meeting ever higher standards cost-effectively would depend very much on managements' ability to develop 'cleaner' processes and to introduce new practices which were more effective in climinating or reducing the problems at source. In a world where more and more companies were expected to meet higher environmental standards, competition would be focused on doing so more cost-effectively and to greater market advantage;
- changes to processes and management practices would need to take place faster than many organisations expected, and be implemented in business functions beyond manufacturing operations. Recent developments, as reflected by the introduction of environmental management systems standards, seem to bear this out. We are now seeing environmental management moving into purchasing, information systems and finance; and
- there would be significant growth in demand for environmental skills in the management and technology fields. We are now witnessing growth in technology and services of the order of 7% per annum. In addition, many organisations seeking to recruit environmental specialists are finding them in short supply.

The business case

In assessing the implications for KPMG, we based our planning on two assumptions. The first was that in most parts of our business we had no option but to raise our staff's awareness to the potential importance of environmental legislation and related developments. In these areas we would need to provide training and develop methodologies for our services to take environmental considerations into account where relevant.

The second was that there were opportunities to attract new clients through added value to existing services, and to develop new services that would result in a new source of fees. It was also recognised that these capabilities and services should not be provided in the UK alone but should be available internationally.

A business plan was therefore developed in four areas:

- environmental management services to private sector clients;
- advice on the development and implications of government policy and legislation;
- integration of environmental considerations into other KPMG services, in particular in the areas of financial audit, corporate finance decisions and corporate recovery;
- provision of services to companies in the environmental technology and service businesses, aimed at helping them to take advantage of the increased demand for technologies and skills;
- development of KPMG's own environmental policy and improvement programmes, in anticipation of client requirements in this area.

The essence of our plans at that time was that businesses, both KPMG and its clients, need to integrate environmental objectives with normal management processes and operational activities. To achieve this, both managerial and technical solutions need to be combined within the business context as part of the business plan.

Since 1989, the National Environment Unit has grown to become recognised as one of the leaders in the field of environmental management, and in dealing with accounting and finance-related issues, particularly due diligence work. The environment also has a profile within the firm which is out of proportion with the fact that the NEU is a team of 15 within a UK staff of over 9,000; most people now know where to go for advice on environmental issues. Financially, as a profit centre, the unit is self-contained; other benefits to the firm are measured in terms of the contribution made towards winning work led by other business areas and in terms of the market profile which is generated through high profile and leading edge projects.

Partly as a result of these initiatives in the UK, other parts of KPMG's international network have put in place their own plans and introduced environmental specialists into their accounting and consulting teams. KPMG now has environmental specialists in 12 countries in Europe and North America. Other teams are setting up in South Africa, New Zealand, Australia and the Far East.

Implications for accountants

Accountants at KPMG are involved in providing a range of investigative and accounting-related services. The most important of these fit into the categories of:

- financial audit;
- corporate recovery/receiverships; and
- corporate finance decision-making.

Although their work can be quite different, the ways in which environmental issues need to be taken into account are based on a common set of issues. These relate to potential legal and financial obligations as a result of environmental legislation, to the investment required to meet regulatory standards and to the revenue implications of market developments.

Financial issues

Current and some of the more recent prospective legislation (shown in Table 9.2) can have considerable financial implications for business, in some cases the impact will be so great as to threaten viability.

Table 9.2 Key environmental legislation with implications for accounting

Current
UK Environmental Protection Act 1990

- Integrated pollution control
- Local Authority pollution control
- Duty of care
- Waste management licences
- Register of potentially contaminated land

UK Water Resources Act 1991

Prospective
EC Waste Directive – Landfill
EC Integrated permitting
EC Eco-management and auditing regulation
EC Incineration of hazardous waste
EC Environmental liability (Green paper and convention)

The implications for business are twofold. First while they continue to operate, businesses could incur additional costs associated with compliance. The potential magnitude of these costs, can mean more than the effects on profits and balance sheet values. These in themselves can cause investors and bankers to take a different view of the business either in terms of earnings potential or business risk. These issues can therefore affect ability to attract finance in its various forms, and the value given to a business in terms of shareholder value. Implications could include:

- current plant may be inadequate to comply with legislation. If operations are to continue both the cost of the upgrade required and the impact on the life, and hence the depreciation charge, of the current equipment could be significant. The purchase of any new equipment and the method of financing will have a resultant impact on the company's gearing;
- raw materials used, or the finished product, may have to be modified/discontinued where usage is either banned (CFC's) or where public pressure and attitudes are such that demand for 'un-environmentally friendly' products declines. Research and development costs associated either with modification of the materials used or the product produced may also be material;

- prior or current operations may have led to contamination of the site. The costs associated with clean-up and site remediation, if this is required by the regulators, could be significant; and
- fines and penalties that will be imposed on businesses acting in breach of the legislation as described earlier are now substantial and there is already evidence that law courts are enforcing these larger fines.

The second business implication is that, should business operations cease, environmental legislation could have the following impacts for lenders and insolvency practitioners:

- as described above, plant may prove to be inadequate to ensure that the operations comply with environmental legislation and this may affect resale value. Resale values of stocks of raw materials and finished goods may also be affected for the reasons outlined above. Both of these will impact on the value of any floating charge over such assets;
- contamination of land could have a significant impact on resale value and hence any charge held. If an investigation of the potential contamination is felt to be necessary, this will delay a potential sale. In cases where there is an inability to give warranties as to the condition of the site this may further inhibit the sales;
- the nature of any potential liability of banks and insolvency practitioners for environmental contamination is currently unclear. However in the light of recent case law in the US, and the EC draft direc-

tive on civil liability for damage caused by waste, it would be prudent to take account of any of the companies activities that are liable to give rise to pollution or contamination, particularly where trading continues prior to sale;

- evidence of a lack of environmental management systems in the business could affect the purchaser's view of potential risks and hence the price to be offered for the company.

Changes to services

The services delivered by accountants, in this section, have not fundamentally changed in themselves as a result of environmental issues. What they have in common, however, is that in each case work and advice depends on an assessment of business risks, and how important these are in relation to the work being carried out (for example, the acquisition of a company) and to other business risks. The environment has become an additional factor in business that can represent a significant business risk. Using the acquisition example this could mean a number of things such as:

- asset values are reduced compared with the latest valuation and compared with book value;
- there may be potential liabilities requiring remediation of environmental damage;
- the market impacts of, for example, competitors producing environmentally-improved products have not been adequately considered in profit projections; and
- the need for capital expenditure to meet legislation has not been adequately provided for.

Professionals in these fields therefore need to rely on being made aware of the key issues, on being able to access specialist information and/or advice, and on having the environment built into normal processes or methodologies.

The National Environment Unit plays a key role in:

- providing briefing seminars and materials that raise awareness progressively and consistently over a period of time. The emphasis is primarily on legislation (current and prospective) and identifying those issues (opportunities and risks) which are relevant to the different types of work and clients;
- building relevant environmental points into existing methodologies and standard documentation;

- building environmental modules into existing training programmes;
- helping to produce relevant newsletters and briefing materials aimed at keeping clients up to date and raising market profile; and
- providing specialist input to client projects.

Environmental management consulting

There is a plethora of different terms in the environmental field, and, specifically, in environmental management. How many different kinds of 'audits' can you think of or have you come across? We have become obsessed with definitions. The term 'environmental management' is so generic, however, that it requires some explanation for the purposes of this chapter. Where environmental management is reasonably well developed in a company we can expect to find that it involves:

- specialist processes, such as environmental review and life cycle analysis, aimed at determining where changes are required to raw materials used, to processes and their control, to products and their effects (including use and disposal) and to management systems;
- normal management processes and systems which have been adopted to include environmental objectives aimed at implementing these changes. These will include product design and development, business planning, procurement, manufacturing, production engineering, finance:
- normal management processes and systems aimed at implementing change, controlling risks and measuring business performance, which have been adopted to cover environmental risks and performance. These will include performance measurement systems and management reporting, internal auditing, training, and external reporting.

Most people who have in some way already been involved in environmental management will accept that meeting the requirements of environmental legislation, and working to improve environmental performance in other areas, involves in some way changing the business and how it is managed, potentially on a broad scale over a period of time. There is also fairly wide recognition that these changes, and their success, depend just as much on attitudes and the behaviour of management and staff (company culture) as on procedures and systems.

Given this, we can see how environmental management consulting needs to address both the specialist areas and those where environmental objectives need to change existing practices, processes and systems. Without both capabilities, or at least the ability to link the two, we believe that our clients may not implement an effective form of environmental management.

Specialists

Both aspects of environmental management require some specialist skills, whether a team of environmental consultants is working to carry out an environmental review, assist in the development of policy and in setting targets, or where an environmental consultant is advising on the requirements for environmental information as part of a new systems development.

KPMG has deliberately sought to recruit consultants who are able to fit into a multi-disciplinary team and who can demonstrate a balance between experience in business, knowledge of management systems and the right kind of technical qualification. The National Environment Unit team is now made up of:

- environmental scientists
- chemists/chemical engineers
- accountants
- business graduates
- economists (including environmental economists
- chartered engineers
- management consultants

Integrated environmental management

In recognition of the need to build environmental objectives and procedures into all areas of management, as a fundamental aspect of environmental management, KPMG's National Environment Unit has worked with other specialists within the Management Consulting division. Most groups of specialists within the firm either have, or are working to adapt, their own services to include relevant environmental features where relevant; notably in manufacturing, banking, marketing, business strategy and financial management. Three areas in particular, however, have been the focus for research and development, and these are used to illustrate how KPMG's management consulting services are developing:

- performance measurement;
- information management; and
- purchasing and supply management.

Performance measurement

KPMG's management consultants specialising in financial management and business strategy development have always assisted clients in performance measurement. In the past the focus for business performance has tended to be on financial and related measures such as product costs, productivity, revenue, market share, budgetary control and so on. Consultants have a long track record of developing and implementing performance measurement systems in these areas.

Recent trends have been towards companies using a range or basket of performance measures – a 'Balanced Scorecard' to use the term that KPMG has given to this approach. Total Quality Management developments since the mid-1980s have been a major factor in influencing managers to recognise that the 'end-result' or 'bottom line' is very much dependent on a wide range of performance factors throughout the organisation, even within departments which have only internal 'clients'. Another development has been the increase in linking performance measures, including those which are non-financial, with performance management and appraisal systems, particularly where a component of remuneration is performance or profit related.

Performance measures at different levels in the organisation, and in non-financial as well as financial areas, are therefore now seen as a key element in management and in introducing desired changes into company practices and behaviour. Management without measurement is nigh impossible; as the teaching profession and its 'customers' are now beginning to appreciate.

Through a development project involving Business in the Environment, and fourteen companies who provided case study material, the National Environment Unit has successfully introduced environmental performance into the Balanced Scorecard approach, to be considered where relevant. The project resulted in the production of a set of guidelines for measuring environmental performance entitled 'A Measure of Commitment'. Companies participating experienced benefits through:

- improvements in environmental performance;
- reinforcement of environmental policy;
- improvements in management systems; and
- greater commitment and motivation for environmental improvement.

These types of benefits are not untypical in other areas of performance measurements.

Purchasing and supply

Setting environmental standards for key suppliers where there are significant environmental issues in the supply chain as a result of materials and services purchased, or where; there is potential environmental risk as a result of the extent to which the customer specifies use of materials and technology, is becoming an essential part of environmental management.

The key business reasons behind the need to build environmental considerations into purchasing can be:

- companies may rely on identifying new suppliers, or new products/materials from existing suppliers to be able to comply with environmental legislation (e.g., elimination of CFC's);
- environmental risks can be important enough within the business so as to threaten the viability of a supplier's company and thereby put at risk the security of supply (e.g., businesses can be forced to cease operations for breaches in licence/authorisation conditions);
- the 'cradle-to-grave' concept is firmly embedded in environmental management thinking and developments, and features in BS7750 as well as the Eco-Management and Audit Regulation. Companies seeking the management, public image, and competitive advantages of obtaining certification to standards, will need to take appropriate actions in purchasing.

Working with KPMG's own purchasing experts in its Centre for Manufacturing Consultancy, with Business in the Environment and with the Chartered Institute of Purchasing and Supply, the National Environment Unit has successfully moved the agenda forward both internally and externally. This work has produced a set of voluntary principles and guidelines for environmental management in purchasing, entitled 'Buying into the Environment'. Together with 'A Measure of Commitment' these projects are examples of how business development works with integrating environmental issues with normal management systems.

Success comes from involving the relevant experts in other fields of management and in demonstrating the market interest (from clients and target clients) and business opportunity. KPMG's manufacturing consultants are integrating the environment with services in Total Quality Management and Purchasing.

Information management

Information systems development, implementation and management are areas which represent the largest part of KPMG Management

Consulting services provision, between 40% and 50%. This fact, coupled with the belief that information provision and management are at the heart of environmental management, led us to do some research. A wide range of external and internal drivers of information demands are illustrated in Table 9.2. For example, consider the requirements of the Environmental Protection Act 1990, and other regulations: a number of actions are likely to be needed to ensure compliance. These could include: changing operator practices, new waste treatment and disposal contractors (and others); investment in new process and/or emissions treatment technology. The list could be longer and include all aspects of manufacturing and process operations. What all these actions have in common, however, is that, for regulatory purposes, a considerable amount of data needs to collected, stored and managed to produce the necessary reports to the regulatory authorities.

This is but one example. Management will be interested in similar information, albeit summarised and in more digestible formats. Similarly, some of this information could be published in corporate environmental reports for consumption by all stakeholders. The availability and quality of information is therefore becoming paramount.

Another example, where availability and quality can be just as important is in the area of product development. Specialists in this field need considerable guidance on which features of the product (materials, performance in use, recyclability) need to be improved. Their work needs to be based on sound information about the relative benefits of one material against another, about the environmental effects of emissions in use, and about the feasibility of recycling. Life cycle analysis is a technique which has been rapidly developed to help with this, and is filling many of the gaps in our knowledge. It can, however, be a cumbersome technique and, in general, is not yet subject to rigorous information management controls.

Environmental management can therefore present, in many companies, a real challenge for information management and systems development. In 1992, we consulted with a number of companies on the scope and priorities for the development of environmental information systems (EIS). The focus for this work, and for a number of discussion groups, is provided by the two main driving forces behind the requirement for EIS:

- the demand for information is growing and multiplying (refer to Table 9.2). Most new pieces of legislation at national and European Community levels require companies to provide or make information available. Particularly heavy 'users' will be related to Integrated Pollution Control, Integrated Permitting, Eco-Management Audit. The development of standards such as BS7750, and eventually the ISO equivalent, will also be information and procedure driven.

There are many groups of different external interests in a company's environmental performance, some for altruistic reasons (such as environmental pressure groups) and some for commercial reasons (such as banks and investment fund managers). Effectively these groups of people are different 'users' of information. Within the company, we see a similarly wide range of interests as environmental management spreads through the organisation to bring at least some change to most business functions; and

- the complexity of information is largely derived from the scale and range of demand. Although much of the data needed is common in basic characteristics (e.g., information on toxic substances, waste), different regulators may set different parameters/definitions (as in 'substances').

  Data must also be gathered from a range of sources which are both external (suppliers, remote sensing) and internal (waste across 20 sites, process control systems, procurement systems). The full range of systems applications will at some point need to be used, including database management, modelling, measuring/monitoring and flexible reporting. These factors will raise many issues.

Emerging from these driving forces the key issues likely to be common to most companies, and which provided a further focus for this project, are:

- *how can such a wide range of information be managed* to ensure effective environmental management? Management will need to cover different geographic locations and different processes throughout different business functions. There is likely to be a need for setting data and reporting standards, and providing some company-wide application tools (such as for managing environmental audits).
- *what information technology systems developments will be needed to meet information needs?* There will at least be a need to develop some new applications, raising the question of how EIS will interface with existing systems (in particular those used for managing Health and Safety). This is likely to require some investment and the setting of technical development parameters/standards.

We concluded that R & D effort is needed to provide a better understanding of our needs for environmental information and how this links with other information areas. To do this we need to consider:

- the full range of environmental management functions, and how other business functions are affected by having environmental objectives;
- the extent to which data across all environmental elements are common, including overlaps with related areas such as Health and

Safety. We will need to determine the lowest common level for data collection and storage;

- the relevance of different applications to the company's environmental management processes, in particular the most specialist areas such as Geographic Information Systems (GIS);
- the availability of suitable data within existing business systems;
- the need for an information infrastructure including parameters and standards in areas such as reporting;
- the business benefits of investing in environmental information systems in a co-ordinated, proactive way.

The results of our work in 1992 led to further work in 1993 which is forming the basis of our service integration work with systems and information specialists in KPMG; we also continue to be interested in hearing from companies who wish to consider R & D work in this field on a collaborative and partnership basis. A model for this kind of work already exists in the shape of the IMPACT programmes which have a proven track record in the information and systems management field.

Conclusion

There is no doubt that the environment presents a mix of opportunities and risks to KPMG's business on a scale which is likely eventually to affect all service areas. Our experience is something that we are happy to share with other businesses. In this rapidly developing field of environmental management, with limited, if any, 'best' practice, learning from experience of others can often be the only way forward or at least can offer the benefit of faster development. This must surely be the value of this *Environmental Management Handbook* and other similar works, and the value, too, of business groups which are forming all over the UK and internationally to explore new solutions and techniques.

From KPMG's experience the key points for service companies are:

- there may be no option but to incorporate environmental elements into traditional services. Failure to do so could lead to market disadvantage or potential liability in some areas;
- there is likely to be growth potential in developing specialist services. There could, however be greater market potential from the integration of environment with some service areas;
- international developments will be important and companies may need to demonstrate international services capability. However, these developments will take shape at differing speeds, with those countries

where there is less 'environmental management' interest or 'infrastructure' (initiatives, professional and business organisations, standards and so on) moving more slowly;

- specialist environmental skills will be required to supplement existing specialisations. The trick lies in being able to identify the right skills and then manage effectively how they fit into the business; and
- the value of bringing in environmental skills and specialist service capability cannot always be easily measured in the traditional financial terms. Companies need to be prepared to, and be capable of, applying a mix of performance measures.

I hope to have given the reader food for thought and to have helped to address the question of 'why is the environment relevant in a service business?' Even if the financial arguments, for and against, don't stand out as being of strategic importance for the short term, we should all ask ourselves whether our businesses are sustainable if we ignore the environment.

Summary

The driving forces for improving environmental performance can be just as important, strategically and operationally, to service businesses as well as to industrial companies. Waste treatment and disposal services, although virtually industrial processes, are directly regulated under most regimes of environmental law. Advisory or contracted-out services are more likely to be affected by customer requirements, often influenced by environmental law and a drive towards quality management.

KPMG Peat Marwick is a service business providing a wide range of accounting and consulting related services on a major international scale. In the UK in particular, but also in several other major countries, the firm believes that opportunities lie in offering environmental management consulting services and specialist advice alongside services. It is also recognised however that in areas such as financial audit, corporate finance, receiverships and management consulting, the environment must be considered as a matter of course alongside other elements of business performance and risk.

KPMG in the UK, has invested in establishing a specialist team – The National Environment Unit (NEU). This chapter has explored the thinking behind the original decision taken in 1989, and examined some of the NEU's work with accountants and in integrating the environment with the day-to-day functions of management.

For *accountants,* whatever their line of work in the firm's different business units, the financial implications of a number of issues need to be considered, notably:

- current plant may be inadequate to comply with legislation;
- some raw materials will be discontinued;
- markets for certain products will grow or shrink/disappear;
- prior or current operations may have led to contamination of the site;
- fines and penalties for breaches in legislation may be punitive.

Environmental management has more to do with integrating environmental considerations with normal management functions and in all areas of business, than with the development of separate management systems. Some specialised processes will, however, become established as part of the emerging environmental profession.

By definition advisers in environmental management need to have a good understanding of all management disciplines, and of business operations – both key factors in the development of the NEU as a multidisciplinary team.

The chapter examined three key areas of management:

- how businesses perform is increasingly being evaluated, internally and externally, on an environmental basis. Using appropriate environmental *performance measures* is therefore a key aspect of reporting, performance management, public relations, benchmarking and decision-making.
- environmental management is no longer only about pollution within the factory gates. Consideration is being given to the products' environmental life cycle – from raw material and natural resource inputs to customer use and disposal. *Suppliers* are an important element in this process. Companies will therefore rely on their working relationships, and information provided by, suppliers.
- effective environmental management, and performance measurement in particular, will not be possible without the availability of the right *information*. Companies need to ensure that information needs are understood and managed, and that steps are taken to put in place processes and systems for its provision.

How service businesses regard the environment will depend on the balance between potential opportunities for new services and differentiation, and the 'no option' need to integrate environmental considerations with existing services. Much will depend on how well businesses understand their markets and specific client needs. KPMG however believes that its business cannot be sustainable if the environment is ignored.

References

1 Gray, Rob with Bebbington, Jan and Walters, Diane; Houldin, Martin (ed. adviser). *Accounting for the Environment*, Paul Chapman Publishing/Chartered Association of Certified Accountants, London (1993).
2 Business in the Environment, KPMG Peat Marwick and the Chartered Institute of Purchasing and Supply. *Buying into the Environment: Guidelines for Integrating the environment with Purchasing and Supply*, HMSO, London (1993).
3 Business in the Environment and KPMG Peat Marwick. *A Measure of Commitment: Guidelines for Measuring Environmental Performance*. Available from Business in the Environment, London, (1992).
4 *KPMG Environmental Briefing Notes*.

10

Environmentally-friendly management in hotels

JOHN FORTE

Director,
Environmental Services, Forte plc, London

- Hotels with successful environmental programmes can communicate the environmental message and influence the thinking of their visitors – and be rewarded by customer loyalty, staff satisfaction, reduction in costs, and increased profitability.

- The first step in implementation of a hotel environmental management programme is to appoint a senior manager to organise a working group of key personnel in all areas of hotel operations.

- Purchasing policy is a central element for action, so as to ensure that the products and commodities purchased by a hotel minimise impacts to the environment; this means that management of the supply chain is of particular importance to hotels.

- Internal audits are used for reducing waste and energy consumption. Hotels can waste up to 40% of total energy used, and much of this waste can be eliminated by sound management practices, staff-awareness programmes, and use of modern technology, such as computerised building management systems.

- Management set targets, based on actual consumption and performance, to make staff responsible for energy, water, and other items of consumption, to monitor progress in environmental performance and to communicate this back to staff.

- Hotels can make judicious use of technology, but should not use gadgets as a means of overcoming bad working practices; training and discipline are far better investments for improving performance.

Introduction

As the environment continues to rise on the public agenda, the hospitality business, the world's largest industry, finds itself ideally in the centre of a huge wheel of activity. Hotels can influence, educate, lead and mobilise many people to join in the one common interest to protect and save our planet.

Each day, hotels are visited by millions of customers. They use the facilities for leisure or work; eating, drinking and sleeping. The staff provide the service, prepare, cook and serve their meals, clean their rooms and respond to all their needs. Suppliers, bring in the materials needed for the operation and contractors maintain the premises and equipment. The hotel with a successful environmental programme can influence the thinking of all its visitors by communicating the environmental message to them. In so doing, the hotel will be rewarded by customer loyalty, staff satisfaction, reduction in costs, and increased profitability, giving its shareholders a good return on their investment.

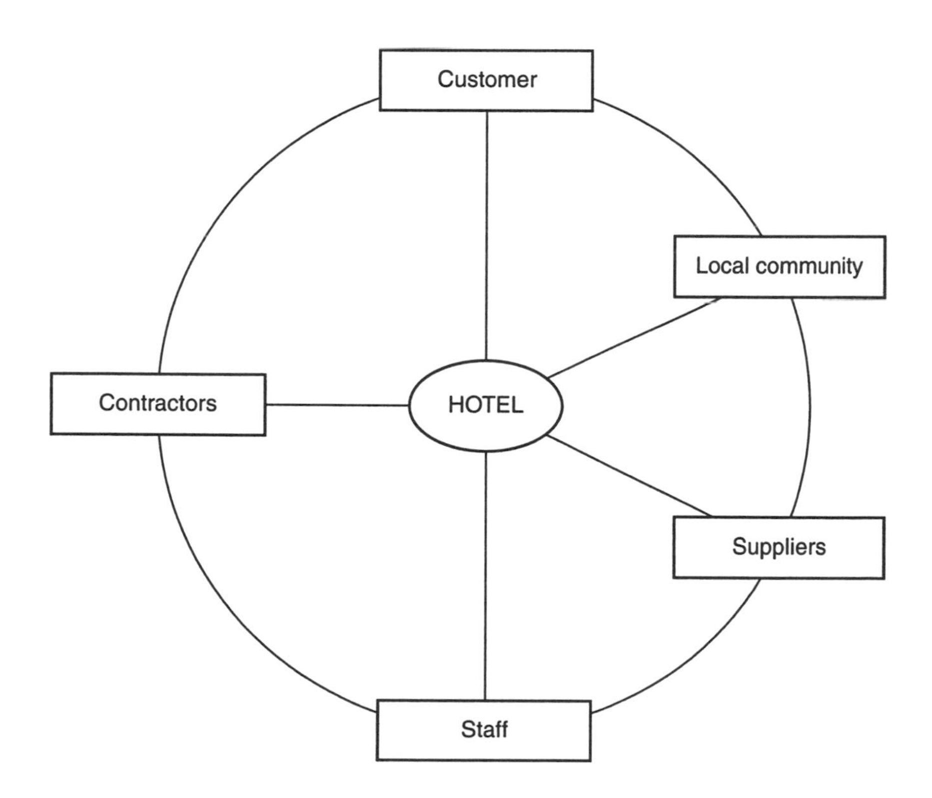

Figure 10.1 The hotel circle

Problem or opportunity

Contrary to popular belief, respecting the environment and adopting environmentally-friendly policies does not necessarily mean having to spend money without a return on capital employed. Far from it; investing in new technologies, like combined heat and power, will give simple pay-backs in under three years. Energy-efficient lighting will reduce lighting costs by at least 60%. Reducing waste will automatically lead to lower operating costs. Clearly, therefore, there are benefits associated with good environmental practices.

It is important to realise, however, that, to succeed, an environmental management system is totally dependent upon the full co-operation and involvement of staff. Relying on mechanical and computerised controls will not achieve the same results.

Staff will need to be made aware of the issues and trained in new methods of work. They also need an incentive in the form of a target to aim for and a simple monitoring system to show how well or badly they have performed against the target set. Monitoring and targeting is one of the useful management tools we discuss.

It is appreciated that many new technologies and systems cannot be retrofitted and the chapter concludes with some thoughts on new designs and developments which must form part of our future thinking, if we are to safeguard our environment and leave our children's children with a world they can enjoy.

Management have to appreciate that issues such as global warming, the greenhouse effect, depletion of the ozone layer and the destruction of the rain forests, are here to stay. Governments world wide are under increasing pressure to preserve the environment for future generations and, unless the hospitality industry recognises the part it can play and take the appropriate action voluntarily, the legislators will impose controls that, in the long term, will prove far more difficult to set up and more costly to comply with. Thus, what now can offer immense opportunities, could in the future, become a heavy and too costly a burden for many businesses to bear.

Creating an environmental policy

Like any other aspect of the business, the environment demands management attention. Therefore, the very first step for the head of the business, however large or small, is to appoint a senior manager who will be responsible for all aspects of environmental management in the organisation. This manager should organise a working group, consisting of key personnel responsible for purchasing, maintenance and engineering, housekeeping, building and design, training, food and beverage, and public relations and marketing.

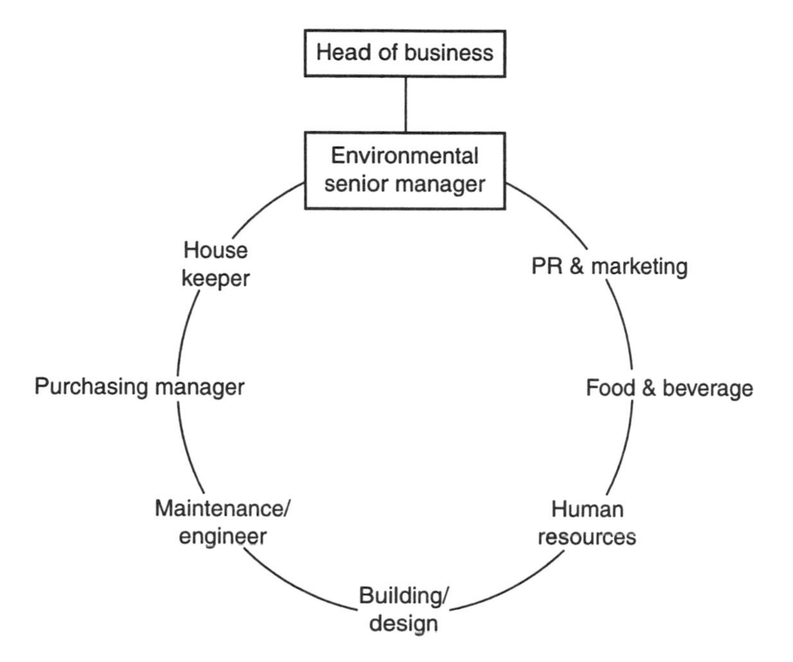

Figure 10.2 Management and key personnel

In the first instance, this working group should establish how the business impacts on the environment, and propose the policy direction that should be taken. The policy then becomes the basis on which all other activities, such as environmental audits, reviews and action plans can be carried out and monitored. It must show that the impact on the local environment has been fully considered, not only in the current operation but also in future developments.

Once the general direction has been accepted and endorsed by senior management, a 'Mission Statement' can be prepared, informing customers, staff and the public at large, of the aims of the business in respect of the environment. The mission statement, signed by the head of the organisation will show that senior management is committed to the policy and that the working group is acting on its behalf, pursuing policies that it fully endorses.

Armed with this commitment from senior management, the working group can now draw up detailed policies for each aspect of the business. Expert help and guidance may be required at this stage, but the group should be able to gain sufficient information, from various sources, to embark on an environmental management programme.

Suppliers and local environment groups, not forgetting members of the staff, all of whom may have some knowledge or involvement in specific areas, can all contribute to this initial exercise. Then, once the first steps have been taken, and provided the policy is kept under review, improvements incorporating new ideas, equipment or technology can be introduced. Thus, knowledge can be acquired with operating experience.

Purchasing policy

Unlike other industries, hotels are unlikely to be major primary sources of pollution. However, to provide the services required by customers, they have to purchase and use many products and commodities that, during manufacture or production, could be causing harm to the environment. Therefore, the next major task for the environmental group is to carry out an internal examination of everything the business needs to purchase in order to function.

Most suppliers will have gone through this process themselves and they will be able to provide all the necessary information on their products and services. Therefore, a simple letter from the purchasing department, making the supplier aware of the hotel's environmental mission statement and policy and, at the same time, requesting assistance from the supplier, to help the hotel to achieve its main aims, is likely to produce a wealth of information on all aspects of the business.

Naturally, to function, the business needs to buy the products and services, but the purchasing policy should specifically look at reducing the amount of waste, for example in the packaging, that comes in with the actual product. By discussing the problem with the supplier, simple changes that could lead to benefits, both for the hotel and the supplier, may be possible.

For example, if it is important to purchase sugar in single-portion sachets, is it necessary to buy the sachets in units of 144 in an inner cardboard box, overwrapped with cellophane? The inners, packed in an outer cardboard case, contain 10 or 20 boxes. Why not, 1,440 or 2,880 sachets in one outer container? Could we not consider going back to loose sugar in sugar bowls? Of course, food hygiene must not be compromised, but all aspects of purchasing and using sugar, or any other product, must be re-examined. By so doing, it will be possible to reduce much of the waste coming in with products.

The purchasing policy must also address other issues such as the life span of equipment and its running costs. For example, a refrigerator cabinet, with 4cm of foam insulation, operating at a design temperature of 37° Centigrade, will cost far less than one built with 6cm of insulation, using equipment designed to run at an ambient of 43° Centigrade. However, the first is likely to be less efficient, will need frequent maintenance attention, will use far more energy than the latter and will have to be replaced sooner.

SPECIFICATION

Therefore, it is essential that, in purchasing any product or equipment, true comparisons between competitive suppliers are made, and that product specifications and data sheets are examined carefully before the purchasing decision is made.

Purchasing must also consider the ease of disposal of waste products and old equipment. The policy should be one of purchasing products only in the amounts likely to be used, with the potential to re-use, repair or recycle as much of the equipment and packaging as possible, keeping final disposal down to a minimum.

Waste management

Having eliminated as much of the 'purchased' waste as possible, the next area to be addressed is that of waste disposal. The major objective of this policy has to be to 'reduce the amount of waste dispatched to landfill'.

Each member of the working group will have to examine the waste generated in each department and decide whether the waste could be reduced or eliminated, re-used, repaired or recycled. A simple internal audit or checklist showing the type of waste it is – for example, cardboard, cans, plastic, glass or food – with the quantities usually collected, can be carried out by each department over a given period of, say, an average week or month. The information can then be collated to estimate the total quantity of each category generated by the hotel over one year. With this basic information, the group can assess the equipment and storage space required to separate and keep the waste in its various categories, and to avoid sending it all to landfill.

By eliminating glass, cans, plastic, paper and cardboard from the main waste-stream the cost of waste disposal can be reduced by as much as 80%.

STAFF GOODWILL

With adequate facilities, separation can be at source, without incurring extra labour costs. However, for the policy to succeed, it is essential that staff are fully aware of the reasons behind the policy and that they are adequately trained on handling the waste materials they have to deal with. Achievable targets for waste reduction and collection of recyclable materials will motivate and give them an incentive to comply with the policy, and regular updates on how they are achieving against the target set will act as a further spur and reminder.

Waste disposal contractors can now be approached to quote for the disposal of the separated waste. Bearing in mind that they will be able to sell the recyclable waste, contractors themselves should either be prepared to capitalise any equipment needed, or offer an amount of money for each category of waste collected. Either way, the amount disposed of to landfill and, therefore, the waste disposal cost, should be reduced.

Food waste

In kitchens, little can be done with the plate waste coming back from the restaurants. However, an examination of this could well reveal that the portions of food are too large. It could also highlight a quality-control problem, either in the processing or initial purchasing of the material. For example, the amount of fat and gristle on a steak or the quantity of vegetables left over can be good indicators.

Customers who have to leave food on their plates, seldom feel they have enjoyed 'value for their money'. Far from it; knowing they are paying for it, they may well feel cheated. It is not unusual for chefs to order far more food than they are likely to use for fear of running out but, whereas this can be acceptable in the case of frozen or canned materials, with fresh or chilled produce over-ordering will automatically lead to waste. It is far better, therefore, to adopt a policy of 'selling out' rather than having waste to dispose of.

Food preparation waste should also be kept under constant review. It is appreciated, of course, that many chefs prefer to purchase raw materials that can be prepared in their kitchens, under their personal supervision, and to their own standards. Nevertheless, most raw materials can be purchased ready-to-use, direct from suppliers. Against the higher cost of purchasing, there are cost savings in energy, labour and waste disposal.

Hygiene considerations, especially in smaller kitchens, are also important. Importing clean vegetables or potatoes into the kitchen, eliminates the 'dirty' operation of cleaning them as well as the risk of importing bacteria with the soil. If the chef does not have to bone and roll the joint of meat or poultry, there is less risk of contaminating working surfaces or utensils. Another important consideration is that preparation waste is thus reduced, making a considerable contribution to the total cost of disposal.

Legislation

Governments, in recent years, have become acutely aware of the significance of disposal of waste in a safe and managed manner. New regulations appear with monotonous regularity and, usually, the legislation seeks to achieve regulatory controls within a given time span. In the

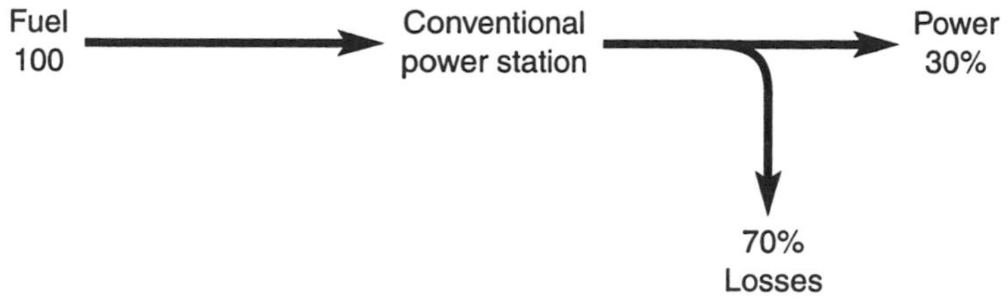

Fuel 100 → Conventional power station → Power 30%; 70% Losses

Figure 10.3 Conventional power station

case of packaging-waste, for example, it is proposed that within five years, 60% of it must be taken out of the main waste stream and, of that, 40% is returned for recycling.

It is important, therefore, to introduce a waste management system into the operation and encourage a culture that seeks to eliminate waste of all kinds in all aspects of the operation. 'Waste not, want not' is as good an adage today as it was yesteryear.

Energy conservation

The benefits that can result from the elimination of waste can be dramatic in the case of energy. The cost of electricity, gas or fuel oil is such that, any savings in consumption are likely to result in substantial benefits in money-saving terms.

The benefits to the environment manifest themselves in a reduction of the emissions of harmful gases into the atmosphere. Conventional power stations can waste as much as 70% of the fuel they use, producing millions of tonnes of 'greenhouse' gases in the process. Therefore, energy conservation is a most significant subject on the environmental management agenda, as any electricity saved by the end-user reduces the amount that needs to be produced by the power station (see Fig 10.3).

By the very nature of the business, hotels are major users of energy. They operate continuously over a twenty-four hour day, usually 365 days of the year. Heating and hot water, air-conditioning, lighting and restaurant services, are taken for granted by customers, regardless of the time of day or night and the industry, naturally, has to respond to the customer' s needs. The comfort of the customer is sacrosanct, and, to be able to respond to any possible needs, many aspects of the business are kept constantly on ' stand-by ', just in case customers call for them. The waste, as a result, can be and is, considerable. It is estimated that hotels waste anything up to 40% of the total energy used.

Much of this waste can be eliminated by sound management practices, staff awareness programmes and, of course, modern technology.

However, computers and mechanical controls are only as good as their operators allow them to be, and senior management must never assume that a good building management system can operate effectively by itself. It still needs a well-qualified engineer to drive the system successfully.

A good engineer is also key to successful energy management, but he or she may need assistance to communicate with the remainder of the staff. Therefore, it is wise to form a sub-group of the main environmental group, which will concentrate on energy conservation.

Energy plan

The group, as their first task, should draw up and agree an energy conservation plan that would seek to look at and review all aspects of energy usage in the hotel. This audit will show the major areas of use and, therefore, of potential waste. It will show where controls can be applied at little or no cost, and where capital investment may be required to eliminate a major fault or waste. It is as well to consider employing an expert consultant to carry out this initial audit and to produce the energy plan for the hotel.

His brief should be to review energy consumption in the building, by examining all the utility invoices and establishing the pattern of use. He should ensure that the utilities are being supplied on the most suitable tariff, matching the business needs and pattern of consumption, and he should produce the energy plan for the hotel. The plan would prioritise all possible areas of savings into 'No cost', 'Low cost' and 'Capital' categories, so that the management group could agree and embark on a structured plan to reduce consumption and save energy.

The choice of a consultant is vital and managers should avoid the temptation to engage tariff consultants on the basis of shared savings as remuneration. It is far better to agree a fee for a specific brief and a day rate thereafter, should the energy group require his or her services.

Monitoring and targeting

The initial audit and plan should show many areas where simple staff action will automatically result in savings: switching off unwanted lights, for example, or delaying turning on equipment until it is needed and turning it off when no longer required, thus eliminating the 'stand-by' mode. For staff to respond, however, it is important to make them aware of the problem and involve them in the resolution.

The problem is the cost of wasted energy, both to the organisation and to the environment. Staff normally have no idea how much the hotel pays for its electricity, gas, oil or water. They never see the bills. They take these commodities for granted; lights are part and parcel of the job!

Hotel:
Address: Company Code: Period End:
Region: Report Date:
Category:

Total Fossil Fuel Consumption Summary

	Consumption (therms/week)		Variance to target	
	ACTUAL	TARGET	therms	%
Period MAY 7	1,755	2,173	419	19
Year to Period MAY 7	125,985	138,631	12,646	9

AVERAGE WEEKLY CONSUMPTION PER PERIOD – therms

Period	Last year to May 88	Weather variance (warmer)	Weather variance cooler	Target reduction	Target	This year to May 89	Variance to target	Future target
Jun 3	1,754	(274)		(74)	1,406	1,545	(139)	1,321
Jul 9	1,389		103	(75)	1,417	1,493	(77)	
Aug 10	1,312	(12)		(65)	1,235	1,293	(58)	
Sep 11	1,569	(4)		(78)	1,487	1,595	(108)	
Oct 12	2,442	(462)		(99)	1,881	2,251	(370)	
Nov 1	3,452		742	(210)	3,984	3,242	742	
Dec 2	4,179	(508)		(184)	3,487	3,461	26	
Jan 3	4,273	(132)		(207)	3,934	3,122	811	
Feb 4	4,379	(71)		(215)	4,093	3,413	680	
Mar 5	4,124	(267)		(193)	3,664	3,003	661	
Apr 5	2,945		519	(178)	3,386	2,922	464	
May 7	2,007		281	(114)	2,173	1,755	419	
An. Tot.	145,925	(7,974)	6,977	(7,296)	138,631	125,985	12,646	

Notes:
1. Targets are based on an overall annual reduction of 5.0% from last year.
2. () on Variance-to-Target denotes performance worse than target.
3. This year on last year total weather variance is (0.7)% i.e. (997) therms
4. This year on last year consumption is 14.3% i.e. 20.939 therms

Figure 10.4 Example of a hotel monthly Monitoring and Targeting report

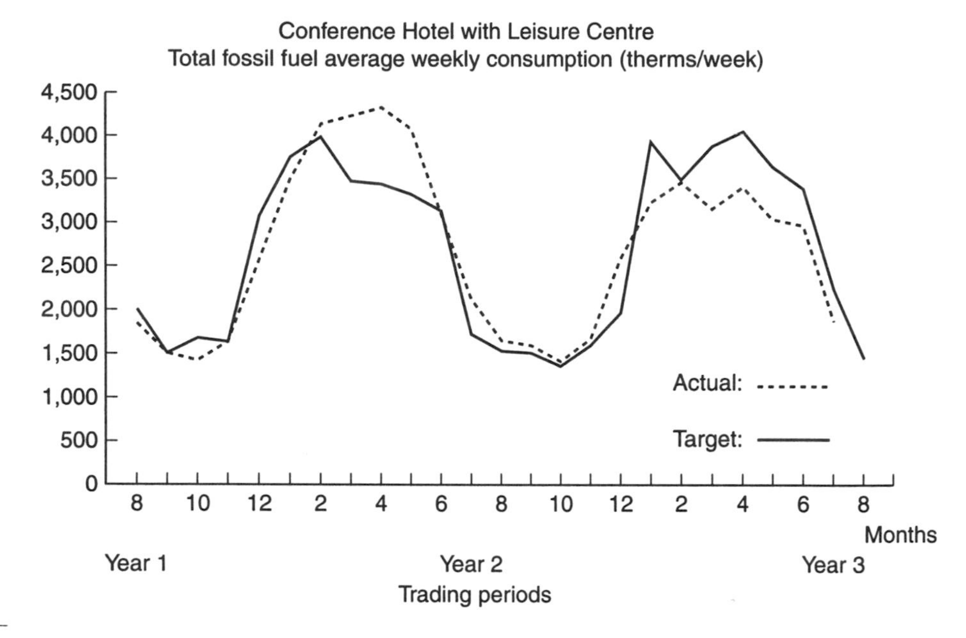

Figure 10.5 Graphs are an effective way of showing staff how they are performing against an agreed target

Unfortunately, managers, too, tend to treat utilities as fixed, unavoidable overheads, rather than as variable and manageable costs.

So, the first priority is to change the management attitude and to treat fuel and energy as an important item on the profit and loss account, to be reviewed every month. At the same time, make staff responsible for the energy they consume in their department.

A fully-fledged Monitoring and Targeting System requires utilities to be sub-metered to each main area of operation, such as kitchen, leisure centre or laundry, but, it is not always practical to install sub-meters. This does not mean that a simpler version cannot be set up.

For example, knowing that, in the average hotel, lighting represents some 15–25% of the total electricity consumed, circulate the electricity bill to all heads of department with a calculation showing the approximate cost of lighting for the period the bill refers to. Ask all heads of departments to run a campaign for the following month, encouraging staff to switch off all unwanted lights and to ask staff for their suggestions on how to save on lighting costs. Review the position when the next bill arrives. If the campaign has succeeded, the lighting costs should have decreased. If the first month shows no savings, try again. Whatever the result, it must be communicated back to the staff and the figure must be published monthly from then on, with staff feedback, ideas and suggestions. Savings can be translated to the amount of electricity that was not used, and the reduction in power-station emissions into the atmosphere that resulted.

Targets, based on actual consumption and performance, can now be set and staff interest can be maintained by publishing monthly figures, actual against target.

The more sophisticated Monitoring and Targeting System is based on actual meter readings, showing how much energy has been used each week, in each department. The target for saving on consumption is agreed with the head of department, having identified the action needed to reduce or eliminate waste. The actual performance is then monitored against the target set.

A good Monitoring and Targeting System will also show up anomalies in the business, such as the cost of a conference or an exhibition, as well as reduction in plant efficiency. The historical data required could be compiled by the consultant as part of the initial audit, and setting up the Monitoring and Targeting System could be part on his brief. Successful schemes can yield savings of anything between 10% and 40% on actual consumption.

Mechanical controls

Once staff are fully aware of the part they can and should play, the judicious application of mechanical and computerised controls can be considered and applied. These can vary from the simplest timing devices and thermostats, to the most sophisticated Building Management Systems. However, managers must avoid the attraction of gadgets as a means of overcoming bad working practices. Training and discipline is a far better investment than money spent on 'engineering' out a staff defect.

Building Management Systems can adjust automatically the heating or cooling levels required, taking account of the climatic condition outside the building at the time. If one side of the building is in full sunlight and the other is in the shade, in a cold airstream, the system can provide heat to the cold rooms and switch it off in the sunny rooms.

The building can be zoned, so that one zone at a time is let. The remainder of the building can be maintained without full heating or air-conditioning until the zone comes into service. Similarly, controls can be applied to conference and meeting rooms, leisure centres and kitchens. Lighting levels can be regulated so that, for example, the lighting levels in corridors during the night can be reduced to the minimum required for safety reasons.

A good system can react to changing conditions far faster than a human being and it will pay for itself within two to three years. However, it must be driven by an operator who understands its technology, is able to make full use of the equipment, and takes full advantage of its capabilities.

Heat recovery

Much of the energy wasted results from heat loss, especially from refrigeration and air-conditioning plant. Most of this heat can be recovered and used, for example, to pre-heat water for the hot water system or for the swimming pool. The reduction in heating costs is substantial and plant tends to become more efficient, needing less maintenance and saving more energy.

In certain climatic conditions, sufficient hot water can be generated without the full use of the boiler systems, which prolongs the life of the existing boilers and, when the time comes to refurbish the boiler house, a much smaller capacity boiler can be considered. Most heat-recovery systems, once installed, need very little maintenance.

If refrigeration for the food areas is carefully planned, so that the plant is centralised and remote from the cold chambers, not only can heat be recovered, but kitchen temperatures will also be reduced. In turn this creates a better working atmosphere and requires less energy to keep the kitchen cool. In new or refurbished kitchens, all such heat-producing plant should be kept outside the cooking and preparation areas.

Energy-efficient lighting

A major waste heat source in hotels is the lighting, especially in the public areas. For it to produce light, a tungsten lamp requires energy to heat the element within the globe to a 'white-hot' state. 90% of the total energy used is therefore wasted as excess heat and only 10% is for producing the light needed.

The opposite is true for energy-efficient lighting. Therefore, far less energy is required to produce the same level of lighting. However, to produce the desired effect required by the interior designer, a lighting scheme needs to be designed by an expert lighting consultant who is well briefed and understands the needs of the scheme and the ambience the interior designer is aiming to create.

Energy-efficient lighting can reduce lighting costs by a dramatic 60% to 80%. Because of its much longer life, lamp-change frequency decreases dramatically, which, of course, means less maintenance. Also the lamps remain cool, so shades or wallpaper are no longer subject to scorch marks. That reduces the need for early replacement or redecoration. The effect on air-conditioning can also be dramatic, as the heat-load in the building is substantially reduced.

The many advantages offered by energy-efficient lighting probably provide the best example of the true benefits that can be gained by adopting an environmentally-conscious policy. However, not all new technology will give the same returns.

New technology

Energy conservation has become a growth industry, with new gadgetry hitting the market every day. Substantial claims of huge savings normally accompany the sales literature of all these innovations and many of these claims can be somewhat exaggerated. However, some can be justified, at least in part. Therefore, the hotel manager should adopt an attitude of caution before commitment.

If the theory sounds good, the product may merit a trial. If the trial shows that savings can be achieved, it becomes a commercial decision, based on return on capital invested. If the simple pay-back is beyond the life of the product, it should not be considered. In practice, actual savings rarely live up to the enthusiastic expectations of the salesperson. Therefore, basic precautions and safeguards need to be taken, especially if the product is new and untested in the hotel industry. Operating conditions often differ to such an extent that, although the product may have proved itself in another environment, it could be totally useless in the hotel application.

Installation in these cases should always be on the strict understanding that, should the product fail, the installer will be responsible for its removal, restoration of the conditions that existed before the installation, and compensation for any damage or loss.

Maintenance

Maintenance of equipment must also be considered at the time of purchase. Much of the equipment will be beyond the technical ability of the hotel engineer to maintain effectively. Therefore due consideration must be given to contract maintenance, using specialist contractors to maintain plant and equipment on a planned basis.

Maintenance on breakdown is a dangerous policy to adopt and one that will result in unnecessarily higher running costs, as badly-maintained equipment becomes highly inefficient and consumes far more energy. Refrigerators with damaged seals, for example, will lose cold air and allow warm air to enter the chamber. This will cause icing-up of the evaporator and, not only will the refrigeration plant need more energy to try to maintain the temperature, but there is also a risk of losing valuable food stock.

Modern boilers need constant checking and fine-tuning to guarantee efficient burning of fuel and, therefore, minimising emissions to the atmosphere. Badly maintained water-cooling towers are a health hazard, spreading legionnaires' bacterium to its surrounding areas. In such cases, by saving on maintenance, the business could be lost for ever.

Water conservation

Water is one of the most precious commodities we possess. Without it, life is impossible. It is also in far shorter supply than most people imagine and, therefore, waste should be avoided at all costs.

Since 40% of the water used in hotels is in the guestrooms and, although some controls to limit water flow on showers or hand-wash basins are possible, there is little that can be done to control the amount of water that a guest uses to fill a bath tub, or the length of time the guest spends under a shower. However, much of the water wastage happens even before the guest arrives. Chambermaids run off far more water in cleaning the rooms than the guests use for their baths or showers. Good staff training and discipline, therefore, can do much to save all the water wasted in guestrooms.

Water leaks can develop in many parts of the system and, unless these are detected and dealt with, thousands of litres can be lost in a short time. Faulty valves on the main tanks or toilet cisterns, for example, will allow water to escape through the overflows, directly to the drain.

It is in such situations that a good monitoring system comes into its own. Regular meter readings and calculations on consumption will soon reveal a problem, for, if the rate of use rises and the business activity cannot account for the increase in consumption, a fault in the system must have developed.

Dripping taps, or taps that are left running for any length of time, are one of the most common causes of waste, as are hose-pipes used for cleaning back areas or kitchen floors or irrigating gardens. Dishwashers or washing machines operating on part-loads, and taps running cold water in kitchen sinks to defrost frozen foods or 'wash' vegetables, are all areas of waste that can be avoided or controlled. The kitchen and back areas account for some 25% of the total water used in hotels, therefore, clearly, any savings that can be made in these areas will be worth while.

Public washrooms and toilets account for a further 20% of water used and much of this is used unnecessarily. Flushing of urinals, for example, happens regularly every few minutes throughout the twenty-four hours, regardless of use. It is now possible to apply controls in these areas by requesting customers to flush after use or by installing mechanical or electronic controllers on the systems. It is extremely important, however, that toilet hygiene is not, in any way, compromised, as a result.

Planning a better environment

So far, this chapter has addressed the major issues that concern the hospitality industry. It has looked at ways in which hotels can take action, not only to contribute to the well-being of planet Earth, but also to improve on their profit performance. Being environmentally conscious does not necessarily mean having to spend more money. However, to adjust to new ideas, technological innovations and changed human attitudes, will mean a measure of investment by existing business.

Many new concepts, equipment or products, cannot be introduced, since retrofitting would cause too much disruption to a seven-days-a-week business. The general appearance and decor would also be affected. The classic example is modifying existing fittings to take energy-efficient lamps. Naturally, under these circumstances, it is better to wait for the refurbishment of the area, rather than attempt the modification. The opportunity to include new technology and thinking in terms of new projects or refurbishments however, must not be missed.

Landscaping

Creating a pleasant environment has always been a prerequisite of our business. How else could hotels attract customers to their establishments? Grounds, gardens and car parks can make a first and lasting impression on the new guest, arriving for the first time. They can also be extremely costly to maintain.

Employing a good landscape architect who has knowledge of the native flora and fauna of the locality will result in a scheme that will blend the building into its surrounding area, will add colour, interest and impact for the visitor, and will provide food and shelter for the wildlife in the area. The architect, knowing the cost of ground maintenance, will use only plants that will thrive in the local climate and will reduce the amount of grass-cutting to a minimum.

Again, replanting an established garden would be impractical and costly, but in a new project for the future, such consideration must form part of the total scheme.

Saving water

Consideration must be given to saving the rainfall that lands on the buildings and its grounds, so that this, rather than treated water coming in via the water mains, is used for irrigation. Not only will the hotel benefit from lower water charges, but precious energy, which would otherwise have been used to treat the water, will have been saved.

Again, consideration must be given to the safety aspects of water storage. Hotels are visited by young children, as well as adults, and water, whether in storage tanks, ponds, lakes or pools, tends to be an attraction for youngsters. Adequate precautions must be taken to prevent accidents.

New technology

As time progresses, new discoveries are made and, in turn, new technology is applied. Old problems may be solved, but new ones are likely to be created. Therefore, at all stages, it is important that careful thought is applied to any new idea.

Disposing of waste is a good example. Most of the waste generated in hotels can be recycled. There would be no requirement to send waste material to landfill, if storage of accumulated waste did not present problems with space, hygiene, and pests. If these problems can be overcome then all waste material can be recovered and re-used, even if it is in a different form.

Food waste and other putrescible material, for example, can be composted. The gas produced can be used to fire a combined heat-and-power unit, to produce heat, hot water and electricity for the hotel. The resulting compost can be used to enrich the soil of the grounds and gardens, and any surplus can be sold. Saving money by not using peat from fast-vanishing peat bogs is another bonus for the environment.

Combined heat and power can normally be justified in hotels with sufficient heat loads throughout the year. A commercial engine, driving a generator, produces electricity. The water, used to cool the engine, together with heat recovered from the exhaust system, is used in the hotel's heating and hot water system. Thus, for the cost of the heat and hot water, a measure of 'free' electricity is also made available.

Such installations can produce simple pay-backs of less than 3 years, but the heat load must be sufficient to allow the system to run for an average of at least 15 hours a day.

Solar and wind power

Energy from natural sources, such as the sun and wind, can be captured to produce heat and electricity, reducing the amount the hotel has to buy and the supplier needs to produce. Such sources are the most environmentally friendly of all technologies, however, the return on capital for the investment is still well beyond the period normally considered viable by accountants in the UK.

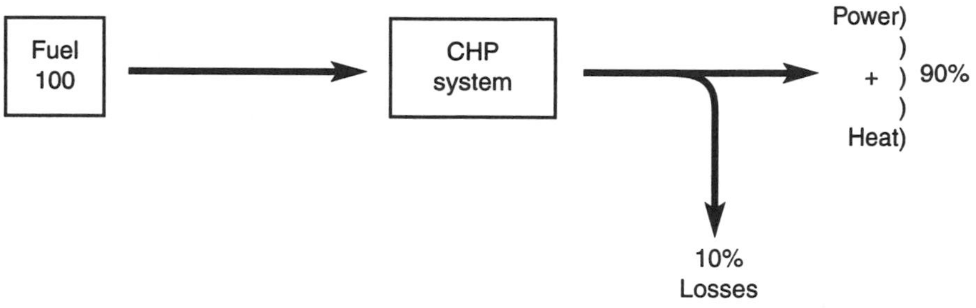

Figure 10.6 Combined heat and power

In areas of guaranteed sunshine, such as the Caribbean or the Mediterranean, solar panels, forming an integral part of the roof, can satisfy a high proportion of the heat, hot water and electrical requirements for the hotel. Similarly, on locations exposed to consistent air movements, wind generators are able to provide a regular amount of the electricity needed.

The viability of these technologies should therefore be considered in each location, always taking local climatic conditions into consideration. It is also wise to keep new technologies under constant review as, very often, costs tend to reduce with increased take-up, while the price of energy will continue to rise.

In any case, incorporating such new technologies in new buildings, as part of the development costs, makes commercial sense and, by so doing, harmful emissions can be reduced appreciably. Then, if proprietors of new hotels undertake to plant one tree in its grounds for every bedroom they operate, the air quality in the local environment will be preserved or even enhanced.

It is obvious, therefore, that there are many areas in which a hotel can make a valuable contribution towards the protection of the environment, if each aspect of the operation is considered with that objective in mind.

The success of an environmental policy, however, depends totally on the full co-operation and goodwill of the staff. It is they who will separate the waste into returnable, re-usable, recyclable, or disposable piles. It is they who will turn off unwanted lights or dripping taps. It is they who will proudly explain to customers the policies in force in their hotel to save Mother Earth.

In turn, what they learn at work they are likely to practise at home and, similarly, the message they give to customers will help to spread the good news and practice to their homes and businesses.

The world at large cannot afford to ignore the environmental issues now facing humanity and, as today's custodians of the planet, this generation will be judged on its contribution towards its safekeeping. Therefore, every opportunity for improvement must be grasped and hotelkeepers, the world over, are ideally placed, at the centre of the issue, to show the rest of the world the way forward.

Recommended reading

Datschefski, Edwin, Landsell, Suzannah and Stewart, Clair (compilers and editors). *The Environment Council Handbook*, London (Looseleaf with monthly updates.)

International Hotels Environment Initiative. *Environmental Management for Hotels*, the industry guide to best practice, Butterworth-Heinemann, Oxford (1993).

Little, B. F. P., Branson, B. and Brierly, M. J. *Using Environmental Management Systems to Improve Profits*, Graham & Trotman, London (1992).

11

Environmental management in a leading bank

VICTOR BRUNS

First Vice President, Deutsche Bank AG,
Corporate Banking (Head Office), Frankfurt

1. **Environmental protection in a bank normally involves:**
   - conservation, recycling, and energy saving;
   - conservation as a means to cost-reduction;
   - taking advantage of the opportunities and managing the risks presented by environmental issues.
2. **Progress to date has included:**
   - promoting the development of environmentally-friendly products and services;
   - accessing the special funds available for the development of 'Green businesses';
   - helping environmental businesses with cash flow, capital investment, information and consultancy.
3. **The obstacles to environmental management in banks are:**
   - the lack of a clear understanding of what 'sustainable development' means in terms of operational goals;
   - a shortage of effective techniques for assessing customers' environmental risks;
   - and it is unclear in lending for environmental projects what is the risk for the bank.
4. **To remove these obstacles will require the development of:**
   - standard methods for assessing environmental risks, and
   - the production of accepted approaches to reporting environmental performance in a form such as the 'ecological balance sheet'.

Introduction

Environmental protection is becoming increasingly important in all sectors of business and industry. For a company in the financial services sector this means tackling environmental problems both in-house – for example by implementing measures to reduce the use of resources – and in its business with corporate customers. For a Bank and its customers environmental protection has become a business factor presenting opportunities and risks. Capital investment in environmental protection offers the bank an interesting new market to tap. To support companies with environmental problems, special products have been developed, such as information services. On the other hand, there are potential risks for a bank in environmental protection which should not be overlooked.

This chapter is divided into four parts:

1. How are banks affected by environmental protection?
2. What environmental protection measures have banks implemented already?
3. What are the stumbling blocks for banks?
4. What measures are needed to remove the stumbling blocks?

The protection of our environment has become one of the most important questions facing our society and this obliges us all to act in a responsible fashion. The current situation is analysed here from the perspective of a bank in Germany.

1. How are banks affected by environmental protection?

ENVIRONMENTAL PROTECTION WITHIN THE BANK

Environmental protection should always begin 'at home' and with a company's own logistics. Although in its operations a service company usually has fewer difficulties than a manufacturing company, which has environmental problems such as the intensive use of resources, emissions and the disposal of special waste, it still has many opportunities to protect the environment. Some banks have approached the problem by testing, and later establishing – in line with conventional financial reporting – a comprehensive system called the 'ecological balance sheet'.[1] Deutsche Bank chose an approach which is based more on individual measures. Responsibility for in-house environmental protection, which has the backing of the Board of Managing Directors, has been assumed by the Organization and Operations Division. The aim is, wherever possible, to introduce environmental measures and to cut back the use of resources in purchasing, building technology and power, waste disposal and the use of vehicles. Some years ago, an environmental protection manager was appointed with a network of contacts in the domestic bank. This group has drawn up a catalogue of over fifty environmental protection measures that are observed in the bank's procurement handbook. The measures include *inter alia* the following:

- environmentally-friendly disposal of shredded microfiche sheets;
- avoidance of products containing PVC; use of environmentally-compatible materials, e.g., cardboard folders;
- Annual report printed on paper bleached without using chlorine (since 1990);
- Reduction of energy consumption by installation of micro-processor-steered heating and cooling plant.

The individual measures are currently being included in an environmental register, which will serve both as a checklist for those responsible for environmental matters at branches, also as a summary of the environmental areas the bank has addressed to date. The register covers eight main areas:

- recyclable waste paper;
- other types of recyclable waste at the workplace;
- recycling of Eurocheque customer and credit cards;
- environmentally-compatible cleaning;
- toilets/water/waste water;
- canteens/kitchens/rest areas;
- recycling/refill of colour ribbon cassettes and toner cartridges;
- energy-saving measures.

The catalogue is constantly being expanded with input from staff and the Staff Council, as the day-to-day behaviour of individual staff members at the workplace is a crucial to the implementation of our thinking on environmental matters. To foster the commitment of our staff, an environmental competition was held in late 1991 under the umbrella of Deutsche Bank's staff suggestion scheme and received an enthusiastic response. Also, the staff magazine, Forum, and Deutsche Bank's in-house office communications system carry regular reports on environmental measures in the bank.

In their relationships with customers many banks are promoting paperless banking by introducing self-service cash dispensers and electronic transfer machines. Home banking programs also help to reduce traffic by saving customers trips to the bank. The use of these systems is likely to become more widespread in the future.

ENVIRONMENTAL PROTECTION AS A COMMERCIAL FACTOR FOR THE BANK AND ITS CUSTOMERS

The German Environment Minister, Klaus Töpfer, has often said that environmental protection does not create costs; it merely reallocates costs which were previously borne by our natural resources.[2] This statement underlines the government's demand that the ecological costs of extraction and waste disposal should be borne in full by the polluter, so that the ecological costs of production will eventually be included in the product price. This objective provides a sound basis for a reorientation of business strategy and social behaviour in the world's economies if we are to achieve sustainable development. It will certainly be a difficult and painful process to reallocate these costs. However, this process is inevitable if we are to protect the quality of the environment for future generations.

Business will have to deal with changes in public buying behaviour and ever-tighter environmental legislation. One example of recent legislation is the new German packaging law introduced in 1991, which allows the customer to return obsolete packaging of goods to the store where they were bought. The store, therefore, has a greater incentive to buy from suppliers who cut out excess packaging.

By using market forces, these and other regulations are encouraging industries to reduce waste. Many companies are finding that pollution control not only reduces waste but can also cut costs. This new cost-saving potential will add a new dimension to price competition in industries which produce a large amount of waste.

ENVIRONMENTAL SERVICES IN CORPORATE BANKING

Once we recognize that we are entering a period of global change in market and regulatory conditions, we must ask: how can banks work with their corporate customers in addressing the issue of environmental protection? They are concerned with environmental protection in their clients' businesses in three different ways:

(i) Banks are seeking to utilize the opportunities generated by new markets which are emerging through ecologically-conscious consumer behaviour and greater environmental regulation. When examining environmental protection from the perspective of our corporate customers, we can see that it affects all industries and extends to all operational areas. Environmental protection is affecting the entire economy and is becoming a decisive factor in competition. Only companies which are quick to recognize the resulting implications can be successful in future. Many companies are therefore adopting an environmentally-conscious style of management and invest in environmental protection.

A study by Ifo, a Munich-based research institute, suggested that in Germany alone, over half a trillion* DM would have to be invested in the construction industry, to protect the water supply and to ensure proper waste disposal, clean-up of contaminated industrial sites, noise prevention and the protection of nature and the countryside. Of this amount, in the new federal states (the former German Democratic Republic) DM 174 billion is needed for the expansion and maintenance of environmental protection facilities. In the old federal states, expenditures of around DM 220 billion are needed for the expansion of environmental protection facilities and DM 187 billion for their maintenance.[3]

Another result of this development is the growth of the 'environmental technology' sector. According to the most recent statistics of 1993, 6,000 companies in Germany are involved in environmental protection technology. In 1992, turnover in this area came to DM 55.6 billion. For 1993, a growth of 4.9% (to DM 58.3 billion) is forecast for the environmental technology market.[4] This presents huge new market opportunities for those banks prepared to do business with these industries.

The media frequently call for banks to examine the environmental implications when financing capital investment projects. However, banks have no statutory duties to act as an environmental police force. Moreover, it is not always clear which individual investment project can be regarded as environmentally friendly. Frequently, such projects are thought to include capital investment in a sewage plant or in equipment for air pollution control. Unquestionably, capital investment in such projects yields benefits for the environmental area concerned (ground, water or air) and it is therefore beneficial. But as an 'end-of-the-pipe technology' for reducing or concentrating pollutants which have already been created, it can only be seen as a component and not as the target of a comprehensive and forward-looking environmental strategy. By contrast, capital spending on a new heating plant which, from the start, emits less pollution into the air should be seen as a better environmental investment.

(ii) In their own interest, banks should recognize and evaluate the risks which their customers are facing as a result of environmental change. There are two kinds of risks to be considered:

(a) There is often an immediate risk resulting from possible damage to the environment, and the resulting reduction in the value of collateral held against a loan. This risk – a major subject of debate in Anglo-American legal literature – is especially pronounced in connection with possibly contaminated land used as security.[5] Rather than foreclosure, which – in the event that it is legally preceded by

* One trillion = 10^{12}

bank ownership – may under certain circumstances impose liability for the clean-up on the bank, it may be better for the bank to completely write off the loan and forfeit the collateral. Another risk under German law is from 'Sicherungsübereignung', a unique legal concept which, upon the extension of a loan, transfers the ownership of movable collateral such as inventories or machinery to the lender.

(b) There is an indirect risk regarding damage to the company's market position which can be caused by a sales reduction through an environmentally-conscious customer base, as well as from new regulations which could destroy a company's entire market. Therefore, in the interest of its own risk management in dealing with companies, the bank needs to develop its own appraisal systems in order to account for environmental considerations. The aim is to achieve a win-win situation and satisfy customers' and bank interests at the same time. Deutsche Bank included the assessment of environmental risks in its risk scoring sheet for the evaluation of loan exposures some years ago. However, it will eventually be necessary to go beyond this and incorporate other environmental aspects into lending risks evaluations, such as the extraction and use of raw materials, components and supplies, through to the production, sale and use of the product, then to its recycling or disposal.

(iii) Banks are, of course, very interested in developing relationships with companies that are well equipped to meet present and future environmental standards. Therefore, they pursue their own interests if they help their customers to manage their environmental risks and to exploit new market opportunities. Apart from these purely economic aspects there is a third reason why banks are concerned with protecting the environment:

In addition to managing the risks and pursuing their own financial goals, banks have a duty as 'corporate citizens' to contribute to the social goal of 'sustainable development'. At the invitation of UNEP (United Nations Environment Programme), several international banks drew up a programme on environmentally-conscious behaviour by banks. This 'Statement by banks on the environment and sustainable development' was signed by 29 major banks on 6 May 1992.[6] In this statement the banks pledged to adapt their policies to encourage 'environmentally-compatible growth'. Protection of the environment with the aim of sustainable development will be one of the most important tasks of business and industry, including banking. It is important now that other banks should join this initiative.

2. What environmental protection measures have banks implemented already?

PRODUCTS

In the late 80s demand for banking products to assist companies with environmental management had risen in line with increased public awareness. To penetrate this new market, many banks have set out to develop specific financial products. Below, some of the products are described which have been designed for four main service areas: conventional financing, cash-flow lending, investment services and information services.

Traditional financing services

For capital spending on environmental projects, some banks offer cheaper loans which they subsidize from their own assets or refinance from assets which depositors provide for this purpose at below-market rates.[7]

In Germany, considerable public-sector financial assistance is available. German banks act as clearing houses for these promotion schemes and provide additional financial assistance from their own funds. These schemes can help companies to cut the mounting cost of capital spending on environmental protection. Users of these schemes need to be familiar with the application requirements and to know how the schemes can be applied and combined, especially as the range of available assistance extends from tax relief, through grants and subsidies, to loans at favourable interest rates, and guarantees.

To provide faster access to more detailed information, two databases, 'db-select' and 'euro-select', are available through the bank, also publicly. The 'db-select' database compiled by Deutsche Bank allows industry and self-employed individuals from Germany to select from around 720 public support schemes offered by the states, the federal government and the European Community, of which *about 50 schemes are aimed at promoting environmental protection*. The database also offers useful information for research and development departments as it contains details of support funding, also forthcoming calls in public support schemes.

The 'euro-select' database operated by several European partners contains the same data as 'db-select' and, in addition, national and regional financial support schemes of six EC partner countries and the European Community. The 2,700 programmes include *around 100 programmes for capital investment in environmental projects*. Users can call up information in their own language about support programmes in other countries.

Cash-flow lending

Some projects qualify neither for conventional bank financing nor for promotion schemes. For these cases there are two alternatives to the traditional credit facility: one is the 'operator model', the other is project financing. With the 'operator model' the bank's financing decision is not based on the borrower's creditworthiness, but on the future cash flow expected from the project financed. For example, the operator model allows local authorities to build sewage plants or recycling facilities where needed by outsourcing the construction and operation to a private contractor. This means that a private, single-property holding company builds and finances the sewage plant – according to the requirements of the local authorities – and reports the property in its balance sheet. Upon completion, the sewage plant is let to a qualified operator. Local authorities must be able to carry out inspections without difficulty so that they can comply with their public duties. One vital advantage of this model is that in many cases the technical expertise of the operator assures greater cost efficiency.[8]

Cash-flow lending enables the investor – often local communities or states – to take pollution control measures without burdening their balance sheet or their investment budget. These financing strategies make environmental projects possible for many investors who lack the liquidity or collateral assets to qualify for conventional financing from a commercial bank.

Ecological deposits and investment services

As previously mentioned, some banks offer special types of deposits and savings certificates at below-market rates; funds pooled in this way are then passed on, together with the interest advantage, to finance capital spending on environmental projects.

Several banks and financial service companies, particularly in the Anglo-American countries and Switzerland, offer ethically and ecologically 'sound' investment funds.[9] However, the difficulty of setting appropriate selection criteria has, to date, kept the German banking supervisory authorities from giving their approval to an ecological label. There have been successful attempts in other countries to examine and certify the environmental integrity of such investment funds. However, at present we still lack a standard evaluation technique.

Information supply

In business, information acts as a trigger for capital investments. This holds true especially where environmental measures are concerned. The supply of information must therefore be an integral part of a bank's efforts to promote environmental improvements. In a study carried out

Results (in per cent) of a survey of small- and medium-sized businesses carried out for Deutsche Bank

Obstacle	Per cent
Difficulties in finding suitable suppliers	30
Time-consuming authorization procedure	33
Lack of know-how	30
Opposition from employees	13
Cost of environmental measures	62

Figure 11.1 Obstacles to the implementation of environmental measures
(Source: Deutsche Bank)

in 1991 – for the second time – on 'Information supply in environmental protection', 90% of the small- and medium-sized companies polled with an interest in environmental protection said that they had implemented environmental measures during the last three years, and 80% planned to do so in the next two to three years. Despite this positive response, day-to-day talks with customers repeatedly show that, particularly for small- and medium-sized companies, environmental protection presents information problems. Two-thirds of the companies questioned specified the costs of implementing environmental measures as an obstacle; however, a third of the companies complained about difficulties in finding suitable suppliers and the lack of available know-how[10] (see Fig. 11.1). Banks must offer their customers support in those areas where they can supply the necessary know-how. In corporate banking, these are primarily the established areas of financing advice and financing itself.

However, at an earlier stage it may be appropriate to provide further services, in particular information and advice on environmental problems.

As a first step four years ago, Deutsche Bank published a handbook[11] containing information about changes in environmental legislation and market conditions, together with recommendations on changes in business strategy and tools for their implementation. This publication was written for small- and medium-sized enterprises and is still being quoted in current literature.[12] Now it is somewhat outdated, which is a sign that we have left the awareness-raising phase and entered a phase of setting goals and taking appropriate action.

Deutsche Bank also hosts 'Mittelstandsforen', briefings for small- and medium-sized businesses. These were originally designed to raise the awareness of environmental issues among business owners. Now the workshops are intended to promote the implementation of preventive or restructuring measures by dealing with more specific topics, such as the new German packaging law.

Satisfying the customers' need for the right information at the right time is a demanding task. To make it easier for them to find suppliers and to improve their know-how on the environmental market, Deutsche Bank, in co-operation with independent partners, has developed an environmental database. It amounts to a 'who offers what?' for the environmental sector, comprising a directory of suppliers of environmental services, products and technologies. Currently it contains around 2,800 companies with about 9,000 detailed offers. A special indexing system makes it possible to link the company's environmental problem quickly with a potential solution. The database is equally attractive to suppliers and users. Suppliers can present their products, services and innovations in a way closely geared to the target group; users can obtain an overview of the suppliers and the solutions they offer for specific environmental problems. Through a co-operation agreement with an Austrian bank, a step has been taken towards internationalizing the database. Since summer 1992 it has contained both German and Austrian suppliers. The database is now also available in a handbook to make the information available to businesses without the necessary computer equipment or expertise.

If we presume that providing the right information to companies at the right time can trigger environmental measures, we must attempt to provide other user-friendly and low-price approaches. In this way, small- and medium-sized companies could be given a further incentive to use environmental protection as a means of ensuring their businesses' success in future. One such attempt is the 'Questionnaire on the environmental risk situation', which was tested early in 1993. This questionnaire is aimed at manufacturing companies with 30 to 300 employees and was developed by an external environmental consultancy company. A checklist is used to make a practical examination of operational areas with regard to environmental protection. The questionnaire is filled in by the companies and then, for a fee, returned to the consultancy company for computer analysis. The report on environmental risks is sent to the company and is treated confidentially *vis-à-vis* the bank.

To summarize, it is important for a bank to provide a range of services for small- and medium-sized companies so that it can act as their partner throughout the decision-making process (see Fig. 11.2).[13]

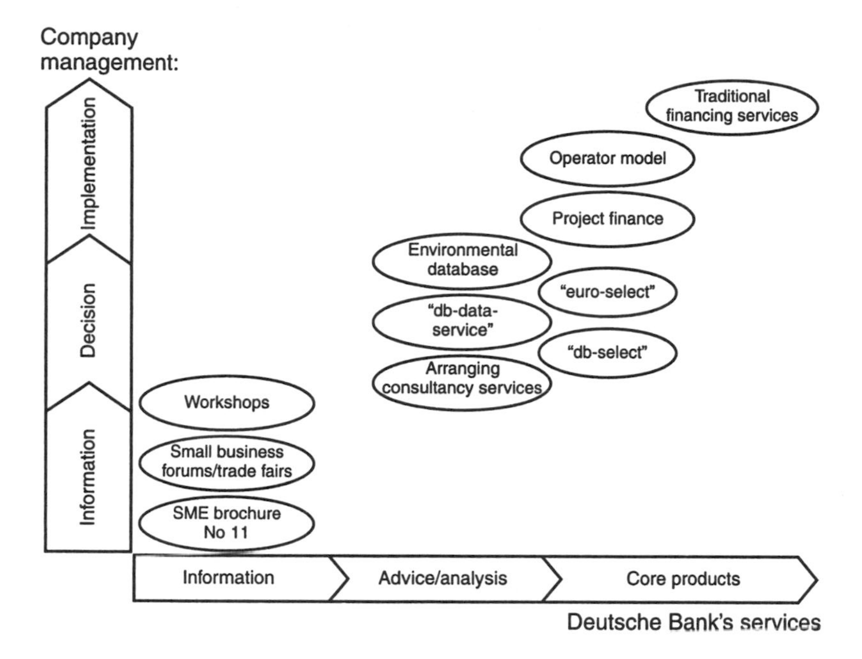

Figure 11.2 Deutsche Bank's environmental services for companies
(Source: Deutsche Bank)

The bank management must recognize that, outside its core areas, it will generally only be able to provide this know-how in co-operation with third parties.

PUBLIC RELATIONS

One task of public relations is to publicize a bank's product range through seminars, contributions to journals and participations in environmental trade fairs.

Another important area is sponsorship – for example, by financing internships in national parks or by offering prizes for environmental achievements in industry.[1]

In partnership with the foundation 'Jugend forscht e. V.' Deutsche Bank annually holds the competition for 'Young Europeans' environmental research'. We invite researchers under 22 years of age from eastern and western Europe who have already won national competitions held by 'Jugend forscht' and their partner organizations. The aim is to bring together committed and creative young people who are concerned about the environment.

3. What are the stumbling blocks for banks?

SUSTAINABLE DEVELOPMENT AND ECOLOGICAL RISKS

Sustainable development is the target as formulated in the UNEP statement. Unfortunately, this term has not yet been defined in detail for banking. Banks are therefore faced with three partly-solved questions:

(a) What specific goal does sustainable development entail for banks?

(b) How does this goal, once defined, translate into day-to-day operations?

(c) What instruments do banks need to develop to take appropriate action?

A number of banks are currently working on the answers to these three questions.

(a) Defining the goal for sustainable development

To answer the first question will require a contribution of the natural sciences and an understanding of the society we live in. The way to the target of sustainable development will be difficult and will call for substantial changes in behaviour, in structures and management systems. In the short term, sustainable development – as the debate about the inclusion of environmental protection in the German constitution shows – is difficult to reconcile with the need not to make the country unattractive as a business location, as not all countries are protecting the environment to the same extent. Also, a bank pursuing 'sustainable development' cannot take up an outside position relative to its business partners as successful and active participation is only possible if the bank works on the basis of a social consensus. The task is to take an active part in industry' s move towards sustainable development. In its business with corporate customers a bank must aim to support companies in their attempts to protect the environment, for the bank's success depends on the success of its customers.

(b) Translating sustainable development into day-to-day operations

When dealing with the second question, we are facing the problem that it would be impossible to take a case-by-case approach in assessing the huge number of loans which are processed at the branches of a large commercial bank. Today, a bank's risk management can only follow the general principle stating that the increased risks of a company – which may arise from environmental or other reasons – will involve an increased cost to the company. Also, each bank must decide for itself which risks it will refuse to accept. Otherwise it would run the risk of earning extra profits from polluters in the short term through increased risk premiums, but in the longer term it might damage the quality of its loan portfolio.

(c) Developing instruments for sustainable development

The third question concerns the instruments. It is not sufficient for banks to define environmental protection as a target if they have no instruments to check individual cases to see that this target is being attained. Banks should have little difficulty in providing their employees with precise guidelines on collateral, since outside experts can be hired to assess the risks linked to potential collateral such as land and buildings. A much greater obstacle arises in assessing the ecological soundness of a business or its production processes, since a product's entire life cycle (that is its input/output requirements at each level of production until its disposal) must be taken into consideration when its overall impact on the environment is assessed. Banks are not in the business of producing goods, developing new technologies or processing materials.

Therefore, bank employees naturally lack competence in production-related industrial technologies and cannot be expected to thoroughly assess the ecological impact of their customers' businesses. Banks do not have the capacity to train all of their corporate banking officers to be environmental experts, and it would be unreasonable to expect an account manager to check his customer's sewage pipes for polluting effluent. A case-by-case approach – as practised by international financial institutions for their large financing projects – simply cannot be applied to a commercial bank's substantial case load.

ECONOMIC RISKS

Assessing the ecological impact of customers' activities is not the only difficulty with environmental loans. Sometimes economic risks are present to an even greater extent than under a credit facility intended for conventional purposes.

A bank's risk in financing environmental projects essentially lies in the lack of creditworthiness of the potential borrower. For example, bank managers are sometimes asked to comment on the financing of soil decontamination projects. Such requests are often based on a misunderstanding about the role of banks in such soil clean-up projects. The question should not be 'Who is providing the loan?' but 'Who is paying the bill'? In other words, who is responsible for the financing costs, and is this debtor able to service the loan?

The financial risk for commercial banks is also a main factor in the tardy implementation of decontamination and pollution control measures in central and eastern Europe, in spite of the substantial and urgent demand for such projects there. A project-financing strategy cannot easily reduce the country-related risks specific to commercial banks, because the typical plant or facility which is financed in most cases cannot earn the currency needed to service a foreign loan.

4. What measures are needed to remove the stumbling blocks?

What can be done to remove these obstacles? If we agree that commercial banks must integrate environmental protection into their daily business decisions, they should not and cannot be charged with a policing function which goes far beyond their usual tasks as banks. If banks are to consider environmental factors in their dealings with customers they will need a decision-making tool which allows them to handle efficiently and safely the large volume of credit decisions in their day-to-day business. The development of a standardized method for ecological assessment would greatly facilitate the quantification of possible ecological damage and resulting risks. This instrument, presumably in the form of a report, should be:

- easy to use, especially for small- and medium-sized companies;
- uniform at a national or, better still, at an international level;
- include an assessment of the ecological impact in terms of the risks and costs involved for all parties; and
- enable the non-scientist to make sound decisions about the risks resulting from the ecological aspects of a project.

A fifty per cent solution would, if it were upwardly compatible, mean a great step forward compared with the current situation. Such instruments have been conceived and recommended by, among others, the International Chamber of Commerce, the EC Commission and the Swiss initiators of the 'ecological balance sheet'.

The European Community could be instrumental in supporting the standardization of such a tool. If we want industry and the banking community to take into consideration the ecological impact of their actions in the light of our common goal to achieve sustainable development, *we should expedite the development and implementation of standardized ecological reporting methods for business*. Equipped with adequate instruments to implement goal-oriented environmental measures, banks and industry can together make their contribution to an incremental, and hopefully global, move towards sustainable development.

References

1 Waeber, M. 'SBV: Erste Schweizer Bank mit Ökobilanz', *Tagesanzeiger*, Zürich, 26/11/92, p. 39.
Gottschall, D. Die Papiertiger, *Manager Magazin*, Hamburg, 9/92, pp. 200–205.
2 Töpfer, K. e.g., Opening address at the ENTSORGA Trade Fair, 9/91.
3 Ifo-Institut quoted in 'Der Baubedarf im Umweltschutz ist riesig', *Frankfurter Allgemeine Zeitung*, no. 133, 10/6/92, p. 16.
4 Press release of the German Federal Ministry for the Environment, Nature Conservation and Nuclear Safety, 7/12/92.
5 Lee, R. G., Sapte, W. 'Lenders, Land and Liability', *International Banking and Financial Law*, Supplement on *Lender Liability*, November 1992.
Schneider, U. H., Eicholz, R. 'Die umweltrechtliche Verantwortung des Sicherungsnehmers', *Zeitschrift für Wirtschaftsrecht*, 1/90, pp. 18–24.
Sweeney, S. 'Lender Liability and Cercla', *American Bar Association Journal*, 2/93, pp. 68–71.
Toft, A. B., Schulte Roth & Zabel, 'US: EPA adopts final rule on lender liability under CERCLA', *International Banking and Financial Law*, November, Supplement on *Lender Liability*, November 1992.
6 Author unknown. 'Commercial banks pledge to support environment', *Wall Street Journal*, 7/05/92.
7 Author unknown. 'Nur verhältnismäßig wenige Anbieter schwimmen auf der "grünen" Welle mit', *Handelsblatt*, Düsseldorf, no. 5, 8/9/93.
8 Rudolph, K.U. 'Einschaltung privater Unternehmen für die öffentliche Abwasserentsorgung', Information brochure complied on behalf of the Federal Ministry for the Environment, Nature Conservation and Nuclear Safety, Witten, 1/91.
Rudolph, K.U. and Gellert, M. 'Das Niedersächsische Betreibermodell – Erfahrungen aus technischer und ökonomischer Sicht', *der gemeindehaushalt*, no. 6, 1988, pp. 121–126.
9 Zaugg, B. *Umfassendes Öko-Rating*, Schweizer Bank, 6/92, pp. 10–12.
Author unknown. 'Umweltfonds: Besser als der Trend (1)', *Ökologische Briefe*, Frankfurt, nr. 32, 5/8/92, pp. 16–18.
10 Roland Berger. *Studie zum Informationsverhalten im Bereich Umweltschutz*, carried out for Deutsche Bank AG, München, August 1991.
11 Deutsche Bank, Mittelstandsbroschüre, no. 11, *Umweltschutz: Fakten, Prognosen, Strategien*.
12 Wicke, L., Haasis, H.D., Schafhausen F. and Schulz, W. *Betriebliche Umweltökonomie*, Vahlen Verlag, München (1992), pp. 572–574.
Günther, K. 'Öko-Bilanzen als Grundlage eines Umwelt-Auditings', in Steger, U. (ed.), *Umwelt-Auditing*, FAZ Bereich Wirtschaftsbücher, Frankfurt (1991), p. 71.
13 Junker, K. 'Umweltschutz - Herausforderung und Verpflichtung', Bank und Markt, no. 7, 7/90, pp. 16–20.
14 Schierenbeck, H., Seidel, E. *Banken und Ökologie*, Gabler Verlag, Wiesbaden (1992), p. 93.

Part III

■

DEVELOPING POLICIES AND SYSTEMS FOR ENVIRONMENTAL MANAGEMENT

'Environmental considerations must become as much an integral part of the managerial decision-making process as financial, legal and human matters. Environmental matters affect every decision.'

Alex Kramer, Board Chairman, Ciba-Geigy,
Business International Research Report, 1990.

'The world we have created today as a result of our thinking thus far has problems which cannot be solved by thinking the way we thought when we created them.'

Albert Einstein.

Introduction to Part III

■

In discussing policy development and the appropriate systems for implementing environmental policies, it quickly becomes apparent that the skill in successfully improving environmental performance is recognising and effectively managing the organisational tensions created. Indeed, some of these tensions also concern the organisation's interface with its external operating environment.

In Chapter 12 Rolf Marstrander, from the Hydro Aluminium a.s., focuses upon the tensions created between the organisation and its operating environment. He perceives the need for the industrial infrastructure to change as well as individual companies changing, if sustainable development is to be achieved. In particular, he cites the need for a better recycling/remanufacturing infrastructure, together with more vertical integration across supply chains. One way of increasing vertical integration across supply chains is for strategic alliances to be developed between suppliers and their customers.

In Chapter 13 Philip Rees considers the tensions created when the construction company John Laing plc decided to formalise its strategy on its environmental performance. He describes the challenge of developing a policy in an industry with little experience of such matters. He also discusses the dilemma of meeting customer needs while doing 'the right thing' for the environment. If the customer does not 'value' environmentally-sustainable features, who pays for such features? What is the benefit of offering them? Clearly, these questions must be addressed within organisations, if an environmental policy is to be successfully implemented.

This tension is further explored in Chapter 14 by Nicholas Reding from the Monsanto Company. He describes how Monsanto has moved through four stages of environmental awareness. In the first stage, compliance was all important; then Monsanto's CEO Richard J. Mahoney took the company a step further on a values basis – Monsanto must improve its environmental performance because 'it's the right thing to do'. This CEO activism turned into institutionalised activism, and finally into the 'era of added value'. At this point, Monsanto is reaffirming its financial obligations to its shareholders. In the future, environmental improvements must have a pay-back for Monsanto's business; costs – such as the cost of the company's programme to reduce air emissions by 90% by the end of 1992, estimated at $115 million – cannot be sustained without business benefits.

It can be argued that some environmental improvements, for example energy-saving product features, have a market value. Others, such as land remediation or reductions in emissions, are costly and difficult to market as a benefit to a specific customer. This begs the question of how to create a truly 'green' market where environmentally-sustainable corporate behaviour is rewarded. Numerous reports have been produced citing the need for government intervention to tackle this issue, such as the development of tax incentives and other fiscal measures.

Karl Kummer describes in Chapter 15 how Rank Xerox implemented environmental policy in a large, functionally-organised business. Based on their commitment to quality, Rank Xerox developed a strategy to involve all parts of the business in the improvement of the company's environmental performance.

In the final chapter Dennis Vaughn from Grand Metropolitan describes how a small centrally-organised department can support the implementation of an environmental policy. He explains the roles and responsibilities of the Grand Metropolitan food sector's Environmental Affairs department. The key activities are: conducting audits; producing guidelines and procedures; providing technical resources, and consulting. He emphasises the need to integrate these activities into the businesses' core activities.

Part V takes the tensions mentioned here and describes some approaches for overcoming them. In particular, it deals with engaging individuals in environmental policy, the role of the environmental-change agent, defining the corporate vision and working with suppliers.

Executive summaries

■

12 Industrial Ecology: a practical framework for environmental management

The change in focus from pollution to sustainability in economic and environmental terms confronts society and industry with new challenges because environment has become a third dimension, integrated with the two traditional dimensions of technology and economics as the critical factors for industrial development. The change in focus from industry to industry as part of society also makes the platform for a systems approach to an environmental management aiming for sustainability. A sustainable development will be one in which we constantly minimise our use of non-renewable resources through recycling, reduced waste, use of renewable resources and environmentally-friendly products. Industrial and societal systems aiming for sustainability will have to co-operate across traditional borders between single companies, different industries, industry and government, and industry and its markets. The pattern of co-operation across traditional borders will exist alongside traditional competition. This pattern of co-operation and competition between subsystems is similar to the pattern of ecology as known in nature. We have given this pattern or concept the name 'industrial ecology'.

The concept of industrial ecology raises new challenges related to technology information as a basis for strategy-forming and decision-making, with new patterns of co-operation between authorities, business and markets being formed. Those challenges are very much a matter of R & D and specialist work, following known patterns of organisation.

Industrial ecology also raises challenges related to eduction, information and the task of getting the organisation of your company started on a route leading towards industrial ecology. Traditional health, environment and safety (HES) work is part of the concept of industrial ecology. It is shown how we, by using the basic principals of HES-work upon other parameters of importance for the systems-oriented concept of industrial ecology, can simplify the first steps towards a better understanding in industrial ecology of our own organisation.

13 Environmental management in a construction business

This chapter covers a four-year period in the development of environmental management systems in John Laing plc from its inception in

1989 to 1993. The Group trades as a general construction and house-building company both in the United Kingdom and internationally. The subject of environmental affairs was seen to fit in well with the overall company ethos of quality, safety, training, community affairs and customer service.

In 1989 little or no work had been completed in the construction industry on environmental management systems and, since then, not much has changed. Much of the work completed by Laing has been of a pioneering nature. The driving forces have been senior management commitment, employee enthusiasm, limited customer pressure and a belief in the need to develop a long-term competitive edge.

From an initial beginning with no information, the author describes the process of establishing a corporate environmental policy, and how practice, procedures and management systems have been subsequently developed. Two further sections review the start made on performance indicators, and the development of environmental education and training. Even after four years, there is still a long list of actions and procedures to implement.

14 Developing and implementing environmental policies and programs in an international corporation

For the St Louis-based Monsanto Company, environmental policy has moved through four major phases: compliance with the numerous laws enacted in the United States in the 1970s and early 1980s; CEO activism, comprising both voluntary initiatives and disclosure of data to the public; institutionalised activism; and environment policy as business policy, or the 'era of added value'.

While each of the four phases is distinct, with its own starting points and characteristics, each also overlaps the others, and all four continue as major influences upon the corporation's environmental policies. This gives rise to inevitable conflicts and tensions within the company, both of which are healthy and vital to both business and environmental goals.

Five factors have been critical in Monsanto's development of environmental policy: a strong sense among employees for doing the right thing; a CEO committed to environmental protection; a determination to break the 'environmental gridlock' in policy debate; a sense of responsibility for disclosing dates about operations; and the growing awareness that environmentalism must add value in ways perceived by customers.

15 The Rank Xerox approach to achieving environmental leadership

Rank Xerox, a leading office equipment company with 28,000 employees, has an integrated environmental policy and programme with a director in charge.

The company has a long record of responsible environmental behaviour and achievement. The Rank Xerox Environmental Health and Safety Policy is recognised as a model which other companies may follow.

The company has set two objectives for its environmental programme:

1. To move from a 'Nobody Cares' to an 'Everybody Cares' situation.

2. Compliance is not enough; Rank Xerox aims for Leadership.

Although Rank Xerox has no big environmental issues, there are many improvements possible which will benefit the environment.

People-involvement and management commitment are prerequisites for a successful environmental programme. The company has established an 'Environmental Network' which includes the functional managers as well as specialists within the marketing operations.

Rank Xerox works to the principles of Total Quality Management (TQM) and the Quality tools are applied to the management of the Environmental Leadership Programme.

16 Environmental management in the food industry

Environmentalism has had an impact on food processing as it has on other industries. Traditionally considered a 'clean' industry, food processing businesses, like service companies and financial institutions, are expected to act responsibly towards the environment. Unlike heavy industry producers, such as petroleum or chemical manufacturers, food processors face an immediate consumer response and reaction from competitors and distributors to 'environmentally friendly' consumer products.

The environmental issues affecting food processors are similar to those affecting other industrial sectors. It has, perhaps, taken regulatory programs a longer time to reach the food industries. This has allowed food companies to learn from the programs and lessons of the chemical and petroleum companies. The management systems developed in response to increased environmental awareness and resulting issues are modelled on those of traditional industries.

The objective of a food processor must be to provide safe food. This must be done while protecting the health and safety of employees, the environment and the communities in which they operate.

Grand Metropolitan (GrandMet) Food Sector operates as one of three business Sectors of GrandMet. The Food Sector is comprised of several businesses and brands, producing such products as ice cream, frozen and canned vegetables, pet foods, prepared dough products, pizzas, baked and frozen dough and pastry products.

GrandMet Food Sector initiated its environmental coordination efforts in 1990, following its acquisition of major food businesses. With the formalisation of its quality management system, a number of management policies were re-issued, among them one for the environment. Recent environmental spills and releases at some facilities also increased employees' awareness and commitment to environmental management. This environmental policy speaks of 'complying with the letter, as well as the spirit, of laws, regulations and operating standards', in protecting employees, neighbours and the environment. The policy places responsibility for environmental performance upon line management. The corporate function, Environmental Affairs, is responsible for technical support, regulatory interpretations, development of company operating standards, and conducting facility assessments to identify issues and risks for company management.

12

Industrial Ecology: a practical framework for environmental management

ROLF MARSTRANDER

Senior Vice President, Strategy, Technology and Ecology, Hydro Aluminium a.s., Oslo

▪ In recent years the focus of environmental management has shifted from the local regulation of industrial pollution to a concern with 'sustainable development' on a global scale.

▪ To put this idea into practice we need to understand how our businesses operate, not just in producing and marketing products and services, but in terms of their impact on the world's ecology.

▪ This involves establishing a new industrial infrastructure for:
(a) recycling and remanufacture;
(b) measuring the economic value of natural resources;
(c) looking at product costs over their life cycle;
(d) assessing environmental risks; and
(e) calculating production costs in terms of energy, etc.

▪ For environmental management to succeed in practice its practioners must agree on the use of simple ratios and measures, like safety managers use 'accidents per million hours worked'.

▪ These factors also need to be agreed between suppliers, customers, and end-users.

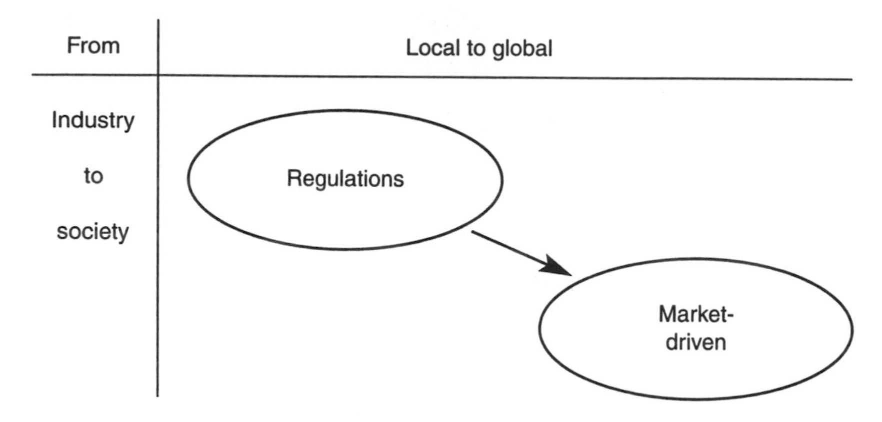

• From unlimited space to limited/closed loop
• From linear thinking to systems thinking
• From economics to sustainability

Figure 12.1 The shift in environmental thinking

During the last years an important shift in environmental thinking has taken place. We see this in the declarations from the UN conference in Rio, where energy conservation, global warming, biodiversity, economic development and technology transfer, world population growth and food supply, became the key themes. We have received the concept of sustainable development, from the UN Commission on Environment and Development. The title of their report, *Our Common Future*, is in itself a reflection of the same trend. We see the same idea in more detail in the industrially developed economies, where life cycle assessments of products, energy conservation, the limitation of waste from households, agricultural pollution and global pollution related to ozone depletion and greenhouse effects, get more attention than direct emissions and discharges from industrial activity.

The shift in thinking

As shown in Fig. 12.1, we have moved from an environmental approach where interest was directed at industry and local pollution, to an environmental view which focuses on global problems related to the way our societies consume and develop. This new perspective confronts industry and society at large with new challenges because the environment has become a third dimension integrated with the two traditional dimensions of technology and economics as the critical factors for industrial development.

In the 'local pollution-by-industry'-era, environmental improvement in industry was initiated to a large extent by government authorities. In the 'society's-global-concerns'-era, which we are entering, this is no longer a viable approach. The solutions have to be found through co-operation between public, authorities, industry and the users of industrial products, and society represented by the final consumers. Industry is needed as a proactive partner to help solve the environmental and ecological challenges and threats which we face globally. Implicit in this conclusion is the belief that technology, and the ways we apply it, will be part of the solution.

The change in focus from local to global and from industry to society, represents a dramatic shift. But more than that, it represents an increased realization of the fact that we are living on a planet which sets its own limits – limits that make it necessary to restrain the amount of waste we produce. This sort of challenge our planet has always faced and it has been solved through ecology. Now that our industry-based culture is faced with a global challenge, the concept of *industrial ecology* has come to the surface.

The focus on global issues raises a series of new challenges:

- If industry is to take the role of a proactive ecology-oriented partner co-operating with markets and authorities, management will require new social insights, new tools for decision-making, new technologies, and a higher general understanding of environmental issues.
- In industry we are used to thinking in terms of single companies competing in the market. That viewpoint will persist, but we need to add new dimensions to our thinking. The car user wants the car to function, but society wants the car to be produced, used and recycled with a minimum use of resources and generated waste. This will require new co-operative alliances which will exist alongside traditional competition.
- In industrial companies we have developed cultures which deal with challenges in the field of economics, technology and an improved environment. We must develop this culture further to achieve a *proactive ecology-oriented participation* from our organizations. Training systems and techniques for organizational development must be created for this purpose.

All these challenges represent possible threats to the individual company if we fail, but much more important is the fact that *any threat is an opportunity*. These questions point to an embryonic situation; we are really at a turning point.

In this chapter we will look more closely at these challenges to industry and possible solutions to them, taking the idea of 'industrial ecology' as our starting point.

We will do this by:

- looking more closely at the evolving concept of industrial ecology. What does it mean in terms of concepts, development of new methods, new technologies and analytical tools?
- discussing how this relates to some of industry's experiences in the traditional field of Safety Health and Environment (SHE);
- finally, we will discuss management practices that combine the SHE experiences with ecology-related approaches.

The concept of industrial ecology

During the last years, we have seen regulators call for life cycle analyses and life cycle assessments of products. We have seen ideas formed around clean technology, environmentally-clean manufacturing and design for environment. All these broadly based ideas point towards a shift in thinking from a clearly process-oriented focus to a more systems-oriented view. Lately, these trends have been brought together into the broader idea of industrial ecology.

Harbin C. B. Tibbs[1] describes an 'industrial ecosystem at Kalundborg' that has very limited losses of energy and waste, because of its integration with a series of industrial activities. These activities include an electric power plant, an oil refinery, a bio-technology plant, a plasterboard factory, a sulphuric acid producer, cement producers, local agriculture and horticulture, and district heating.

A collection of cases from companies around the world shows how business has embarked on a path to sustainable development. The book by Williams and Golüke[2] discusses the role of 'The business charter for sustainable development' drafted by the International Chamber of Commerce, and the actions needed to fulfil Agenda 21 from the UN Rio Conference. It presents a series of cases from different companies, the actions that were taken and the results achieved. The majority of cases demonstrate the capacity of organizational dedication and technical ability to improve processes, utilize waste and to co-operate with customers, suppliers and the surrounding community. An important message is *the need to formulate clear goals related to identified areas for improvement.* Examples of this are DOW's 'Waste reduction always pays' program (WRAP), and Du Pont's targets in the same area. Both companies have publicly committed themselves to targeted reductions of waste. And both companies reflect these targets in their inter-company programs for improvement.

The proceedings of the National Academy of Sciences of February 1992 has references to a colloquium entitled 'Industrial Ecology'. In the introduction to the colloquium, industrial ecology is described as '*a new approach to the industrial design of products and processes and the implementation of sustainable manufacturing strategies. It is a concept in which an industrial system is viewed, not in isolation from its surrounding systems, but in concert with them. Industrial ecology seeks to optimize the total material cycle from virgin material to finished material, to component, to product, to waste product, and to ultimate disposal.*'[3] The concept of industrial ecology stems from the definition of ecology and the similarities we face between ecology and an industrially developed society in the obvious limitations which exist if we want our industrial system to become really sustainable. A biological ecosystem is a natural system in balance, but dependent on energy input from the sun. Some organisms in this system use water and minerals to grow, others consume these organisms along with minerals and gases. The waste of these organisms again are food for others etc. The whole system is a network of inter-related processes in which everything produced is used in another process. This balance is dynamic. We can see dramatic changes in the system, but the system as a whole adapts.

In the nineteenth century, we allowed ourselves to run industry as if we had unlimited resources and we could produce unlimited waste without doing any harm. The smoke from the plants was a good sign; it meant healthy business. Today we are learning by experience that landfills are limited, and long-range pollution like acid rain sets obvious restrictions to what we can emit from our smokestacks, etc. We are entering a period in which we want to base our industrialized society on a model which has limited energy and limited resources as inputs, and limited waste as

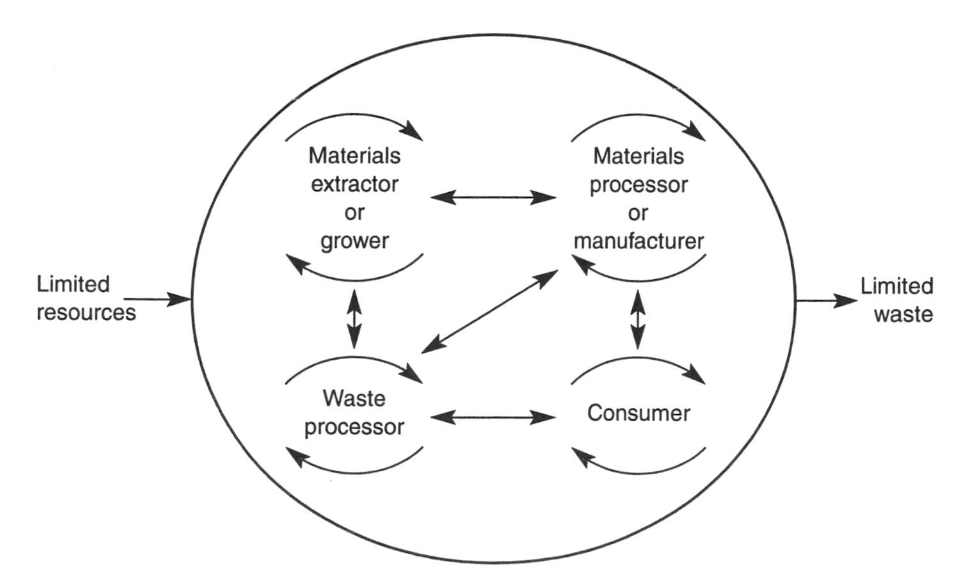

Figure 12.2 A model of an industrial ecosystem

its final output. This is described as an 'industrial ecosystem' as represented in Fig. 12.2. The main areas of activity in an industrialized society are identified as the raw-material producers, the processing and manufacturing industry, the consumers, and the waste-handling industry. In each of these four areas we will find activities where we can improve environmental performance. We may also find activities between the areas as shown in the Kalundborg example, or as in all sorts of activities related to the recycling of used products.

For it to function properly, the industrial ecosystem will require a major improvement in communication between the different areas of activity, and between the often competing actors in each of the four areas. If we add to this the fact that society at large, often represented by the public authorities, also has its role to play, it is clear that the biggest challenge will be one of communication and education – showing individuals and groups how different activities can each contribute to the end-result by limiting the consumption of non-renewable resources and reducing the generation of waste. The rewards will be seen in the increased economic activity which will be developed to support the new interchanges in this industrial ecosystem.

A new industrial ecosystem

The challenges we will face in establishing this new industrial ecosystem have been listed by Jelewski *et al.*[3] They concentrate on the following topics:

- The development of an industrial infrastructure for recycling and remanufacturing. We see this developing today for some categories of products and for valuable metals.
- Evolving models and methods that will enable us to measure the value of natural resources in economic units. Without this development, we cannot use our normal economic and financial techniques to take decisions in an industrial ecology system.
- The development of methods and rules for the use of life cycle analyses, life cycle assessments, or systems studies of the application of alternative materials to specific products. These systems studies will be based on the engineering and scientific data which is needed to assess environmental effects.
- Logistics and transport studies, including risk assessments, related to an industrial process.
- The analysis of production systems and their use of energy. This will include energy-efficiency measures in a single process, also the effi-

ciency improvements which can be achieved by re-designing the complete industrial system.

All these topics will have to be studied by the different actors in the industrial ecosystem, from a practical point of view. Universities and research institutes will also have an important role to play in the development of an industrial ecosystem. John R. Ehrenfeld[4] describes some of the R and D topics he sees for Massachusetts Institute of Technology (MIT) which are needed for the education that is necessary for supporting the development of industrial ecology:

- *Data acquisition* – Develop data networks that can access and combine relevant information from different sources.
- *Analysis and transformation of data* – Develop approaches to indicate and quantify the material flows across industrial sectors. Research on life cycle assessment to provide a systematic framework for analysing the product and for processing the implications of LCA, will be part of this development.
- *National accounts* – Development of 'sustainable' performance measures, including standard account systems for national economies and the integration of these into management information systems for the public service.
- *Policy analysis* – New and improved processes for negotiating and purchasing upon ecological values and criteria. Processes that help introduce scientific analysis into policy discussions and processes to deal with uncertainty will be included.
- *Implementation studies* – Studies of organizational behaviour in choosing among technologies. Many alternative 'clean' technologies exist, but are not used, in spite of the potential cost savings and product quality gains. Research into reasons behind these decisions, including studies of innovation and design practices, is needed.

This brief summary of the concept of industrial ecology indicates some of its characteristics. Industrial ecology is a *systems approach* to guide the development towards, a more sustainable industrial system. The concept, and the solutions it points towards, are very much in an embryonic state, but it is a proactive concept. It can act as the framework for strategy which will embrace technology development, the positioning and development of relationships between industries, and communication with public authorities and potential markets for individual companies as well as industries.

The concept of industrial ecology is complicated because we must deal with a very complex system that we cannot yet fully describe. Also industrial ecology is a system in dynamic adaptation. The management of a company that wants to be 'environmentally proactive' in line with the concept of industrial ecology will be faced with challenges in two dimensions:

- *Expertise* – One dimension is related to the exploration of concepts and the development of the in-house expertise which is needed to be able to handle the systems and solutions required. This is the same task we face in any R and D effort.
- *Communication and education* – The other dimension is the task of building an organizational understanding of industrial ecology, including all the employees and communicating this to our customers and to society. This is a matter of communication and education.

The challenge related to development of expertise needs to be seen in a longer time horizon. The development of expertise does not solve any immediate problem. It is related to the strategic choices about future market positions, future products, technologies and services.

The challenge of building understanding in our organizations is also long term, but it can give more immediate feedback. An important factor in any ecology-oriented development will be our present efforts on safety, health and environment. We will discuss in some depth how we can relate these efforts to the larger concept of industrial ecology.

Some characteristics of safety, health and environmental work

So far we have been occupied with the general concept of industrial ecology without being company-specific. Seen from the individual company's point of view, industrial ecology is a system like the one indicated in Fig. 12.3. The 'value' of the end-product is defined by the sum of the processes needed to make it, its properties in use, and the possibilities of the end-product being re-used or recycled. By 'value' we mean its economic, technical and environmental characteristics. When discussing the role of Safety, Health and Environmental work (SHE) in this system, we see that SHE work is part of all the processes and so must also be represented in the end-value of the product.

It is important to remember that the authorities' approval of SHE-work has been related to the regulation of discharges, the use of dangerous substances, etc., for each particular process or plant. Similarly, each company has been concerned with SHE-work related to its own processes and based on its own organization. This also reflects the fact that the product's performance is not directly linked to the emissions and safety hazards related to making it. But the SHE-value may be an important factor in marketing, or may be demanded by the authorities as we see in demands for LCAs or eco-labels. The experiences and philosophies behind SHE-work, are discussed elsewhere in this book.

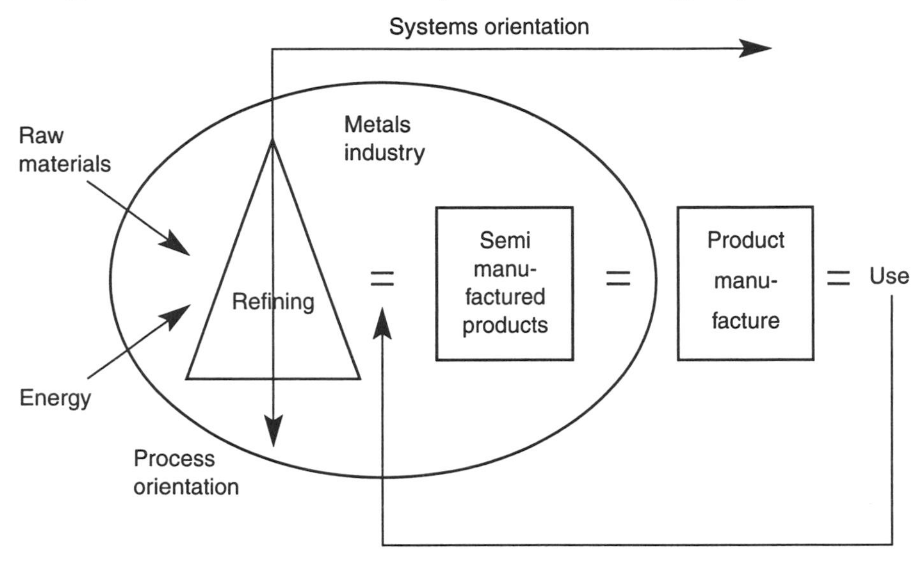

Figure 12.3 SHE is an integral part of ecology-orientation

Here we will focus on one characteristic of SHE responsibilities as seen in safety work.

Most major companies in the western world have achieved impressive improvements in safety performance during the last ten years. Most of these gains have come as a result of *a very simple but effective management approach. It has been a focus on a single common denominator, accidents per million hours worked, and the responsibility has been defined as part of the line manager's job in all departments of the company.*

The use of a common denominator has made it possible to establish a simple control loop as shown in Fig. 12.4. This control loop is traditional and well understood. It gives us the basis for a systematic approach to continued improvement. The definition of line responsibility ultimately leads to every employee from the CEO down through the company. But it is obvious that the safety performance of a sales representative spending most of his time in a car is different from that of a mechanic using his spanner tightening nuts, or an operator watching his control panel. For the design engineer, safety in the processes he designs, is different again. The important lesson from this analysis is that the safety denominator has, combined with line responsibility, helped us to identify and explain to every employee the complex field of safety improvement. Also, the experience gained in safety is helping companies to implement and develop an understanding of the wider concept of Total Environmental Quality Assurance.

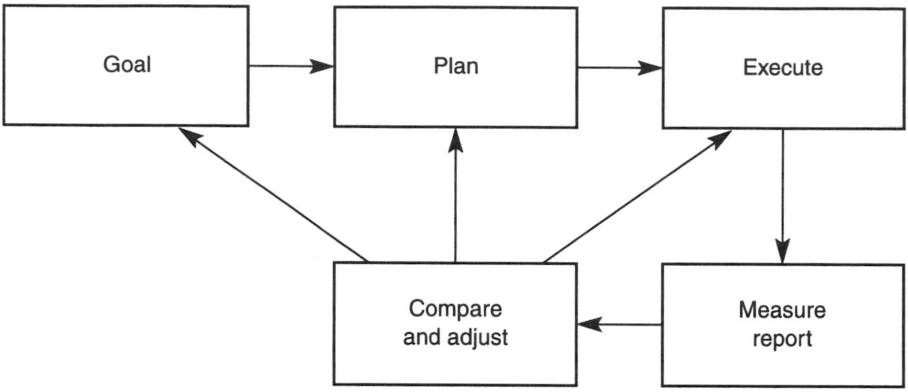

Figure 12.4 The control loop

An example of this can be taken from the refining of aluminium oxide into primary aluminium. In this process fluorides are needed and they are costly. Emissions of fluorides as particulates, and as gas, can cause harmful effects in the environment. The refining process has therefore been linked to cyclone batteries and a dry-scrubber system, using aluminium oxide to catch as much as 97% of the fluorides in the off-gases coming from the reduction cells. The aluminium oxide from the dry scrubber is used as raw material in the reduction process. This is an example, as indicated in Fig. 12.5, of a closed-loop-process reducing the loss of a costly raw material that can also be harmful to the local environment. The process is dependent upon a minimum loss of off-gases containing fluorides before they have been recaptured. With the overall goal of reducing the amount of fluorides lost, we get a common goal for the total quality in the closed-loop system consisting of

oxide silo ⇒ reduction cell ⇒ gas transport system ⇒ dry scrubber system ⇒ oxide silo

Building organizational understanding to ecology-related indicators

Implementing the concept of industrial ecology depends very much on the relationships the individual company has with its customers, and with the end-users of the product. If we want to make an analysis of an end-product in a supply-chain like the one shown in Fig. 12.3, we can use the kind of a matrix illustrated in Fig. 12.6.

That sort of matrix presented in combination with the systems-based concept of industrial ecology, can be complicated and difficult to communicate to the different organizations in the supply chain. The complete concept of industrial ecology will be useful mainly in strategic analyses, and in the development of concepts for new products and new processes.

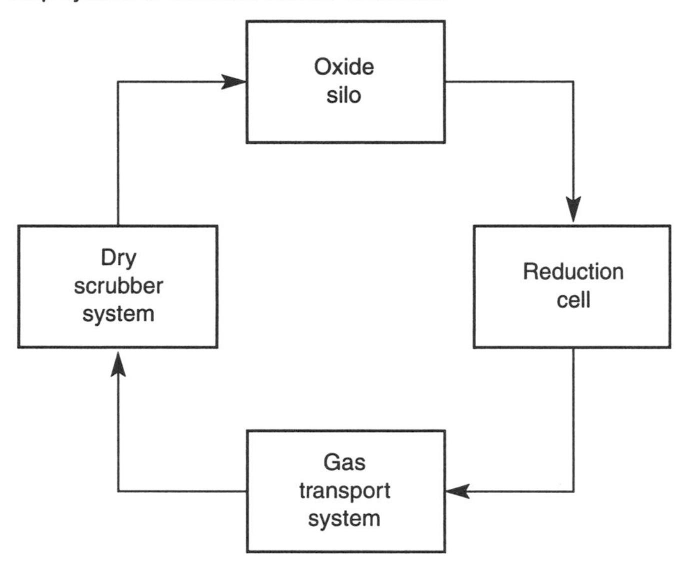

Figure 12.5 The refining of aluminium

In an operational setting on the shop floor, or in our communication with a specific customer, our local community or to the public at large, we need a simpler vehicle to get our message across.

If we take the lessons learned from the safety and environment-related activities about goal-setting, we should look for simple indicators which relate to the whole system and use them to reduce the complexity of the industrial ecosystem. Experience in metal fabrication and other material-producing industries, suggests indicators like:

- energy use and energy efficiency;
- waste reduction;
- recyclability; and,
- transportation of materials and products.

All of these relate to all the processes in the system and to the use and handling of end-products.

We have seen in *Business and Sustainable Development,*[2] in Dow's WRAP-program and Du Pont's goals for waste reduction, examples of big corporations using similar indicators in their environmental policies. The policies also involve goal-setting and line responsibility as we found with regard to safety. In preliminary studies of energy use and logistics in Hydro Aluminium, we found that these parameters could be used in the same way.

LCA-* data for each material and product	Energy efficiency	Recyclability	Environmental factors waste/emissions	Safety	Food preservation
Process-related • • •	●			●	
Recycling-related • • •		●	●		
Transport-related • • •					
Product use characteristic • • •	●		●	●	

* Life Cycle Assessment

Figure 12.6 Ecology measurements for a product

It is reasonable to assume that for any company we could define a set of relevant ecology-related factors that we should measure; the raw materials, the manufacturing processes, the end-products, the recycling possibilities and the links between them. Applied as we described in our safety example, we would produce:

- similar approaches to those used in Safety-Health-Environment work;
- access to indicators that can help reduce the complexity of the system,
- a basis for communication with customers and suppliers on ecology-related matters.

The management challenge will be to give the process some priority in resource allocation, and to provide the necessary expertise and status for the ecology programs. The industrial ecology programs will require systems studies, communication and educational activities, outside and inside the company.

Practical implications

Our focus in this book is on environmental management. The challenges we face in this are derived from the challenges we also face in the environment. In recent years we have seen a change of emphasis away from local issues to global challenges, and a focus away from industry-specific concerns towards a view of industry as an integral part of society. This change-over is a consequence of taking an ecological perspective, and seeing industry as a total system. This has led to discussions around the concept of 'industrial ecology'.

Our discussions clearly demonstrate that we are at a very early stage in the development towards an industrial ecosystem. But we can also see that the concept which is rooted in industrial ecology is likely to raise some new challenges related to:

- a new industrial infrastructure for recycling and the re-use of all sorts of waste;
- increased co-operation between industries, and between companies;
- a more clearly defined role for the consumer in the industrial ecosystem (e.g., in recycling waste and conserving energy).

These challenges require us to develop approaches which will enable us to include performance measures for environmental factors in our established process of techno-economic planning, and strategic analysis.

All these challenges are identified and discussed in the literature, but the concept of industrial ecology can provide the basis for a more comprehensive and systematic approach.

Ecology is a concept that implies dynamic change. Dynamic change has two different dimensions:

- the strategic long-term dimension; and
- the immediate dimension of education and increased awareness.

The strategic long-term dimension contains the traditional R and D elements. The management challenge is very much one of giving vision and support to the development of technology, and providing techniques for planning and analysis. The concept of industrial ecology will provide a conceptual framework for co-ordinating these strategic efforts.

The dimension of education and ecological awareness is different. From our experience with traditional Safety-Health-Environment work, we can see that the principles of environmental management, as expressed in setting goals on identifiable parameters and giving responsibility to line management, can also be applied to the concept of industrial ecology. This concept suggests that we might choose to set goals in terms of elements that bind the industrial ecosystem together like:

- energy use and energy efficiency;
- recycling and waste; and
- transport.

Experience indicates that each of these parameters can be used as a measure to reduce the complexity of the message for involving employees, customers and suppliers; defining clear goals, policies that can be understood, and results that will encourage further improvement.

Developed and used as indicated in this chapter, the concept of industrial ecology is one that can help management to build on the foundation of policies and practices already established for environmental management.

References

1 Tibbs, Hardin B. C. 'Industrial Ecology – An Agenda for Environmental Management', *Pollution Prevention Review*, Spring 1992.
2 Willums, Jan-Olaf and Golüke, Ulrich. *From Ideas to Action: Business and Sustainable Development,* ICC report on the Greening of Enterprise. ICC Publishing and Ad Notam Glydendal, Oslo (1992).
3 Jelewski, L. W., Graedel, T. E., Landise, R. A., McCall, D. W. and Patel, C. K. W. 'Industrial Ecology : Concepts and Approaches', *Procedures of the National Academy of Sciences*, USA, vol. 89, February 1992, pp. 793-797.

4 Ehrenfeld, John R. 'Implementing Industrial Ecology/Design for Environment: Roles for the University', *MIT Program on Technology, Business and Environment*, MIT, Cambridge, Massachusetts prepared for the National Academy of Engineering Workshop on 'Industrial Ecology and Design for Environment', July 13–17, 1992.

13

Environmental management in a construction business

PHILIP K. REES

Group Director, Environmental Affairs
John Laing plc, London

- John Laing plc is a major construction, housing, mechanical engineering and technology-related business, employing 12,000 staff in the UK and 800 overseas. The company has 600 construction sites and operates in Europe, the Middle and Far East, and has housing interests in the USA.
- In the past 4 years Laing have pioneered the development of environmental management systems in the construction industry, driven by a strong commitment from senior management, support from employees and pressure from customers.
- Management believe that these environmental initatives will help to develop a competitive edge for the company.
- This chapter describes the process of establishing a corporate environmental policy, management systems and procedures. A start has also been made in setting performance indicators and establishing training and education.
- As the systems become more comprehensive and sophisticated, environmental management will be incorporated into the normal way of running the company.

The beginning

August 1989 marked the beginning. The previous three years had seen the consolidation and development of the Laing Group's external trading capabilities in the environment field, particularly in the areas of ground contamination, asbestos removal, water treatment and energy efficiency.

The time had come to look inwards. The simple instruction from the Group Chairman, Martin Laing, was *to develop an Environmental Policy and the procedures and management systems that go with it.*

There were a number of pressing business reasons why the time was right. The Group, in existence since 1848, is a family-run, and largely family-owned, company. It has a culture and ethos based on customer service, safety, training, quality and community involvement. Environmental affairs was identified as a missing key component, highly compatible with these other corporate values. Customers were pressing hard for environmentally-acceptable solutions to their business problems. Although these customer perceptions were not yet articulated within their building programmes, that time would surely soon arrive. Employee pressure was such that everyone in the company wished to see the Group provide a lead and direction in standards of environmental performance in their everyday working life. The considerable work in progress to reinforce the quality culture within the Group at that time had led to the introduction of quality accreditation and measurement procedures, which had identified in turn the potential for measuring and managing performance in a wide range of business areas including the environment. Finally, there was strong personal interest and leadership provided from the top of the Group to ensure that the environmental credentials of the Group were based on sound foundations for the future.

That there was no-one to talk to became immediately apparent, in the sense that for the construction industry in 1989 environmental affairs were at the pre-nascent stage. A rapid review showed that initial contacts would need to be with organisations outside the construction industry or from within the Group itself.

The remainder of 1989 was therefore spent establishing contacts, deciding what the environmental issues were that were peculiarly important to the construction industry, and how these should be translated into a feasible and workable policy for the Laing Group.

Policy development

Although the construction industry itself had not yet woken up to the importance of internal environmental good housekeeping, there were clearly a number of organisations that were looking in from the outside and forming views and opinions as to what should be happening.

Through Martin Laing's chairmanship of WWF (UK), it was possible to develop contacts with a wide variety of organisations and people, to understand the issues progressively and develop a perception of where the Group should stand. The blank piece of paper was rapidly replaced by a welter of paper, and the need for selectivity in reading, plus a large waste paper basket, became essential.

Within the Group, an Environmental Committee was formed in early 1990. This comprised senior line managers (not environmental specialists) from each of the Divisions who would become responsible for ensuring the implementation of environmental policy down to grass-roots project level. At the first meeting of this Committee a final draft of the Group's Environmental Policy Statement was presented for discussion and agreement.

This policy document had been developed by the Group Director of Environmental Affairs and two internal advisors who had sifted through the issues, culled from them those of importance, then balanced the need for environmental action with the economic cost and commercial feasibility of achievement. Very early on, a number of issues were identified as especially related to the construction industry.

- *The housing sector was already in recession and the construction sector was rapidly heading that way.* Any actions with a significant overhead cost, or site costs, attached to them would be unlikely to be implemented, unless a customer could be found who was willing to pay for them.
- *There was clearly a considerable gap between customers' desires for good environmental performance and their willingness to pay.* Even in the area of housebuilding, virtually no premium existed (or exists now) for environmentally-acceptable products, which contrasts with the 'green' consumer market.
- *The majority of construction projects are designed and specified by others directly employed by the building's owner.* Unlike other industries, therefore, the builder's role in these circumstances is to persuade for change rather than to eliminate environmental impacts directly. This aspect has had a considerable effect on policy statements.
- *For the remainder of the Group's businesses, where the total process is within the Group's control, more specific and absolute statements can be made and implemented.* Therefore this, in environmental management terms, splits the Group into two parts.

In addition, the Group discovered that many actions that were appropriate for environmental reasons were already in place for other good company cultural reasons, thus reinforcing the logic of integrating environmental affairs into the overall Group ethos.

LAING AND THE ENVIRONMENT

Throughout its long history Laing has endeavoured to maintain a philosophy of caring, conscious of its responsibility towards its clients, its shareholders, its employees and towards the community at large.

In the current age of global communication however it is becoming increasingly evident that the pollution of the Earth and its atmosphere which started with the Industrial Revolution, coupled with further unchecked pollution of all types, could endanger the continuity of an acceptable quality of life for future generations.

Public pressure, with Governmental support in many countries, is increasingly demanding that products and services are environmentally sensitive, with calls for steps to reduce pollution in an endeavour to rebuild a healthy Global environment. This needs to encompass a balanced approach to the use of land, materials and resources.

In recognition of this background, and taking into account Laing's key role in contributing to the built environment the Company is formalising its position on environmental, matters by the publication of the following Policy Statement:

JOHN LAING PLC
ENVIRONMENTAL POLICY STATEMENT

1. LAING will ensure that in the use of land, both reclaimed and greenfield, its developments are sympathetic to the environment; paying close attention to the treatment and disposal of any hazardous or potentially toxic materials to avoid environmental problems.

2. LAING will, wherever practicable, utilise building materials and products which originate from sources which can be shown to be sustainable and which are re-usable or can be recycled, and which it has established are environmentally sensitive.

3. LAING will pay particular attention during construction activities to the emission of pollutants, reduction in noise, dirt and to the careful use of pesticides, herbicides and toxic treatments, taking the most stringent precautions to avoid health hazards and ensuring that environmental impacts are minimized.

4. LAING will continue to develop an environmentally aware approach to the management of the Company recognising also that sound management of energy and resources in land and materials cuts cost and creates competitive advantages.

5. LAING will continue its philosophy of seeking to employ caring people promoting a sense of responsibility towards the Environment by Management and Staff within their working role and in their home situations.

6. LAING will continue to develop and practice 'in house' environmentally caring policies covering its use of premises, sites, plant, and other assets. The Company will also continue to promote energy efficiency and sound environmentally sensitive practices.

7. LAING will achieve this policy by establishing a clear set of environmental objectives at Group level with each Division appointing a senior manager responsible for ensuring comprehensive implementation of these objectives.

Figure 13.1 The Laing Environment: Practice Note No. 1

The crystallisation process was lengthy and, on occasions, painful – particularly for what, in retrospect, looks like a short and simple document. Inevitably, the balancing act between saying something positive, achievable and not too costly, and being bland, pulpy and open to criticism for inadequate commitment, was the most difficult issue. In this, the Group was helped and advised by the UK Centre for Economic and Environmental Development (UK CEED) which was able to give an independent view on that balance.

Spring 1990 saw the publication of the Group Environmental Policy; its Practice Note No. 1 is reproduced here as Fig. 13.1.

Practical implementation

As the policy statement was being finished, work began on fleshing out what the policy actually meant in practice. Strategically, it was decided that the Group would establish a list of topic headings where real and practical initiatives could be introduced to improve environmental performance. The resulting 'practice notes' gave a brief introduction on the issues and then a succinct statement on Laing's policy. They are printed on distinctively coloured yellow paper in a form suitable either for external distribution or for posting on office or site notice boards. All the points are covered on one side of A4 paper.

So far the Practice Notes have covered a comprehensive range of topics, and a typical example is shown in Fig. 13.2. The complete list to date is as follows:

- Environmental policy statement
- Recycled paper
- The use of CFCs – Chlorofluorocarbons
- The use of timber – Hard- and Soft-wood in construction
- The control of substances hazardous to health
- Noise
- Company cars
- Company catering and health
- Landscaping and the use of peat
- Dust
- Recycling paper
- Cleaning of company premises and property under construction
- Temporary lighting on site
- Temporary office accommodation
- Trees
- The protection of water courses and aquifiers
- Energy effciency programme for business accommodation

LANDSCAPING AND THE USE OF PEAT

Landscaping forms an important element of many of the Company's contracts and projects and it is a key feature of housing development where planting can mature to benefit the environment.

From the 1960s Horticultural Peat Products have been in common use by Landscaping Contractors and others to improve the condition of soil and as a growing medium.

There is now concern over the loss, through peat extraction, of unique wild life habitation and the long term damage caused to Sites of Special Scientific Interest.

LAING POLICY

The Company will cease using peat in landscaping operations carried out by Laing Homes and John Laing Developments, and where possible will seek to promote the use of alternative materials in its contract operations.

Figure 13.2 The Laing Environment: Practice Note No. 9

These Practice Notes provided a comprehensive framework to allow each of the Divisions to begin to develop practical procedures and management systems for implementing and controlling our environmental practice. Some actions were taken at Group level. For instance, a comprehensive review of the company car fleet led to the decision in 1990 to switch totally to cars with catalytic converters as soon as manufacturers could deliver and, late in 1992, to switch to diesel models for all cars with smaller engines. In the first decision, environmental considerations overrode cost issues, while in the latter, cost considerations swung the scales on what is a balanced environmental argument between diesel fuel and lead-free petrol.

The majority of Policies are implemented at Divisional and site levels. Initially the practical procedures were fairly *ad hoc*, particularly as BS7750 was not in place. In particular, the difference initially identified during the policy-development phase between in-house control of design and specification, and external control of design and specification, began to have a major impact. Laing Homes was therefore able to begin a comprehensive Environmental Audit and to identify and implement a number of changes to eliminate adverse environmental impacts. The Company already employed timber-framed, low-energy use, construction methods and was producing a house that is highly energy efficient.

In addition, Laing Homes had in place a policy of using *brown land*, i.e., land that had previously been developed normally for non-housing uses, and over 70% of housing developments were on this type of land. The management then reviewed the whole of their procurement policy to eliminate the use of tropical timbers, CFCs and peat, and expanded their subcontractor control procedures to ensure that environmental requirements were incorporated into subcontracts and their execution.

Furthermore, design and planning criteria were reviewed which led to the introduction of recycling stations on selected sites and a further increase in the insulation standards of all house- and flat-units. Regular six-monthly checks have been instituted to ensure compliance with procedures.

This comprehensive and controlled approach contrasts considerably with the Group's construction activities. The construction businesses rapidly discovered that, although site-control procedures could be, and were, implemented in the areas of dust and noise prevention, of avoidance of water course pollution, etc., in the key areas of design and specification, their role was very different. Frequently irrevocable decisions which might have adverse environmental effects had been taken before they became involved. Thus, for example, they have been faced with having contracted to install boardroom panelling and joinery work from unsustainable forests, halon-based fire fighting systems, landscaping full of peat, and air-conditioning systems full of CFCs.

Responsibility has been placed with the Purchasing Departments to identify these potential environmental hazards with suppliers and subcontractors, and to evaluate and cost suitable, more acceptable, alternatives. On a case-by-case basis, the customer or, more likely, his designer, is then approached and the issues discussed with a view to eliminating environmental impact. There is no contractual basis for these discussions, therefore the Laing team has to decide how far it can push the environmental argument, especially when the alternatives may incur additional costs for the customer. Generally it has been possible to find a solution, although it has to be recognised that environmental Luddites still exist, and that the sensitivities of a designer can be ruffled by the inference of lack of knowledge on the subject.

In these circumstances, when persuasion has not been possible, regrettably it has been necessary to knuckle down and perform the contract obligations, despite the impact that this might have on the environment.

Environmental management systems

Inevitably at the start the practical implementation of the policy was not always easy. During the period 1989 to 1991, the majority of the Group's businesses had registered for and obtained BS5750 (quality management systems)[1] and were increasingly familiar with the concept and implications of formal management systems. As an understanding developed of what was needed in the environmental field, a debate began about what management systems would be required and in particular, where did BS7750[2] (at that time under development) fit? The Group has now accepted the philosphy of Total Quality Management and believes that quality, health, safety, and environmental issues should progressively be integrated into one overall management system.

The debate hinged therefore on what BS7750 might offer that was not already available through BS5750. Subsequently the Group has been represented as an observer on the Construction Sector Code development of BS7750, in order to gain an insight as to how other environmental experts, both inside and outside the construction industry, feel the standard should be implemented.

From this group there appears to be an increasing consensus that the key factor which will provide BS7750 with added value is pan-industry agreement on the environmental impacts caused by construction and agreed solutions to minimise or avoid these impacts. To date there has been no progress in generating this pan-industry agreement, and it is difficult to see how it will satisfactorily be resolved in the future.

Until this agreement is reached, there appears to be no significant benefits in obtaining separate registration and accreditation. During the past nine months, therefore, a small working party comprising representatives of various parts of the Group has been evaluating methods of formalising environmental management systems within our overall management approach. Pilot work began on the Second Severn Crossing project, a project to develop, design, construct and operate a further bridge over the River Severn between England and Wales.

A full environmental impact assessment was required in order to obtain legislative and planning approval. An early decision was taken to involve interested local environmental and community groups proactively, and to ensure that the Group could respond to the concerns of these parties. Landscaping, noise, air quality, protection of ecology (particularly on two Sites of Special Scientific Interest), archaeology and agricultural procedures, were identified as the key long-term environmental impacts. The protection of the Sites of Special Scientific Interest, impact on local communities, loss of commercial fisheries, and permanent entrapment of archaeological finds, were identified as the key impacts during construction.

A considerable amount of investigation, modelling and technological evaluation was undertaken providing a range of solutions to both the long-term and construction issues. Environmental management procedures were then developed in order to ensure that environment matters were automatically integrated into general site practices, and that solutions were implemented with their success or failure clearly measured and evaluated. This information has formed the basis for communicating, discussing where appropriate, and agreeing, future actions with local interest groups.

The procedures developed on this project have been successful in controlling and managing the impact on the environment of construction processes, and in providing good information for internal and external debate. They are being used to form a sound basis for development and use elsewhere in the Group.

Legislation has begun to have an impact on the construction industry. The first piece of purely environmental legislation that has required a strong procedural approach has been the Controlled Waste (Registration of Carriers and Seisure of Vehicles) Regulations 1991. This legislation controls the removal and transport of all waste material (including earth and soil excavation) from site to approved tips. Most importantly for Laing, having established licencing and control procedures, it imposes a duty of care upon the producer of waste, i.e., Laing, to ensure the regulations are obeyed and enforced. This duty of care is backed up by the threat of considerable fines and even imprisonment for breach of the regulations. Fortunately it is relatively straightforward to incorporate this type of obligation into existing management systems, and satisfactory procedures were already in place before registration was required on 1 April 1992. More difficult to implement has been the progressive reduction and banning of ozone-depleting substances under the Montreal Protocol. The law, such as it is, is aimed at controlling production, and the Laing Group took a decision in February 1993 to avoid the use of all these substances with immediate effect, wherever it can be established that they are being used. This will be achieved through our design briefing, and procurement procedures.

Nevertheless, despite the initiatives outlined above, we anticipate that it will be at least five years before environmental management systems are implemented on a Group-wide basis, unless there is a significant shift in customers' demand for such procedures.

Measurement and performance indicators

The development of practical management systems began to raise the need for measurement and performance indicators to allow the Group to determine what progress was being made. Over the past two years considerable work has been undertaken on quality performance indicators, measurement and improvement procedures. This work was expanded in 1992 to start to review environmental performance indicators. In this, the practical problems mentioned earlier between internally and externally controlled design arose again. Thus Laing Homes was able to demonstrate the elimination of tropical hardwoods and peat, the minimal use of CFCs and, by using the National Homes Energy Rating system, they were able to measure the progressive improvement in the energy efficiency of the houses they sold.

In construction, because of the unique nature of each project, measurement generally is project-oriented. Thus the Group decided to work with BiE (Business in the Environment) which had established a small task force to look at environmental performance indicators. In appraising the areas where the Group might make a contribution by developing a case study, it became clear that historically there had been little or no value in cost comparisons between projects because each project is different.

We chose the topic 'energy efficiency in the use of temporary site accommodation' because approximately half the staff and all the operatives are based on site, and energy use is a significant and controllable expense.

The work proved invaluable in developing procedural performance indicators based on ensuring good practice, such as air-testing hutting each time it was moved, making use of thermostats and good office management controls, as opposed to measurements such as costs per unit of area or costs per employee. The exercise emphasised the benefits of performance indicators even when there is no mechanism for, or value in, consolidating information across differing sites or businesses. This will enable us to demonstrate the progress the Group is achieving in minimising the impact its work has on the environment, even though we do not expect to have available the high level of detail achieved by other companies.

Education and training

The thrust of most of the work described thus far can be defined as *top down*. During the policy-development phase, it was clear that out in the field the problem was not going to be one of winning the hearts and minds of staff, operatives and subcontractors, as there was generally as much, if not more, commitment to environmental ideals at the workface than there was among senior and middle management. The important issues were ones of direction and consistency. There was a need to provide a simple guide to the environmental issues most important to the construction and housebuilding industries: what the Laing Group view on these issues was, what the Group was striving to achieve over the coming period and, as far as they were definable, what the goals were.

The macro issues were dealt with by way of an induction video, which was produced during late 1990, shown to all staff during the Team Briefing sessions in early 1991 and now forms part of the formal induction of new starters. About 12 minutes long, it is in two parts, initially covering the impact of the construction industry and buildings on the environment and, secondly, highlighting the key points from the Environmental Policy Guidelines. This induction video is intended to give no more than a general pointer to staff that the environment is *important*.

Considerable thought was given to the need for general environmental education and how to train those whose work was likely to have the greatest impact on the environment, particularly site staff and operatives. In this there are unique aspects to the construction industry which separate it from the factory or the office workplace. Each construction project is different and tailor-made. The location is variable with an average duration of 12 months. The construction team is assembled project by project. Stability, consistency and repetition are unusual in the construction process.

The industry has responded to this challenge by developing sophisticated project-management systems, and the key to solving the environmental training problem lay in the Project Plan. At the commencement of every project a Project Plan is prepared which identifies all the site activities, programme, logistics and resourcing. This plan clarifies the way in which all site operations will be executed. It therefore allows the consideration of environmental impacts to be incorporated into the procedures in the same way as health, safety and productivity measures are included.

As environmental demands develop, therefore, the solutions are incorporated into operating procedures and the management systems, and automatically become the way the job is done. Thus environmental training occurs through the structuring of how a task is performed rather than the specific training of an individual in environmental matters.

Although this implies a continuance of the *top down* approach, in reality site staff and operatives are closely involved in the development of site planning procedures, particularly through quality improvement teams. Networking of good solutions is an automatic process within the Divisions and in this way good environmental practice is becoming part of the overall drive for Total Quality Management. As the environmental management systems become more comprehensive and sophisticated, it is anticipated that the development of environmental training will be achieved through the systemising of the job, not through the separation of environment as a discrete and unique issue.

Conclusions

As with many other issues in the service sector, environmental affairs should be driven by customer demand. In this sense, perhaps the construction and housing industries are different from other service sectors because customer demand for environmental excellence is still muted, and certainly is not yet forcing constructors and housebuilders to take action. This is the reason for inertia that exists across the industry not only in the UK, but also in many other developed and developing countries. Nevertheless the Laing experience shows there are many actions that a service company can take to improve its environmental performance, provided management recognises the need to be flexible and responsive in the way they tackle the issues.

Any manager reading this chapter will have seen that the demands upon his management skills are no different in the environment area than in the other areas of service provision. Although some of the issues are more abstruse and complex to understand, integration of environmental issues into other already established management systems will form a solid foundation that will allow the service company to respond effectively to reduce the impact that it makes upon the environment.

References

1 BS5750 is the British Standard for Quality Systems and was first published in 1979 and revised in 1987.
2 BS7750 is the British Standard for Environmental Management Systems and although a separate standard to BS5750 when it was published in 1992, it makes it clear that it is modelled on the procedures outlined in BS5750.

Recommended reading

A Measure of Commitment, guidelines for measuring environmental performance, produced by Business and the Environment, London, September 1992.

Construction and the Environment, report of the Environment Task Force of the

Construction Industry Employers Council, published by the Building Employers Confederation, London, May 1992.

Schmidheiny, Stephan, with the Business Council for Sustainable Development. *Changing Course: A Global Perspective on Development and the Environment*, Massachusetts Institute of Technology Press (1992).

Your Business and the Environment: A D-I-Y Review for Companies, produced by Business in the Environment, London, June 1991.

Your Business and the Environment: An Executive Guide, produced by Business in the Environment, London.

14

Developing and implementing environmental policies and programs in an international corporation

NICHOLAS L. REDING

Vice Chairman,

Monsanto Company, St Louis, Missouri, USA

■ Monsanto's response to environmental concerns has evolved through four different phases: compliance, CEO activism, institutional activism and added value.

■ The critical factors in this development have been: a strong concern for employees to do the right thing; a CEO fully committed to environment protection; a determination to break the 'gridlock' which sometimes characterises public debates on environmental policy; a sense of responsibility for disclosing information about Monsanto's operations to those potentially affected; and a growing awareness that *environmental issues must add value in ways that customers understand.*

■ The environment is becoming more than a cost of doing business; it is becoming *part* of Monsanto's business.

Introduction

At a meeting of the National Wildlife Federation in Washington, DC, in January 1990, Richard J. Mahoney, Monsanto Company's chairman and chief executive officer, announced the Monsanto Pledge.[1] This seven-point statement of commitment to environmental protection and sustainable development essentially codified Monsanto's definition of what was coming to be known as 'corporate environmentalism', and it did so in a very public way.

Publicizing the commitment served two purposes: it gained a favourable response from environmentalists, employees and other important constituencies; it also made the company publicly accountable for environmental protection and progress. The Monsanto Pledge demonstrates two approaches which had come to characterize the corporation's environmental programs: (1) voluntary initiatives to reduce the pollution created by the company's operations, and (2) voluntary disclosure of data about the company's operations to the public.

The Monsanto Pledge is not only a commitment in itself; it also stands as a reference point for understanding how Monsanto has evolved its environmental policy, and the direction in which that policy is heading.

Environmental policy at Monsanto has moved through four major phases: (1) compliance, (2) CEO activism, (3) activism institutionalized, and (4) the environment as business. For ease of explanation, each of the four will be discussed as separate entities.

However, to gain the best understanding, each should be seen as overlaying another. This means that, while the four are distinct in defining new policy directions, they are not distinct in day-to-day operational practice, and indeed, all four continue as significant influences on policy development and implementation.

This gives rise to inevitable conflicts and tensions within the corporation – conflicts and tensions which are healthy and vital to both business and environmental goals. The successful resolution of these conflicts will ultimately mean that business and environmental goals are not mutually exclusive but actually one and the same.

The era of compliance

Founded in 1901 in St Louis, Missouri, Monsanto is a worldwide manufacturer of agricultural products like Roundup herbicide; chemical products like nylon fibre and plastics; pharmaceuticals; and food products, including NutraSweet brand sweetener.

Beginning in the 1970s, in response to growing public concern, a virtual tidal wave of regulation began to sweep the United States and other industrialized nations. Major US laws included the creation of the Environmental Protection Agency; the Clean Air Act; the Clean Water Act; the Toxic Substances Control Act; amendments which strengthened the laws governing pesticides; the Resource Conservation and Recovery Act; and Superfund. Similar laws were enacted at the state and local levels.

Business and industry, including Monsanto, responded predictably. Each new proposed law was greeted by trade associations and individual companies predicting the apocalypse – the loss of jobs, the loss of competitive position, and enormous costs to the economy. As each proposal moved toward law, there was acceptance and a focus upon compliance. Economic dislocations did happen – but there was no apocalypse. Business cried 'wolf' so many times that their concerns, even when valid, lacked credibility.

Compliance became the environmental paradigm. And compliance was often a stretch, costing money and resources. And compliance remains a costly stretch – with the annual bill for Monsanto about $250 million annually.

Because compliance was a stretch, it became a hallmark of performance. The objective was to comply with the law at the lowest cost possible. But while business viewed compliance as a hallmark, the public viewed it as a benchmark, the minimum performance expected.

At Monsanto, to cope with the rising tide of environmental regulation, an Environmental Policy Staff, headed by a senior vice president, was created. This staff, in turn, created a series of worldwide operating guidelines, to ensure that all of the company's operations were in compliance with all the laws and regulations. These guidelines were periodically updated, to accommodate new regulations as well as new company policies and programs. However, the guidelines were maintained as internal company documents, and were not disseminated outside Monsanto.

Meeting the goals of the guidelines was the responsibility of Monsanto's operating divisions, but an oversight function was provided by a corporate Environmental, Safety and Health Committee.

Environmental compliance as the operating paradigm contained several ultimately fatal shortcomings.

First, compliance focused on yesterday's public concerns – the enactment of a particular law was the end-point of the policy process. By definition, it does not demand tomorrow's solutions, does not encourage new technology, and does not anticipate new problems. Instead, it focuses on taking care of yesterday's problem today.

Second, compliance worked only to the degree that the public trusted the federal and state regulatory agencies – only to the degree that government standards and regulations were credible. As time passed, the public began to believe that the regulators were not doing their jobs. This perception was underscored in the late 1970s and early 1980s by scandals in the Environmental Protection Agency and at independent testing laboratories, the listing of waste dumps as Superfund sites, the contamination at Love Canal in New York, and dioxin contamination at Times Beach, Missouri.

Third, compliance mandated a heavy-handed, uniform approach, one that was highly prescriptive, what today is called 'command and control'. With both the goal and the steps to achieve the goal mandated by regulation, little room was left for industry to use its creativity and innovation.

Fourth, the focus on compliance, with the political wars fought over each piece of law and regulation, left a legacy of mutual mistrust and suspicion among the parties involved in environmental issues.

By the early 1980s, the paradigm of compliance was unravelling, but it was dealt a death blow in 1984 with the tragedy at Bhopal, India. With more than 3,000 people dead, assurances of compliance rang hollow. To continue to earn the right to operate, companies like Monsanto would have to move beyond compliance, toward something else. That 'something else' was vague and unknown.

The era of CEO activism

While Monsanto had nothing to do with Bhopal, the corporation moved swiftly in response. CEO Mahoney appointed a senior management task force[2] to examine all operations worldwide in an effort to determine whether any Monsanto operation posed a similar risk to any of its communities.

This 'High-Hazard Materials Task Force' generally determined that there was no comparable risk, but it did identify hundreds of ways that potential risks to employees, contract people and local communities could be reduced.[3] This program has now been institutionalized into routine high-hazard audits of all Monsanto manufacturing facilities, and has resulted in changes in storage, distribution, transportation and manufacturing.

When he appointed the task force, Mr Mahoney also changed existing company policy to open Monsanto plants to the public and to make available technical data on materials used and manufactured at company plants. These Material Safety Data Sheets (MSDSs) had long been made available to emergency services: local fire and police departments, regulators and those in the transportation and distribution networks. However, they had not been made available to the public. Monsanto plants placed these MSDSs on file at local libraries. Interestingly, very few requests for the documents have ever been made, but the company was given high marks for taking this step toward a broader concept of the public's 'right to know'.

This response by Monsanto inaugurated the era of CEO activism. CEO activism essentially consisted of two policy aspects. First, *a voluntary initiative* by the company to respond directly to public concerns. In the case of Bhopal, this voluntary initiative was the high-hazard task force program.

The second policy aspect dealt with *disclosure*: disclosing information to the public, information which was previously available only to Monsanto employees or to a relatively small group of technical, regulatory and emergency people. This aspect consisted of both opening the plants to the local community and making MSDSs publicly available.

The US Congress also responded to the tragedy of Bhopal. In 1986, the Congress passed, and President Reagan signed, the Superfund Amendments and Reauthorization Act (SARA). In addition to reauthorizing the federal Superfund program for waste sites, Title III of the law also broadened the definition of the public's right to know.

Two sections of SARA Title III had the effect of fundamentally changing manufacturing. Section 312 required *the annual reporting of inventories of toxic materials* stored at manufacturing and warehouse locations. This data is filed with the local emergency planning committees established by the law and with appropriate state officials.

Section 313 required *the annual reporting of emissions* and releases of toxic chemicals to air, water, land, local sewage treatment plants, and underground injection sites, as well as transfers to others. This information would be reported to the EPA in Washington and to the states. Ultimately, it would be made publicly available, although how that was to happen was unclear at the time the law was passed. The law would go into effect for the 1987 reporting year, which meant that the first data would be reported to the government by 1 July 1988.

A simple requirement to report data – simple in concept but complicated in execution – had a profound effect upon industry, including Monsanto. The reason for this effect was that the data (the numbers to be reported) would be large – millions and tens of millions of pounds. It might all be legal and under permit; it might not be causing any health or environmental effects. But the size of the numbers would be staggering.

Monsanto grappled with two main issues of SARA Title III: how to communicate the information to all important audiences, and what to do about the information itself. The law didn't require anything except the reporting of data, but was some response by the company called for?

Through the winter and into the spring of 1988, Monsanto gradually concluded that the company itself would make its data public. An extensive program utilizing the principles and practices of risk communication was adopted.

In early June 1988, the data on emissions and releases which was to be filed with the EPA and the communications program surrounding that data, was presented to Monsanto's CEO. Mr Mahoney looked at the data, said it was unacceptable to him and that it would surely be unacceptable to the public; he also said that the company would focus on those emissions of greatest potential exposure to the public – emissions to the air – and reduce them by 90 per cent by the end of 1992, and then work toward the ultimate goal of zero. And this would be done despite the fact that all of the company's studies conclusively demonstrated that there were no risks to human health or the environment from these air emissions. The announcement said that comparable programs would be developed for Monsanto operations outside the United States, which were not governed by SARA Title III.

The era of CEO activism was reaching its zenith.

When Monsanto announced its 90 per cent program on 30 June 1988,[4] at a press briefing in St Louis, the voluntary initiative far overshadowed the release of all of the Title III data. Yet this public disclosure of the data – a year in advance of when the EPA finally made the data available – was a significant step in Monsanto's environmental policies. Releasing the data made the company publicly accountable, and provided an extraordinarily easy way for the public to track the company's progress in reducing waste from operations.

Again, this particular version of Monsanto's 'CEO activism' included both voluntary initiative and public disclosure components.

Shortly after the announcement on the 90 per cent program for air, Mr Mahoney asked Monsanto's operating divisions and subsidiaries to develop individual plans for waste reduction across all environmental media. Each unit developed a specialized, and broadly different, plan. Monsanto's chemical group, for example, developed a tiered list of priority wastes, the first tier of which would be reduced by 70 per cent. These programs were developed by early 1989, but did not have the consistency of the corporate program for air emissions.

In the public policy arena, Monsanto's 90 per cent program gained broad publicity. Numerous other companies adopted similar programs. Environmental organizations, ranging from small, local groups in Chicago to the National Wildlife Federation,[5] began to call upon business and industry to 'take the Monsanto Pledge' and similarly pledge to reduce air emissions.

Mr Mahoney was invited to speak at the annual meeting of the National Wildlife Federation's Corporate Conservation Council in January 1990, in Washington, DC.[6] He used this speech to broaden the company's own definition of the Monsanto Pledge, and in so doing redefined the development of environmental policy at Monsanto.

In this speech, Mr Mahoney explained the Monsanto Pledge, which reads as follows:

- It is our pledge to reduce all toxic and hazardous releases and emissions, working toward an ultimate goal of zero effect. It may take time, but we will not be satisfied with anything else.

- It is our pledge to ensure that no Monsanto operation poses any undue risk to our employees and our communities.
- It is our pledge to work to achieve sustainable agriculture with the lowest inputs feasible, through new technology and new practices.
- It is our pledge to ensure the safety of groundwater. If our products are found to pose a problem, we will solve it by whatever means necessary. We will make our technical resources available to farmers who may have concerns about groundwater contamination, even if none of our products is involved.
- It is our pledge to keep our plants open to our communities, bringing the community into plant operations. Our employees and our communities will be kept fully informed of any significant hazard. If we can't eliminate the hazard, we'll work to eliminate the source of it.
- It is our pledge to do our part to halt deforestation and declining biodiversity. All corporate real estate, including plant sites, will be managed with the benefit of nature as a serious operating factor.
- It is our pledge to search worldwide for technology that will reduce and eliminate waste from our operations, with the top priority being not making it in the first place.
- Our commitment is to achieve sustainable development for those aspects of the environment where we have an impact. Our commitment is to achieve sustainable development for the good of all people in both developed and less-developed nations.

Monsanto employees were immediately enthusiastic: the CEO received more mail from employees on the Pledge than on any other topic. But the Pledge raised questions and created considerable tensions within the corporation. How would it affect Monsanto's Worldwide Guidelines? Would new guidelines have to be written? What did terms like 'zero effect', 'sustainable agriculture', and 'sustainable development' mean? How did it fit with the Chemical Manufacturers Association's Responsible Care program? Did each statement of the Pledge imply a new voluntary initiative?

Activism institutionalized

With the questions surrounding the Monsanto Pledge, with a CEO who kept pressing the company for more and greater environmental progress, with ambitious yet widely different operating unit programs for waste reduction, with the issue of cost and impact on businesses and products, it was becoming increasingly clear that Monsanto needed to institutionalize its version of environmental activism.

The decision was made to replace the Environmental, Safety and Health Committee with a corporate *Environmental Policy Committee* (EPC).[7] Instead of committee members drawn exclusively from the environmental and manufacturing functions, members of the EPC would broadly represent corporate Monsanto and its operating divisions, including business managers. For example, the current membership of Monsanto's EPC includes five representatives from environmental functions, five from public affairs, four business vice presidents, two from manufacturing and four from various corporate staff functions like research and law.

Since its creation in 1990, the EPC has been focused on a number of initiatives. For example, a team led by the vice president of Monsanto's Fibers Division researched and designed a program to drastically reduce and ultimately eliminate most discharges of wastes to underground injection wells. This program affects only three manufacturing facilities, but this waste disposal method accounts for some 70 per cent of all wastes discharged by the company in the United States and 60 per cent on a worldwide basis.

Also at the direction of the EPC, the individual waste reduction programs adopted by the operating units were consolidated into one overall program to reduce releases and transfers of toxic chemicals to all environmental media, including air, water, land and underground injection wells by 70 per cent by the end of 1995, compared to the base year of 1990.

Further, a waste elimination/pollution prevention technology program has been initiated. A series of meetings have been held for scientists from across the corporation, including corporate research and the operating units, to examine specific waste problems and recommend possible solutions. This has already resulted in the identification of promising technologies to reduce waste. Additionally, a small technical team has been examining ideas for remediation of waste sites.

The EPC, which meets quarterly, establishes policy direction. On a day-to-day operational basis, the operating companies assisted by the Environment, Safety and Health (ESH) organization, a corporate staff function, are responsible for policy implementation and compliance with all laws and regulations.

In this era of institutional activism, communications with important constituency audiences has assumed more formal characteristics. Beginning in 1991, and repeated in 1992, Monsanto published an *Environmental Annual Review* to report on progress being made under the Monsanto Pledge.[8] It is modelled after a business annual report, with a chairman's letter, narrative section, and extensive data section. The review now serves the purpose of publicly disclosing the data reported to the EPA under SARA Title III, along with other waste generation and reduction statistics.

The public response to the review has been extremely positive. The chairman's letter has been given high marks for candour, for it not only describes successes but also discusses shortcomings. Environmentalists have particularly noted the data section, praising the company for disclosure – while also suggesting the inclusion of additional data in future reports. *The Economist* magazine cited Monsanto's review as setting the pace for industry,[9] and in fact several companies have published similar reports.

As part of this 'institutionalized activism,' the corporate ESH staff has been focused in two areas. The first, and largest, area is *compliance* – to ensure that the corporation complies with all laws, regulations and internal guidelines worldwide, and that company policies and programs are aligned with the Monsanto Pledge. This is the traditional role of the ESH function.

The second, and newer, area of focus is *future policy options and directions*, and it is clear that this area is already assuming all the characteristics of the next era of environmental policy at Monsanto: the environment as our business, or adding value to the corporation's businesses through environmental protection.

The era of added value

Monsanto's program to reduce air emissions by 90 per cent by the end of 1992 had a number of positive effects, both in the public policy arena as well as inside the corporation. But it also cost money – approximately $115 million, with a virtually non-existent return on investment. Shareowners ask questions about investments with little or no return, and rightfully so.

Competitive concerns are raised as well, because those companies which focus strictly on compliance and forego voluntary initiatives can find themselves with a competitive advantage over those trying to do the right thing, at least in the short term. In a highly competitive marketplace, corporate environmentalism can impose a severe penalty.

Corporate environmentalism, if it is to endure, must be sustainable in the business sense. In other words, it must add value to the corporation's businesses.[10]

For a company like Monsanto, added value means examining the full gamut of our operations, and making decisions and creating programs which protect the environment and do so economically, by improving a cost position, or increasing operating efficiencies, or making a difference for customers. And this process includes supplier audits, finding a workable framework for examining the life cycles of individual products, and determining one's own responsibility for the actions of one's customers.

Monsanto's focus on future policy options and directions fully embraces the concept of adding value. Focus groups have been held with customers to understand their environmental problems, and what value Monsanto's programs provide to them. Programs are also ongoing in competitive benchmarking – comparing the environmental programs of competitors to determine whether or not Monsanto's programs provide a competitive advantage, or could provide one.

And because Monsanto employees play such a critical role in the success or failure of ambitious environmental initiatives, ways to motivate our people are being found and implemented.

For example, in 1992, Monsanto announced the first winners of the *Monsanto Pledge Awards*. Employee teams from all parts of the corporation, and from all world areas, submitted projects in four categories: pledge performance, innovation, marketplace, and community service. The winners designate an environmental group or organization of their choice to receive $25,000. In the two years that the Pledge Awards have existed, some 300 project submissions have competed.

The awards program is one way we are building environmentalism into the corporate culture, by empowering employees and making environmentalism 'pay its way.' There are others, including rotating business managers into environmental jobs as part of their career development and making managers accountable for environmental improvement.

In this era of added value, government has an extremely important role to play as well.[11] While government has always had the ability to carry a big stick, particularly to ensure compliance, government has not been as much involved in offering the carrot. Given compliance, what can government do to encourage companies to go beyond compliance and achieve truly superior environmental performance? Are there positive incentives which the government can create to encourage manufacturers to clean up faster – beyond what the law requires?

In the United States, ideas along these lines were considered by the EPA under Administrator William Reilly, and hopefully will be studied seriously by the EPA under Administrator Carol Browner. Possibilities might include moving a product registration application by a 'volunteer' company to the head of the approval line, a tax credit, green labels, reduced permitting requirements or a patent extension. The US Clean Air Act Amendments of 1990 included a provision for exempting companies from certain costs if they achieved a 90 per cent reduction in emissions by a certain date.

The goal of positive incentives is clear: find ways to encourage companies to go beyond the letter of the law.

Another role for government to play in this era of added value concerns priorities. Consider the US 'Superfund Program'.

Designed to remediate old waste sites, Superfund has become everyone's favorite nightmare. It has no clear, agreed-upon purpose. Since the enactment of the law in 1980, more than $14 billion has been spent on, or committed to, the program. One study, by the University of Tennessee,[12] suggests that current policy will require between $300 billion and $750 billion to be spent. If the sites identified by individual states, and sites for which the US Departments of Energy and Defense are responsible, are added, the total cost soars toward $1.7 trillion. With incineration the best available technology, we could be spending $1.7 trillion to burn dirt! And that's just for addressing site remediations. Also with Superfund liability provisions, there are potentially huge legal and natural resource costs.

But the government could do much to reduce the cost and speed the programs:

- For high-priority sites, move quickly to protect health and environment. If technology to do so is available, use it. If technology is not available, use interim methods to contain and control risks. And invest money in the development of cost-effective technology to achieve permanent protection of health and the environment.

- For those sites which do not pose a threat to health and the environment, action can be deferred until appropriate technology is developed, recognizing that such sites can be re-examined and rated based upon new information.

It is critical that liability standards be modified – to hold parties responsible for their fair share of the cleanup costs and not burden them with costs of the entire site. This may be one of the most important steps, because it would foster a sense of voluntarism among all parties working together, instead of all parties trying to saddle each other with as much of the costs as possible.

Offering positive incentives and setting priorities in programs like Superfund are two ways government can make a tremendous contribution to this era of added value.

Summary and conclusion

Monsanto's response to environmental concerns, and how it has developed and implemented policy, has evolved through four distinct phases: compliance, CEO activism, institutionalized activism, and added value.

This evolution through the four phases has not been marked by dramatic breaks with previous approaches to environmental concerns and issues. Instead, it has much more closely resembled a process of one phase overlaying another, for, in a very literal sense, all four continue as major influences upon the corporation's environmental policy and how it is determined and implemented.

The critical factors have been: a strong sense among employees for doing the right thing; a CEO fully committed to environmental protection; a determination to break the 'environmental gridlock' which so often characterizes public policy debate; a sense of responsibility for disclosing information and data about Monsanto's operations to those potentially affected by those operations; and a growing awareness that environmentalism must add value in ways perceived and understood by customers.

The environment is becoming far more than a cost of doing business. It is indeed becoming Monsanto's business.

References

1 'A Brief Candle or a Splendid Torch: A Corporation's Commitment to the Environment', remarks by Richard J. Mahoney, Monsanto Co., at the annual meeting of the Corporate Conservation Council of the National Wildlife Federation, Washington, DC, 30 January 1990.
2 'A Special Message to Employees,' *Monsanto Company General Bulletin*, 12 December 1984.
3 *Monsanto Backgrounder*, report on the High Hazards Materials Task Force, October, 1987.
4 'Monsanto Announces Program to Reduce Air Emissions by 90 per cent', *Monsanto Company General Bulletin*, 30 June 1988.
5 National Wildlife Federation, '*Toxic 500 Report*,' Washington, DC, August 1989.
6 'A Brief Candle or a Splendid Torch,' ibid.
7 'Corporate Environmental Policy Committee Formed'. *Monsanto Company General Bulletin*, 16 November 1990.
8 *Monsanto Company Environmental Annual Review*, 1991 and 1992, St Louis, Missouri.
9 'Environmental Protection: Goody Two-Shoes'. *The Economist*, 2 November 1991.
10 'The Environment is Our Business', Remarks by Richard J. Mahoney, at the annual meeting of the International Institute of Synthetic Rubber Producers, Washington, DC, 16 May 1991.
11 'From Compliance to Added Value: The Environment and Competitive Advantage', remarks by Nicholas L. Reding, Monsanto Company, at St Louis University Conference on Business Ethics, St Louis, Missouri, 10 April 1992.
12 *Hazardous Waste Remediation: The Task at Hand*, study by the University of Tennessee Waste Management Research and Education Institute, Knoxville, Tennessee, December 1991.

Recommended reading

Cairncross, Frances, *Costing the Earth*, The Economist Books and Harvard Business School Press, Boston, Massachusetts (1992).

'Focus Issue on Corporate Environmentalism', *Columbia Journal of World Business*, vol. XXVII, nos. III and IV, Columbia Business School, New York City, New York.

Deloitte Touche Tohmatsu International, *Coming Clean: Corporate Environmental Reporting*, DTTI, London (1993).

Easterbrook, Gregg, 'Cleaning Up Our Mess', *Newsweek*, July 24 1989, pp. 26–42.

Easterbrook, Gregg, 'Everything You Know About the Environment is Wrong', *The New Republic*, April 30 1990, pp. 14–27.

Matthews (ed), Jessica Tuchman, *Preserving the Global Environment: The Challenge of Shared Leadership*, W. W. Norton & Co, New York City, New York (1991).

Smart, Bruce, *Beyond Compliance: A New Industry View of the Environment*, World Resources Institute, Washington, DC (1992).

Smith, Emily T., 'Growth *vs* Environment,' *Business Week*, May 11 1992, pp. 66–75.

15

The Rank Xerox approach to achieving environmental leadership

KARL KUMMER

Environment Director, Europe,
Rank Xerox International Headquarters, London

- Rank Xerox is one of the top 150 companies in Europe, employing 28,000 people, and earning revenues of £2.8 billion per year through the sale of photocopiers, workstations, laser printers, and electronic printing systems.

- In 1990 Rank Xerox Management designed the Environmental Leadership Programme with the aim of not simply complying with environmental legislation, but becoming a leading company in the field of environmental management.

- The Programme has involved the following developments:
  (a) an Environmental Health and Safety Policy;
  (b) a list of Environmental Principles;
  (c) a survey of attitudes to environmental issues;
  (d) the establishment of 'benchmarks' for company performance; and
  (e) the implementation of a wide range of programmes for change.

- Rank Xerox has had the help of its parent company Xerox Corporation which, in 1993, received the World Environmental Centre Gold Medal for Environmental Achievement.

- The company has learned that from research to the customer office, from the legal to the communications department, every function is affected to some degree. The environment is everybody's concern.

Overview

The document company Rank Xerox is one of Europe's leading high technology companies, providing a wide range of products, systems and services to handle office documents. Starting with the first office photocopier in 1956, the Xerox Corporation pioneered reprographics, workstations, laser printers, electronic printing systems and colour copiers – in short the systems needed to capture, create, store and distribute information more effectively. Today Rank Xerox is one of the top 150 companies in Europe, serving 500,000 customers, employing 28,000 people, working with 400 component suppliers, and earning revenues of £2.8 billion per year, 70% of which is in the European Community (1992 figures).

We employ 4,000 people in our manufacturing operations in France, the Netherlands, Spain and the UK, and a further 800 people work in product development at our site in Welwyn Garden City, UK. There are also research facilities at Welwyn, specialising in paper and printing technology, and at Cambridge where our EuroPARC laboratory investigates the uses of interactive media technology. In 1993 Rank Xerox opened a new laboratory in Grenoble, France, with the aim of researching processes for effective multilingual document handling.

Rank Xerox is organised into seven business entities, supported by an international headquarters based in Marlow, UK.

Rank Xerox was formed in 1956 as a joint venture between the Rank Organisation in the UK and Xerox Corporation in the US.

The environmental leadership strategy

In 1990 Rank Xerox management designed the environmental leadership programme. News of catastrophic events was alarming the world:

- the destruction of forests by acid rain;
- irresponsible reduction of the rainforests;
- the destruction of the ozone layer which protects the earth against dangerous UV radiation;
- the 'greenhouse effect', was producing global warming, with unpredictable consequences.

Major accidents involving oil tankers and the disaster in the chemical factory in Bhopal, India, created increasing awareness among people in all countries of the world. The Club of Rome, founded in 1968, had already delivered stern warnings about environmental developments in the coming years. It had now become clear that global resources had to be protected.

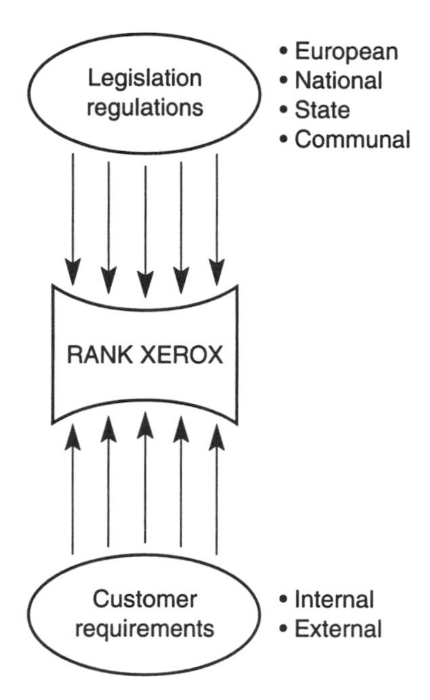

Figure 15.1 Basic reasons for Rank Xerox environmental leadership strategy

The Rank Xerox Board of Directors decided, therefore, to make a situation analysis and identify the areas where the company could contribute to helping the environment.

Many people, both inside and outside the organisation, could hardly believe that Rank Xerox, in the document management business, had an environmental problem at all. However, the directors and others in management felt that even if the company did not have a big environmental problem, many improvements were necessary and possible. The pressure 'to do something' was coming not only from legislation but, increasingly, from customers and from several other sources (see Figs. 15.1 and 15.2).

Employees want to work for a company which cares for the environment, and customers make environmental demands. As a multinational company Rank Xerox has customers in the UK, in continental Europe and outside Europe. Customers in Germany, for instance, expect that we should collect the packaging material from their premises and that we take back their machine for recycling when it becomes unserviceable, or when the customer decides to replace it.

> 'Governments increasingly insist in their documents of tender that environmental criteria such as the EC Eco logo shall be met; they want machines with a stand-by-mode to save energy; and they require suppliers to take care of waste e.g. packaging material and defective spare parts.'

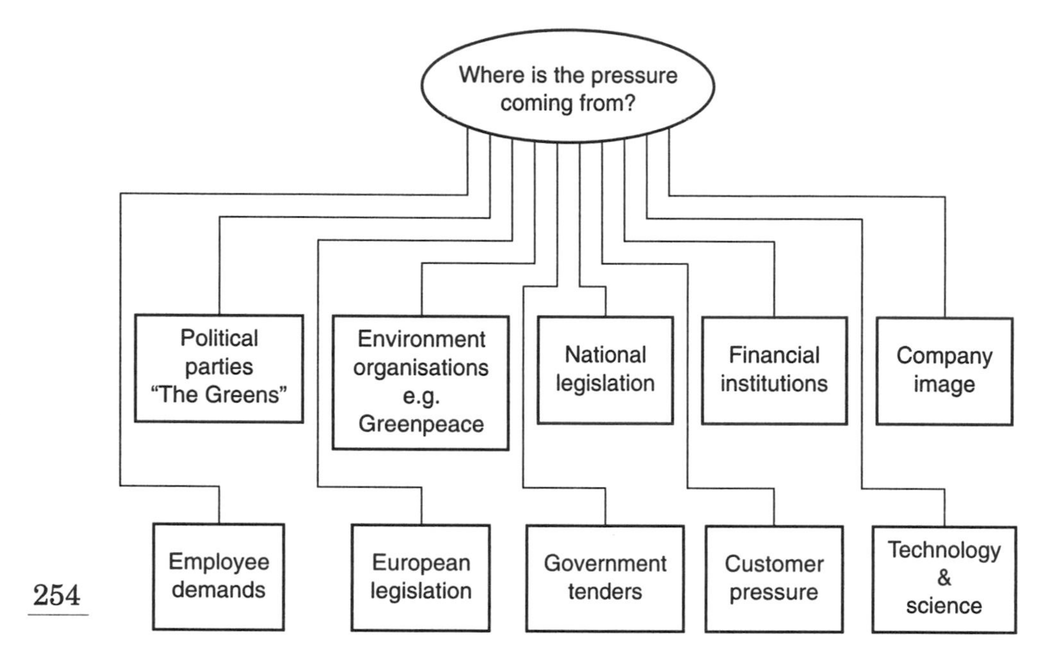

Figure 15.2 The pressures for change

'The polluter pays' principle raises the question of liability for pollution. Financial institutions have to safeguard their investments and they are increasingly interested in the environmental performances of their customers.

We at Rank Xerox realised that we did not have much time. Legislation gives us a reasonable time to react; the competition gives less; and the customer gives no time at all. There was only one conclusion: Rank Xerox had to have an Environmental Leadership Strategy. The Rank Xerox Environmental Health and Safety Policy was published in 1991 (Fig. 15.3) and the Environmental Principles were then communicated to the employees (Fig. 15.4).

The Rank Xerox commitment was given further emphasis when Bernard Fournier, the Managing Director, signed the Charter for Sustainable Development with the International Chamber of Commerce (ICC). For Rank Xerox, the job of implementing the Business Charter and its sixteen principles of environmental management implied that we should continue to follow the corporate policy which we had already been applying on a worldwide basis. Rank Xerox had recognised environmental management as one of the company's highest corporate priorities and as a key factor in sustainable development. A Director for the Environment was appointed in January 1992.

Rank Xerox is committed to the protection of the environment and the health and safety, primarily, of its employees, customers and neighbours. This commitment is applied worldwide in developing new products and processes.

- **Environmental health and safety concerns take priority over economic considerations.**
- **All Rank Xerox operations must be conducted in a manner that safeguards people's health, protects the environment and conserves valuable material and resources.**
- **Rank Xerox is dedicated to the continual improvement of its performance in environmental protection and resource conservation.**
- **Rank Xerox is dedicated to designing products for maximum conservation of resources, and to taking every opportunity to recycle or re-use waste materials generated by its operations.**

Figure 15.3 Rank Xerox environmental health and safety policy

Our Environmental Leadership Strategy has two fundamental elements:

- The move from 'Nobody Cares' to 'Everybody Cares'.
- To achieve a 'leadership' position instead of a 'compliance' status (Fig. 15.5)

In order to achieve an 'Everybody Cares' attitude, a great deal of work is required. We are creating environmental clubs in all major locations, we use our internal communications network to create awareness, and we provide training and education wherever necessary. Design engineers, especially, need to know how to design products for the environment.

We are committed to leadership in environmental protection: a commitment which is applied worldwide in developing new products and processes.

The standards we set ourselves go beyond the controls of even the most stringent EC regulations and our policies have set benchmarks for our industry.

There is fundamental realisation in the company that we can change the way things are made and the way business works, in order to protect our world.

Figure 15.4 Rank Xerox's environmental principles

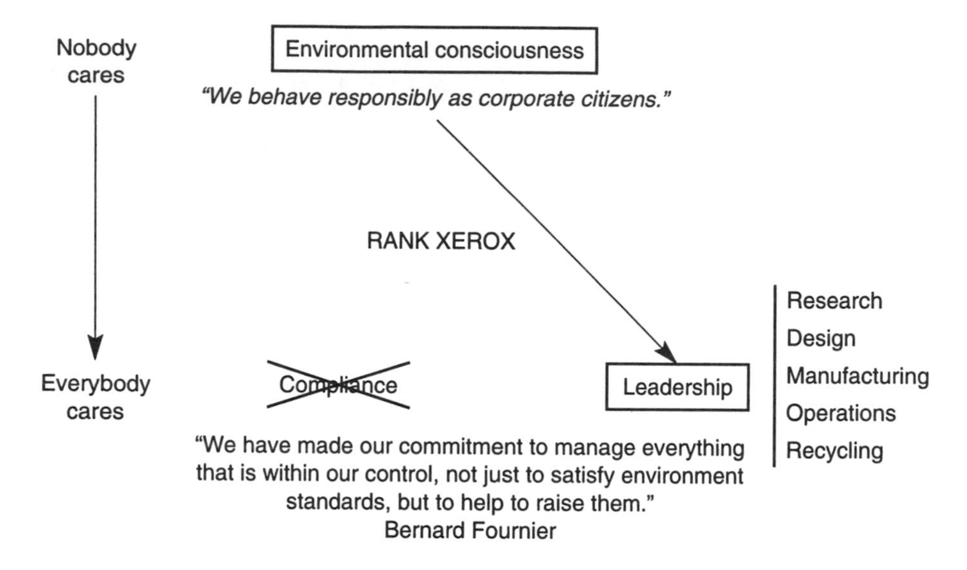

Figure 15.5 The management of change

How can Rank Xerox help the environment

An analysis of the situation showed that Rank Xerox had an outstanding track record but that we should also help to solve global environmental problems. In the past we had focused more – but not exclusively – on the health and safety of our employees and customers. The analysis now showed us that we had to expand our work to much wider environmental issues. The areas of concern are:

- *Global warming:* CO_2 was produced as a result of energy consumption: from the energy usage of our own products, from company cars, material transportation, and the manufacturing processes.
- *Ozone depletion:* Ozone depleting chemicals were used in our production and service processes.
- *Rainforests:* Our impact was restricted to the use of hardwood furniture. Our copier paper is exclusively produced from sustainable non-rainforest resources.
- *Air, water, oil and noise pollution:* Our products create a small amount of noise and dust emissions.
- *Global resources:* Any product which uses materials contributes to the reduction of global resources if it is not re-used or recycled. Rank Xerox products were no different from any others.

- *Waste problem:* Substantial amounts of production and office waste as well as 'end of life' products were being disposed of in landfill sites.

NIne out of ten respondents agree that:

'All organisations should have an environmental policy';

'Quality of life is a key issue of the 1990s'; and

'Environmental action should have a scientific base'.

Over eight out of ten respondents agree that:

'Environmental issues should be seen as an opportunity rather than as a problem'; and

'Organisations should investigate the environmental impact of goods and services they use'.

Figure 15.6 Survey of British Managers

Environmental customer requirements

Environmental awareness and the attitude to 'do something about it' vary from country to country. Even within countries the differences are huge.

Industrial pollution	83%
Energy conservation	74%
'Greenhouse' effect	60%
Ozone depletion	60%
Waste disposal	57%
Cutting out waste	55%
Recycling of materials	39%
State regulation and control	37%
Quality of design	32%
Congestion	32%
Environmentally-friendly purchasing	18%
Improved working environment	17%
Use of recycled paper	15%
Noise reduction	14%
Consumer lobbies	12%
VDU screens	3%

Figure 15.7 Environmental issues in order of importance

SURVEY RESULTS

There are plenty of environmental attitude surveys published in the UK. Dr Colin and Susan Coulson-Thomas produced a survey *Managing the Relationship with the Environment* for Rank Xerox. The respondents were senior managers of UK-based organisations and there was substantial agreement in their opinions about the environment (Fig. 15.6).

We also received clear message concerning the priorities in people's minds, and found that top of the list were: industrial pollution, energy conservation and the greenhouse effect. The overall ranking of environmental issues in order of importance is given in Fig. 15.7.

Most of the environmental priorities are in addition to the traditional Health and Safety requirements which people take for granted. The safety of our employees, customers and neighbours has top priority in the company, and our safety performance record in the production plants has benchmark character (Fig. 15.8). Our products meet, or exceed, European safety standards.

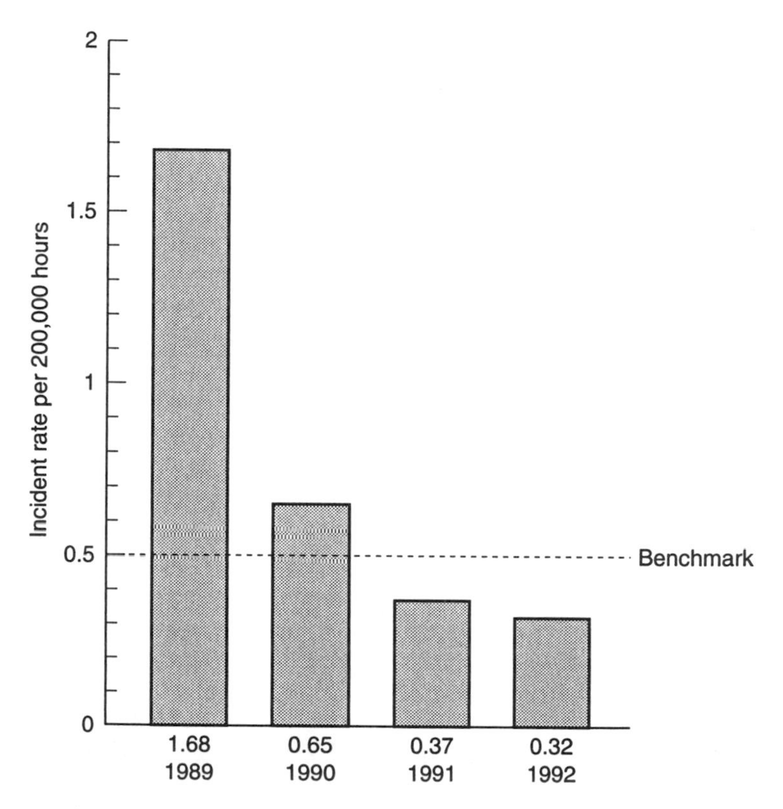

Figure 15.8 Reduction of health and safety incidents: Rank Xerox Ltd Manufacturing

EC ENVIRONMENTAL LEGISLATION

In 1991 Rank Xerox made a study of the impact of EC environmental legislation on Rank Xerox's logistics and marketing operations.

There is a flood of laws coming from the European Commission relating to the environment, however national laws are of prime importance and are often ahead of EC laws. Much of the new legislation is crucial and needs to be understood, but in this chapter it is possible to mention only a small percentage of this legislation. The EC regulations and directives which are especially important to Rank Xerox are:

- the EC Eco-management and audit scheme;
- the proposal for a council directive on packaging and packaging waste;
- the proposal for a directive to improve energy efficiency;
- the various directives concerning waste and the shipment of waste; and
- the regulation on a Community Eco-labelling award scheme.

About 300 environmental directives and regulations have been issued by the EC alone. Many are still in draft form, and often disputed.

The environmental performance record

Rank Xerox began practising 'environmental leadership' before it became fashionable and before good environmental behaviour became a legal necessity. Initially Rank Xerox, as part of the Xerox Corporation, had the benefit of the US example, especially in dealing with hazardous waste, on which EPA rules are strict.

In the mid-1980s Xerox discovered solvents in the groundwater at its Webster, New York, facilities. This led Xerox and Rank Xerox to survey all its facilities worldwide wherever solvents were being used or had been used in the past. Where solvents were found in the soil or the groundwater, a remedial work plan was developed and cleanup activities were begun. This was done in all the countries in which we operate, even in countries that did not have any legal requirements for such a cleanup.

The list of achievements is indeed long. Some of the key facts are listed below:

- In the 1970s Xerox products already had energy-saving modes, as well as automatic two-sided copying which reduces paper usage by 50%.
- As early as the 1970's we eliminated PCB-containing components (i.e., capacitors and transformers) from Xerox products.

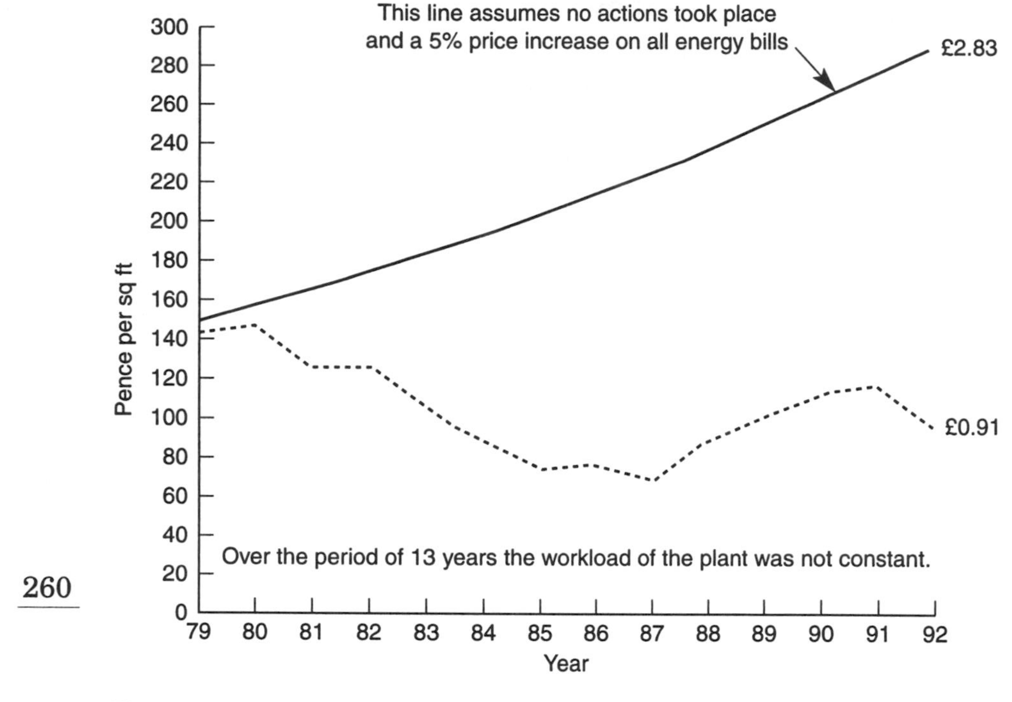

Figure 15.9 Rank Xerox Mitcheldean Plant energy management progress

- In the 1980s we began the introduction of water-based cleaning and water-based painting.
- We have been recovering machine components for re-use and recycling since the 1960s. In 1987 the Asset Recovery Operation was founded in Venray, in the Netherlands. This operation deals exclusively with customer-disposed machines and parts of machines. It is second to none in Europe and in 1992 received the ISO 9000 certification. This is unique for an operation which deals with 'waste' and indicates how seriously Rank Xerox takes the protection of global resources.

In early 1993 Xerox received the Gold Medal for International Corporate Environmental Achievement awarded by the World Environment Centre (WEC) in New York. The WEC Gold Medal is awarded annually to a multinational corporation which has established a worldwide corporate environmental policy, has demonstrated the implementation of the policy and has shown evidence of continued leadership in the environmental area. The award is open to corporations from all over the world.

The Rank Xerox Energy Saving Project is a well-documented quality project. It was not started for environmental reasons but it demonstrates how effectively energy can be managed by giving it enough attention and priority. It is an example of how care for the environment can result in cost avoidance (Fig. 15.9).

Managing the environmental leadership programme

There are different ways of managing a company. It is important that environmental leadership principles and goals are included in the objectives set by the CEO, the board, or whoever has the policy responsibility. *The environment is a vital business need and an opportunity.*

Products, processes and behaviour have to be assessed, environmental issues need to be identified and action plans must be worked out. Rank Xerox works on a global basis and the areas to be addressed are global as well. From research to the customer office, from the legal department to the communications department – every function is affected to some degree (Fig. 15.10). *The environment is everybody's concern.*

Rank Xerox has adopted the Total Quality Management Concept (TQM) and we cannot think about managing the business in a different way. Our processes are well defined and maintained. We won the first European Quality Award, sponsored by the Foundation of Quality Management in 1992.

'This reflects the efforts of 28,000 employees across Europe,' said Bernard Fournier, the Managing Director. 'They have increased customer satisfaction and regained market share by employing quality principles and practices.' To make the environmental leadership process effective it is vital to make the environmental requirements part of the normal work processes. This is necessary in all business stages:

- research, development and design;
- manufacturing;
- logistics;
- maintenance;
- asset recovery (recycling); and
- administration.

Because of the complexity of managing this task effectively, Rank Xerox decided to establish an Environmental Network.

The environmental network

Rank Xerox has four environmental networks with different purposes:

- the Manufacturing Environmental Steering Committee
- the Environmental Champion Team
- the Operating Company Network
- the Safety Network.

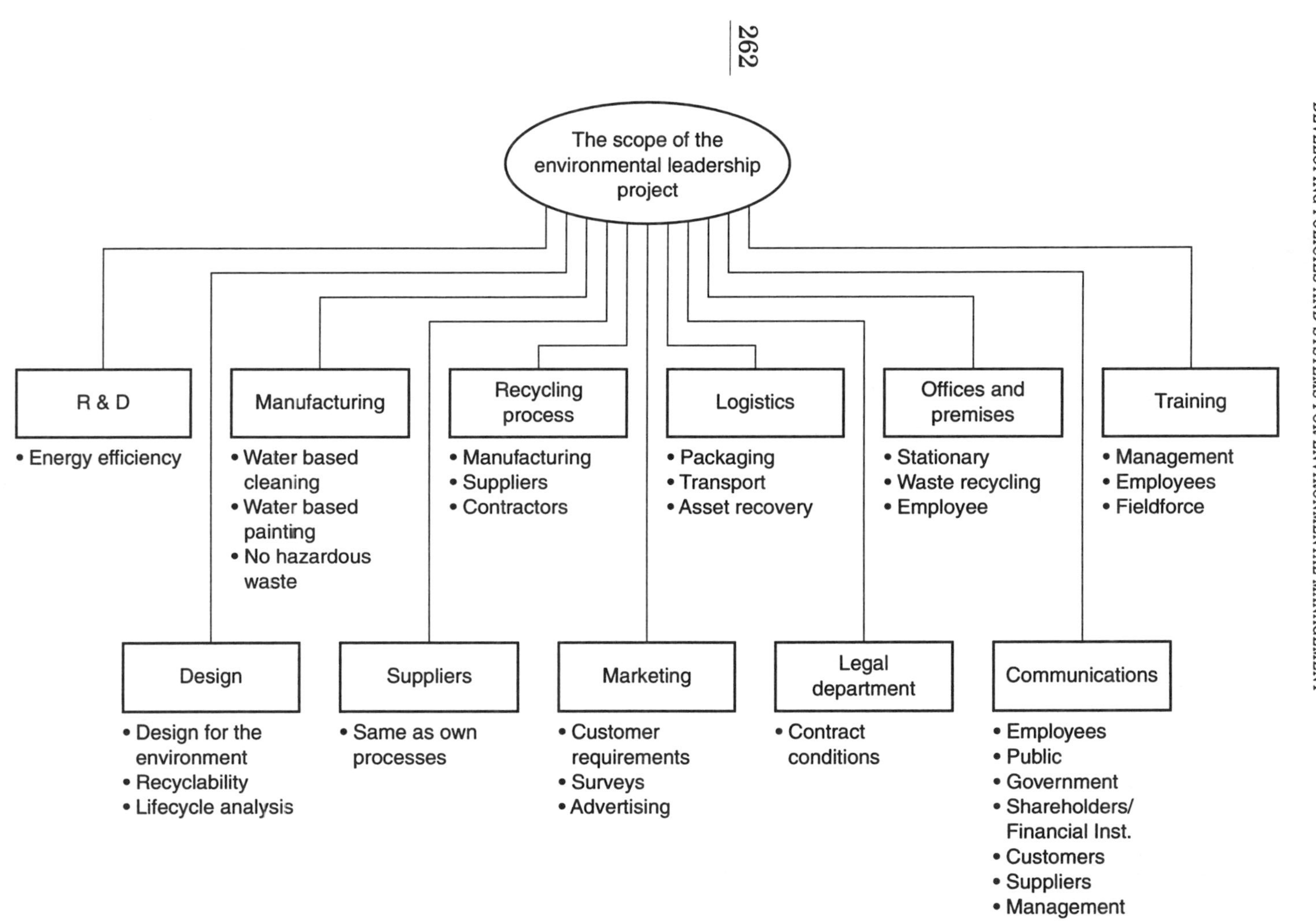

Figure 15.10 The scope of the environmental leadership project

The network organisation is shown in Fig. 15.11. In addition to the networks, which are co-ordinated by the Environment Director, several operations in the UK and in other European countries have formed environmental groups and clubs which provide a forum for employees to participate in the attempt 'to save the earth'. There is a broad willingness to help the environment and not only among the young generation. The clubs are managed by the employees and we are planning to have this forum for employees in all major locations.

THE ENVIRONMENTAL CHAMPION TEAM

The members of the 'champion team' are functional managers who can make decisions on their own. The team has been made responsible for achieving our strategic goals for which we have defined target dates. The goals are set for the following areas:

- reduction of waste;
- energy efficiency;
- switching to recyclable and less polluting alternatives; and
- making our products and process designs environmentally sound.

The elimination of CFCs and other ozone depleting chemicals is one of the champion team's projects and good progress is being made. In June 1993 Rank Xerox eliminated CFCs completely from its manufacturing processes.

Quality techniques and processes are a 'way of life' in Rank Xerox and the Champion Team has adopted a quality approach to its work. For example visitors will find 'fishbone' diagrams in many of the members' offices. A team of skilful people always achieves more than individuals do. (See Fig. 15.12.)

The elimination of CFCs requires substantial investments in cleaning, painting and other areas. However, modern, water-based processes are often investments which bring a fast return and they can reduce costs, especially if the cost for disposing of hazardous waste is taken into account.

The champion team is also working hard on the goal to reduce waste disposal by landfill to nearly zero. We are confident that this goal is achievable over a couple of years. Space for landfill is running out and consumers and industry have to help to reduce waste by different means.

Rank Xerox has adapted the 3R's approach:

REDUCE – RE-USE – RECYCLE

When we changed our packaging from expanded polystyrene (EPS) to paper-based corrupad materials and the exterior packaging to Kraft Brown carton, our packaging engineers were able to reduce the cost by 5%. In 1991 we received a Silver Award from the UK Institute of Packaging.

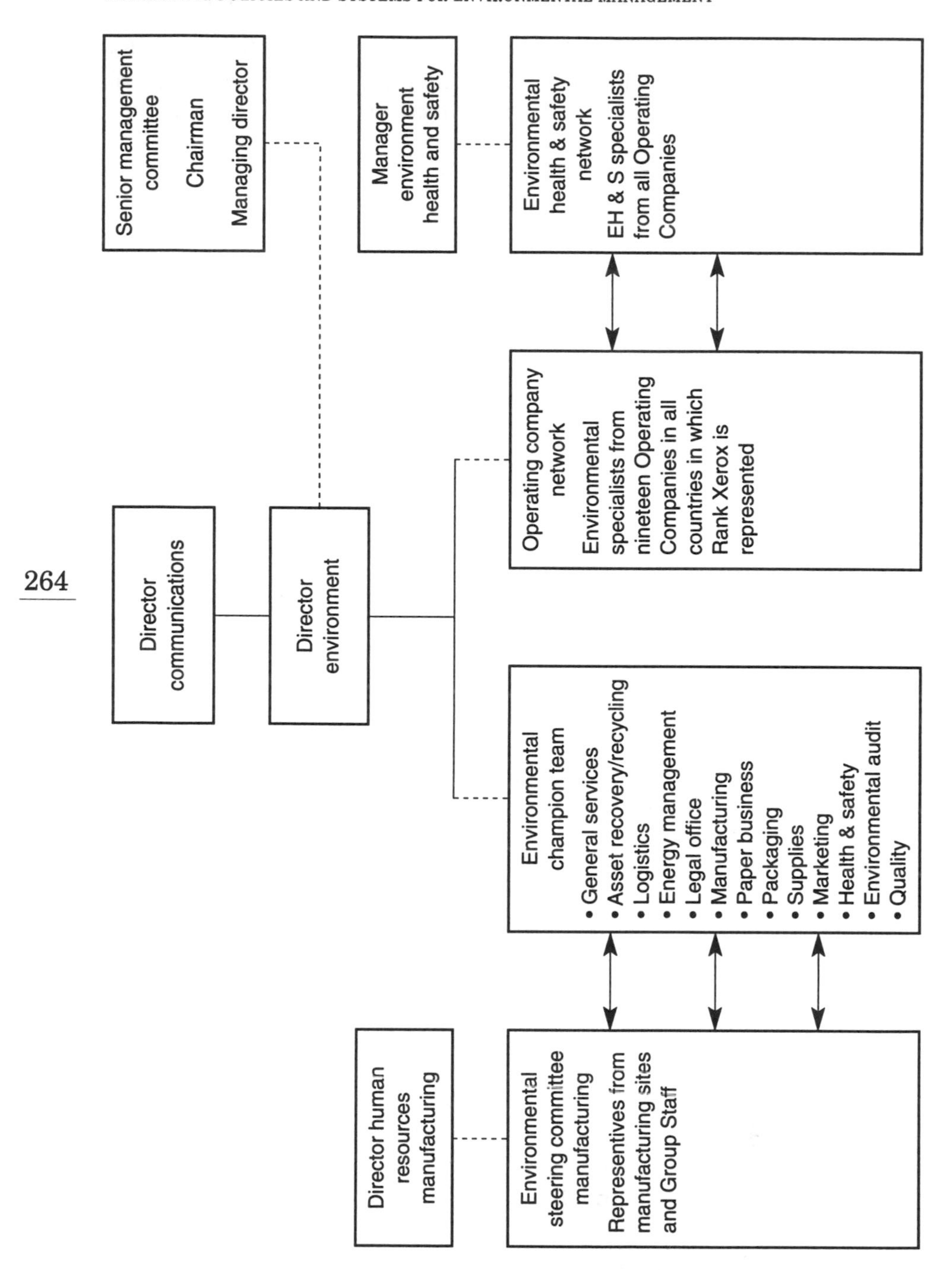

Figure 15.11 Environmental leadership: the functional organisation

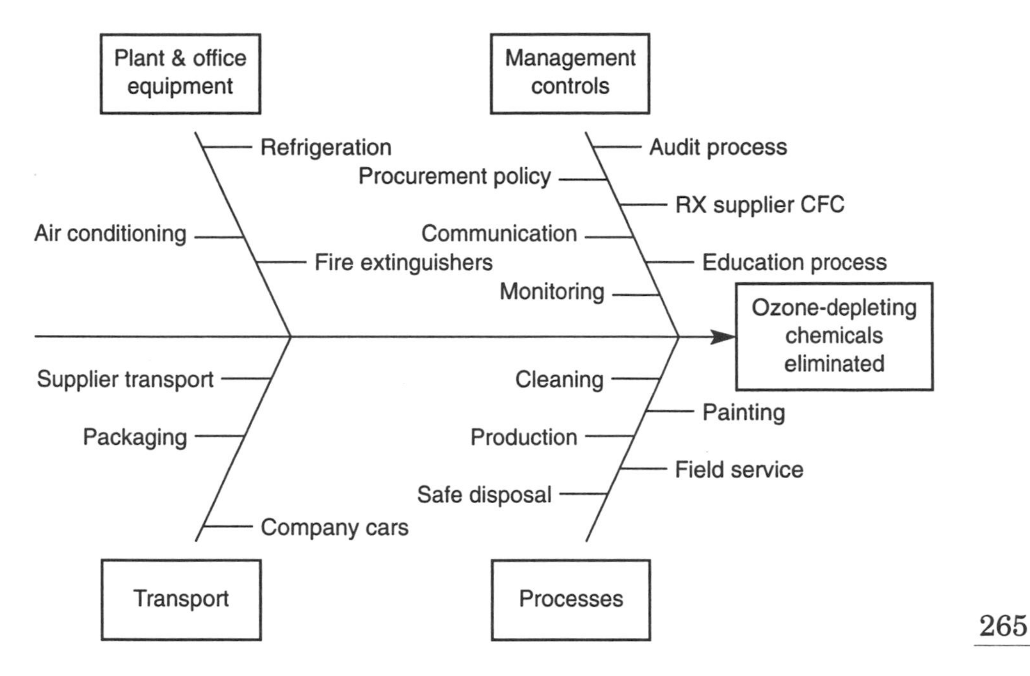

Figure 15.12 A fishbone diagram for eliminating ozone-depleting chemicals

THE ENVIRONMENTAL HEALTH AND SAFETY NETWORK

In a company which works on a multinational basis it is vital to know the national legal requirements concerning environmental health and safety in each country concerned. There are environmental health and safety experts in all our Customer Business Units, from Norway in the north to Italy in the south. Most of the experts have other jobs as well. They have to keep their knowledge up to date by working in national health and safety organisations, by reading literature and attending local training courses. Their role is to advise the local management on questions about employee health and safety and to advise the customers on product safety and the impact of the products on the working environment. Such questions could include items like noise, dust, ozone-depleting emissions, etc. The corporate standards on environmental health and safety are set by the Manager, Environmental Health and Safety, in the Rank Xerox Systems Centre in the UK. He also co-ordinates his network.

THE ENTITY NETWORK

Environmental requirements are moving more and more towards customer requirements which are in their turn often the fulfilment of new national laws. The demands are different in all countries. There are, however, certain requirements which appear to be common:

- The supplier is expected to take back the packaging material, when the machine is installed, for re-use and recycling.
- The same 'take-back' commitment is expected for old machines, defective spare parts, empty toner bottles, copy cartridges, etc.
- An energy-saving mode, a double sided copying capability, and the capability to operate with recycled paper are requirements for copiers.
- Emissions like noise, dust, ozone-depleting and electromagnetic emissions must be reduced far below acceptable health and safety standards.

The team members of the Environmental OPCO Network have their normal jobs in marketing, service, personnel and communications. The team produces valuable material through surveys and customer contacts.

THE ENVIRONMENTAL LEADERSHIP STRATEGY TEAM

The manufacturing and logistics operations have a powerful environmental management organisation. Any environmental policy is in the long term only credible if the products and services are perceived by the customers to be environmentally sound.

The bulk of the environmental changes have to take place in the production area, the distribution system and the design offices. To ensure that the necessary actions are taken we have a steering committee in which the actions are agreed, funding provided and the changes are implemented through the manufacturing network.

In this network our people have to understand what the Product Life Cycle, the Life Cycle Analysis, and the Cradle-to-Grave approach, really mean.

Product recycling and disposal

In the past the product life cycle was finished when the product was sold to the customer. After the product life the customer was normally responsible for the disposal which was usually by landfill.

This 'easy' way is now closed. Upcoming legislation in Germany, and also in the EC, requires that all products have to be taken back for recycling (Fig. 15.13).

The design of environmentally-friendly products, or 'green products', is one of our current challenges. It will take years before international environmental standards and life cycle analysis will be established. In Rank Xerox we are not waiting for this is to happen. We have defined the environmental goals for our future products, also the requirements and checkpoints in the Xerox processes. We call it, in our jargon, the 'Product Delivery Process'.

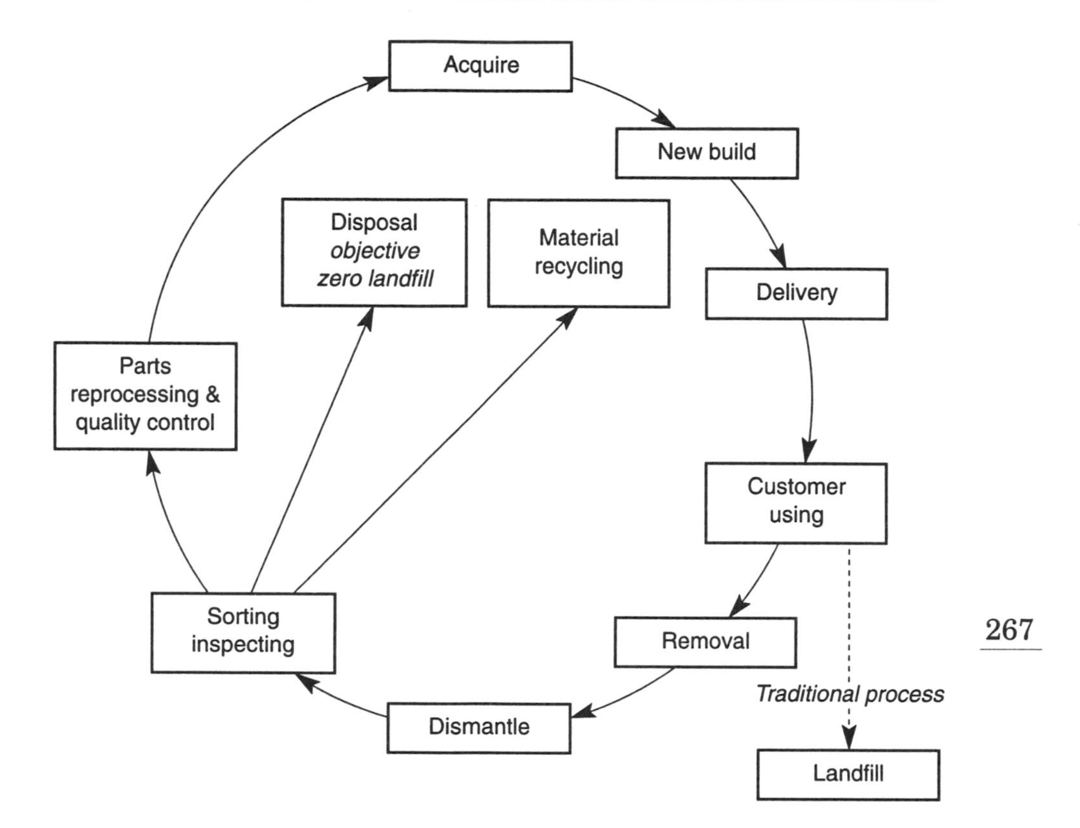

Figure 15.13 The product life cycle is a closed loop

The chief engineer decides during the design phase what has to happen with the product at the end of its life. The re-use, the recycle method and, if necessary, the disposal instruction is decided on the drawing board for new products. Materials are selected to ensure that the recyclability and disassembly of products is easy.

Among product environmental criteria, the design for re-use and recycling is the most important because

- it helps to protect natural resources;
- it contributes to energy savings and therefore reduces CO_2 emissions;
- it avoids excessive landfill;
- finally, it reduces the cash outflow.

For many materials recycling is not a big issue. For plastics, however, the recycling processes and capacities are not yet established. However, we have for some time, been successfully using recycled plastic material in our copy cartridges. The recycled material is exactly the same as the original and we are sure that a certain level of recycled material has no impact on the copy cartridges' quality and reliability. Regrinding the plastic is carried out in our asset recovery plant in Venray, Netherlands. The manufacture of new products in which we can use recycled material is just around the corner. Increasingly, this kind of 'closed loop recycling' is the solution which the industry has to aim for.

In all our manufacturing locations we have established waste management systems which have resulted in a 50% reduction in waste disposal by landfill. In the UK we have decided to co-operate with a waste management company to reduce our office waste.

Environmental audits

Environmental audits have long been in discussion. The EC has issued the eco management and audit scheme (Council regulation no. 1836/93), in the UK, BS7750 also includes an audit. But it will be some time before these policies will be internationally common standards. However, companies can start their environmental audit internally without waiting for the legislation which, in any case, will only be voluntary.

Rank Xerox has been auditing its manufacturing sites in Europe, India and Egypt since 1991. The environmental health and safety audit is measured against internal and external standards. For hazardous materials we adopted the tight Dutch regulations.

Each deficiency identified for corrective action is assigned a priority level as follows:

Critical Corrective action for the deficiency must begin immediately. The status of the corrective actions must be reported to the Environmental Health and Safety Department within one week. A shut-down of the processing plant may be necessary.

High High risk of injury to people, damage to property, or the environment or regulatory non-compliance fines.

Moderate Moderate risk of injury to people, damage to property or the environment; possibly regulatory non-compliance fines.

Low Low risk of injury to people, damage to property or the environment; possibly regulatory non-compliance fines.

The site director is responsible for taking corrective action.

What does environmental leadership mean to us?

We believe that a company which is an environmental leader needs to audit its performance on a regular basis. We are also committed to the following aims:

- Company employees should know that we are caring for the environment.
- We want to meet or exceed all statutory requirements in all countries where the company operates.
- The company helps to raise the standards in the EC and in European member states by supporting environmental commissions and associations.
- Our production plants and office facilities should be 'role model' examples in their neighbourhoods.
- Our products and services should be perceived by the customer as environmentally superior, or at least equal, to our competitors.
- Our products should qualify for environmental labels whenever a label is available (Blue Angel, Swan, EC Eco logo).

The 'Blue Angel' and other awards

The Blue Angel is a German environmental label which is awarded to products with advanced environmentally-friendly characteristics. There are specifications for a large number of products, though not all. Two specifications are relevant for Rank Xerox; they are *copiers* and *copier paper*. The specifications are reviewed after a period of three years and tightened if better 'environmentally friendly' products come on the market.

The requirements brief is given in Fig. 15.14. Other labels are the Swan of the Nordic countries and the Eco logic of the European Community. The EC Eco logo will probably replace the national logos in the long run.

Rank Xerox had received the Blue Angel award for ten of its copiers against the specification valid until 31 December 1993 and for a recycled paper brand. We received the Swan for a recycled paper and a chlorine-free paper.

The environmental logos play an increasing role in the buying-decision process. Large organisations already have environmental requirements included in their documents of tender. The Blue Angel is highly respected in Japan and the US.

Communication

It is not sufficient to be good or even to do something significant. The interested stakeholders have to be informed about it.

Recycled paper - RAL-UZ14
The paper must be produced from 100% recycled paper of which 51% of the pulp must be lower grade paper quality. The final product must be of similar quality to normal copier paper.

Example of "Blue Angel" logo for copiers

Copiers - RAL-UZ62
The specification defines maximum levels for the following emissions:
- Dust
- Ozone
- Noise

There are other criteria, e.g.
- Photoreceptors must be taken back by the supplier for recycling or professional disposal.
- The user manual must contain the company's commitment to take the photoreceptors back after use.
- Waste toner has to be dust free packed and incinerated or recycled.

Figure 15.14 The German 'Blue Angel' label: requirements for copier paper and copiers

For environmental leadership we have identified seven target groups which need to be informed about our strategies. These groups are as shown in Fig. 15.15.

The options for communication are numerous and they are all effective:

- Annual Report, Environmental Progress Report, product leaflets, advertising campaigns, videos, communication meetings, exhibitions, annual Environmental Day, environmental brochures, lectures to public audiences, articles for handbooks, and inhouse newsletters to employees.

Positioning

Environmental communication should not stand alone. It should be an integrated part of the corporate communication strategy. The same logic applies to the whole environmental leadership programme. All activities should be in support of the company's objectives. For Rank Xerox they are: Customer Satisfaction, Employee Motivation and Satisfaction, Market Share, and Return on Assets.

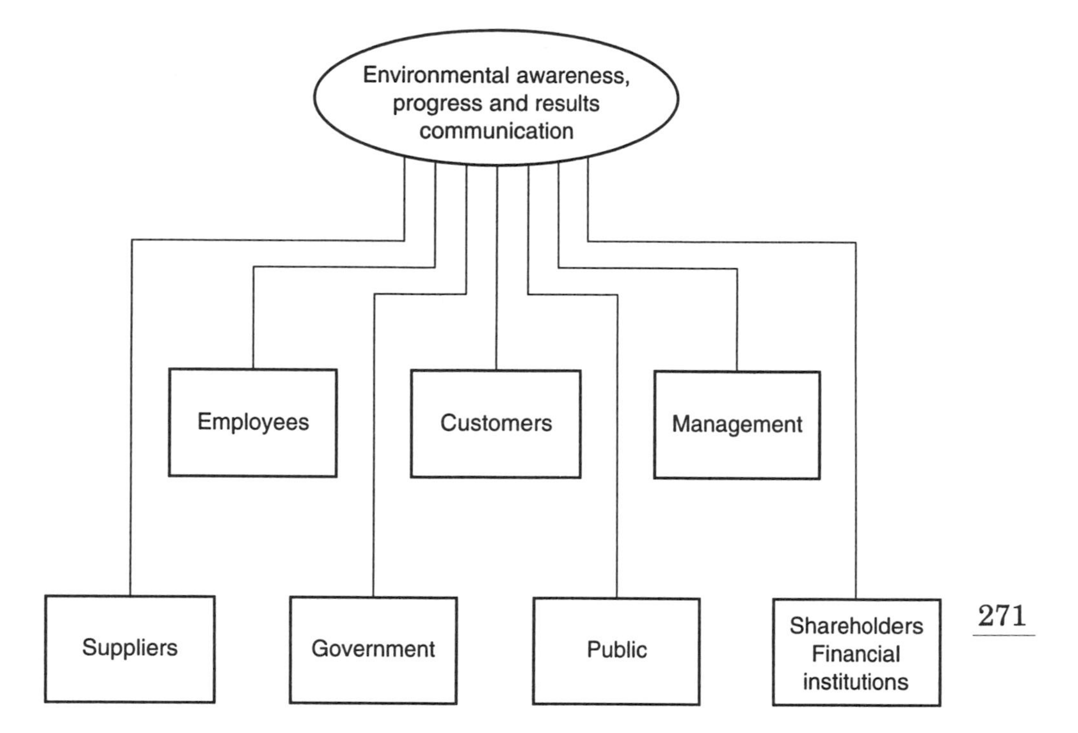

Figure 15.15 Informing the 'stakeholders'

Effective environmental management is vital for a company's success, therefore, it is important to ensure that environmental responsibilities form part of an organisation's regular policies and processes.

Where do we go from here?

There is still a lot to be done. The ultimate 'green products' have still to arrive and our training has not reached all the people who need to be informed. There is no lack of awareness, but what employees, customers and management want to see in the future is *action*.

Recommended reading

British Airways Annual Environmental Report, British Airways PLC, Environment Branch, London, Hounslow, September 1992.

Specification for Environmental Management Systems (BS7750), BSI, Milton Keynes, MK14 6LE, 1992.

Coopers and Lybrand Deloitte, *Your Busines and the Environment*, Business in the Environment, London.

Coulson-Thomas, Colin and Susan, *Managing the Relationship with the Environment*, Adaption Ltd, London, 1990.

Green Rights & Responsibilites, Department of the Environment, London, February 1992.

Report of the Financial Sector Working Group, DTI The Department for Enterprise, Department of the Environment, London, February 1993.

IBM, *IBM and the Environment – a Progress Report*, IBM USA, October 1992.

Monod Jéȓome, Gyllenhammar Pehr and Dekker Wisse, *Reshaping Europe: A report from the European Round Table of Industrialists*, The European Round Table of Industrialists, Brussels, September 1991.

Umweltzeichen Produktanforderungen-Zeichenanwender und Produkte, RAL Deutsches Institut fur Gütesicherung und Kennzeichnung e.V. Bonn, Germany, January 1992.

Environmental Policy and Practices, Rank Xerox Ltd, Parkway, Marlow, Bucks, SL7 1YL, 1991.

Several Internal Documents, Rank Xerox Ltd, Parkway, Marlow, Bucks, SL7 1YL.

Renger, Michael and Nathanson, Nabarro, *Environmental Audit*, The Institute of Chartered Accountants in England and Wales, ISBN 1 85355 3549, 1992.

Health of the Planet Survey, Social Surveys (Gallup Poll) Ltd, 307 Finchley Road, London, NW3 6EH, May 1992.

Smart Bruce, *Beyond Compliance*, World Resources Institute, ISBN 0 915825-72-2, April 1992.

Spiegel Spezial, *Bericht des Club of Rome 1991 Die Globale Revolution*, Spiegel-Verag, Rudolf Augstein GmbH & Co KG, Hamburg, February 1991.

Spiegel Spezial, *Europa ohne Grenzen – Alarm für die Umwelt*, Spiegel-Verlag, Rudolf Augstein GmbH & Co KG, Hamburg, January 1992.

The Council of the European Commission, *Draft – Proposal for a Council Regulation Establishing a Community Scheme for the Evaluation and Improvement of Environmental Performance in Certain Activities and the Provision of Relevant Information to the Public (ECO Audit)*, Official journal of the European Communities, December 1991.

The EC Environment Guide, The EC Committee of the American Chamber of Commerce in Belgium A.S.B.L.V.Z.W, 1992/93.

The United Nations Conference on Environment and Development Rio de Janeiro 1992, *Earth Summit 1992*, The Regency Press Corporation, London, 1992.

Willums, Jan-Olaf and Golüke, Ulrich, *From Ideas to Action – Business and Sustainable Development*, ICC Publishing, 1992.

Winter, Georg, *Business and the Environment*, McGraw-Hill Book Company, GmbH, Hamburg, 1988.

16

Environmental Management in the food industry

DENNIS J. VAUGHN, P.E.

Vice President, Environmental Affairs,
Grand Metropolitan Food Sector, Minneapolis, USA

- Environmentalism affected food processing much later than it came to the petroleum and chemical industries, but the food industry is now subject to strong environmental pressures, and intensive regulation.
- At Grand Metropolitan Food Sector in Minneapolis, the Environmental Affairs function was established in 1990.
- In the last three years environmental management at Grand Met has evolved from a 'command-and-control' approach which relied on instruction from the centre, to the implementation of management systems using participative approaches similar to those established for safety and quality.
- Food Sector Environmental Affairs coordinates environmental initiatives for all Food Sector facilities on a worldwide basis. This includes 54 facilities in North America, and 31 facilities in the UK, and continental Europe.
- Environmental Affairs is responsible for providing assistance to line management in the form of audits, guidelines and procedures, technical information and consultancy.
- The Sector sets targets, and monitors progress towards them on a whole range of environmental issues including: toxic releases, environmental incidents, packaging and chemical storage and handling.
- The aim is that each facility should have an environmental committee, and that these committees should provide grassroots support for compliance programs, training and awareness. Affiliations are also in place with outside environmental interest groups to provide support materials, project ideas, and awareness training to these plant committees.

Introduction

The Grand Metropolitan Food Sector is comprised of GrandMet Foods–Americas, headquartered in Minneapolis, USA, and GrandMet Foods–Europe, based in Paris, France.

GrandMet Foods–Americas includes Pillsbury baked goods, Green Giant vegetables, Alpo petfoods, GrandMet Foodservice and the world's leading superpremium ice-cream, Haagen-Dazs.

GrandMet Foods–Europe produces and markets baked goods, vegetables and ready meals for the European markets. Its brands include Brossard, Erasco, Jus-rol, Pillsbury and Green Giant.

Brossard is the number one European cake manufacturer, and in the Americas GrandMet holds number one positions in prepared dough, frozen pizza, frozen vegetables and superpremium ice-cream.

Operating profit in 1992 approached 190 million pounds sterling.

This is a review of environmental management programs for Grand Metropolitan Food Sector. During its first year (1990), the Environmental Affairs department developed baselines and established measurement systems to track its environmental progress. Key components established included: issuance of an environmental policy; development of an incident-reporting system; development of energy and solid waste tracking systems; and implementation of an environmental assessment program.

Our environmental initiatives emphasize compliance with the spirit as well as the letter of the law, avoidance of incidents and protection of the company's employees, property and reputation. The Environmental Affairs function has added value to the Grand Metropolitan Food Sector business units in pursuit of these initiatives. With limited resources, corporate programs have been developed to enable facilities to comply with both company and regulatory standards. Development of single procedures at the corporate level reduces repetitive efforts by facility staff and insures a consistent response to regulatory or company initiatives.

Environmental initiatives are increasingly recognized as a key component to our continuing businesses. Business objectives now consider environmental programs in determining fiscal year incentives. Capital projects are routinely reviewed for their environmental impacts and requirements. Business teams have recognized the necessity for considering environmental issues in product and process designs. Increased regulatory enforcement has prompted development of detailed requirements and procedures for the facilities, in order to reduce both personal and corporate liabilities.

Organization

During its inception under GrandMet in 1990, the environmental function reported to the Operations Engineering department. In late 1990, the Health and Safety responsibility for Food Sector was incorporated into the environmental programs, creating the Environmental Control and Safety department. This change allowed closer and appropriate coordination of elements common to both programs. In 1992, a majority of the Health and Safety duties were reassigned to the Loss Control department, in recognition of escalating workers' compensation costs. The newly named Environmental Affairs function was then aligned under the Senior Vice President, Technology, Food Sector. This structure permitted closer coordination on several key programs, such as chemical labelling and handling, crisis response, and sanitation procedures affecting wastewaters.

The primary focus of implementing environmental initiatives remains at the facility level. Each facility is required to designate an environmental coordinator to serve as the liaison with Environmental Affairs. This system has proved to be very effective in responding to plant-specific issues.

Food Sector Environmental Affairs coordinates these environmental initiatives for all Food Sector facilities, on a worldwide basis. This includes 54 facilities within North America, and 31 facilities in the UK and continental Europe.

In fiscal year 1992, the Environmental Affairs function had a staff of four. Fifty percent of the budget is for salaries, with 20% allocated to travel. Current corporate resources are limited, with each technical staff member supporting an average of 28 facilities.

Key functions

The key areas of responsibility for Environmental Affairs are:

- audits;
- guidelines and procedures; and
- technical resource and consultant.

Our auditing system is more properly called an 'environmental assessment'. These assessments are conducted on an annual basis for each manufacturing facility of the Food Sector. During these site visits, the physical, protective systems at each plant are assessed, in addition to conducting a review of the permit and paperwork requirements under environmental regulations. Corporate standards are also assessed during these reviews. At the conclusion of these site visits, a report is issued to both plant and corporate management identifying the key concerns and recommendations, along with a request for a corrective action schedule. This assessment program evaluates compliance, identifies key risks, and specifies budget needs for major improvements.

Based on the needs or issues identified at the facilities, corporate standards or procedures are sometimes developed. These identify the program requirements, and also serve to simplify and compile a variety of standards or requirements into a single document. These guides may be regulatory in nature, providing instructions on completing permits or paperwork for environmental regulations, or the guides may be standards for physical improvements needed at a majority of the facilities. Guides issued include: propane storage and dispensing; above-ground storage tank requirements; asbestos management procedures; ammonia operator training requirements; and chemical labelling systems.

Environmental Affairs also serves as a technical resource or consultant, responding to the day-to-day operations of the business units. Questions of interpretation and implementation are often raised by the facilities, and addressed. Pollution control designs are routinely reviewed and major capital projects assessed for treatment system needs and technical support. Close coordination is maintained with plants, consultants, and regulatory agencies to identify and respond to both short- and long-term issues and opportunities.

These three key program functions provide a focal point for efforts in other areas, such as solid waste reductions, worldwide standards, and long-term strategy. By visiting and maintaining contact with the facilities, specific issues are addressed, and overall common needs are identified. By tracking environmental regulations and trends, long-term strategies can be identified and budgeted, for incorporation in the business unit's strategies.

Toxic releases

Government agencies have identified specific toxic chemicals, that, if released, could pose a potential threat to people in plant communities. Companies who use these listed chemicals above specified amounts must submit annual reports to regulatory agencies, listing the amounts used and released.

Companies submitting this toxic release data are subject to the compliance requirements and associated penalties. In addition, public relations can be affected by being 'tagged' as a pollution source, and by being closely watched by environmental groups.

GrandMet facilities use several of the toxic chemicals, and about one-third of all US facilities are required to file the annual forms. GrandMet has established a goal to reduce the listed toxic releases by 90% by 1995.

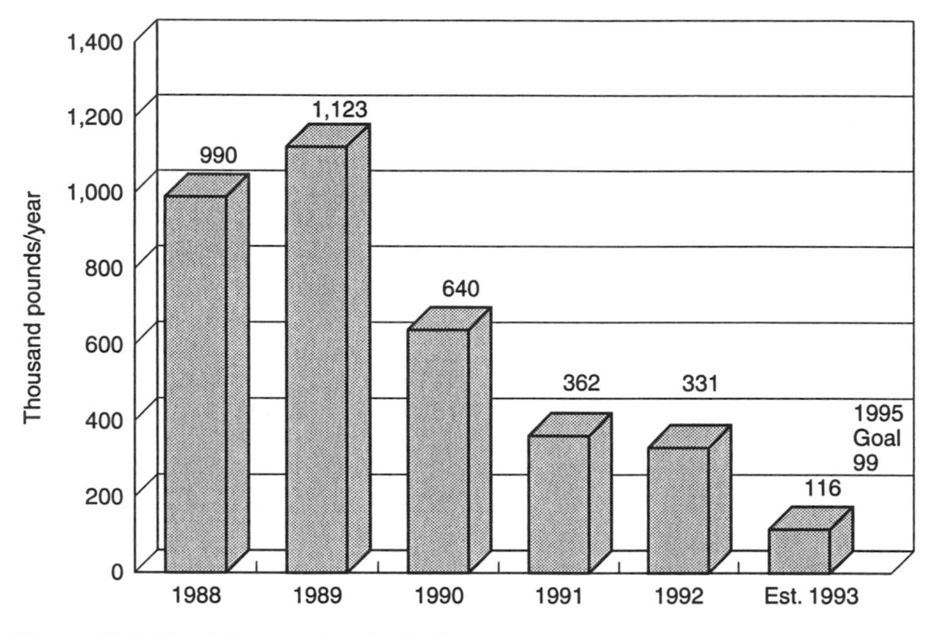

Figure 16.1 Food Sector chemical releases

Submission of the required toxic release reports is tracked by Environmental Affairs. Guidance is issued to all facilities to assist in report completion. Who reports and quantities of chemicals released is compiled into an annual summary. Comparisons are then made between these annual results to track progress on reductions.

In 1991, the total releases of regulated chemicals were 362,000 pounds. This compares with 640,000 pounds in 1990, or an overall reduction of 43%. The number of facilities exceeding the reporting threshold dropped to 13 from 19 in the prior year. In 1992, these total releases dropped to 331,000 pounds, based on reports from 13 facilities.

Environmental incidents

A variety of regulations govern the timing and responses made when an abnormal release of contaminants to the environment occurs. These releases can occur as part of the normal safety functions of ammonia refrigeration systems, for example, or may be due to a waste-water treatment system upset. In addition to correcting the source of the release and conducting necessary cleanup activities, a variety of reporting obligations are triggered. These reporting requirements can be confusing, even sometimes forgotten, during the incident situation.

Environmental releases that may be reported to outside agencies can also affect the perception of the company. It is therefore important that internal communications should be triggered to ensure a prompt notification of corporate management and key functional resources. A variety of requirements also exist for the proper conduct of enforcement agency inspections. Assistance can sometimes be provided to facilities in determining proper actions and responses during these inspections.

Because these types of incidents can have a significant affect upon both the facility and GrandMet as a whole, Food Sector has developed an internal crisis-response plan. This plan identifies a procedure for communicating incidents within corporate management. The communication network allows GrandMet to provide any necessary resources to the facility in response to the incident, and ensures that any reporting obligations are met.

This incident-reporting system has evolved into the use of a toll free 800 telephone number; the single-number system eases reporting by the facilities and ensures consistency for all plants. Postings identifying and promoting the use of this toll free number have been developed and issued to the facilities.

Specific environmental incidents are subject to this reporting requirement. An environmental incident is any abnormal release or spill into the environment; any release that causes injury, evacuation or death; or any visit from an environmental regulatory agency.

Once a reportable environmental incident occurs, facilities are required to notify the toll free number within one hour of discovery. This notification is then routed to Environmental Affairs for response back to the facility. Facilities are also required to provide a written follow-up notification to Environmental Affairs within 24 hours of making the toll free call. All responses are logged and tracked to assess causes and reporting compliances.

The attention to these releases has significantly reduced the quantities of materials being released. Also, Environmental Affairs has been able to assist plants in making the required calls and reports to regulatory agencies. Proper reporting to agencies is a factor in avoiding or reducing penalties. Information gained from these releases is shared between plants.

Chlorofluorocarbons

Chlorofluorocarbons (CFCs), typically referred to as freons, are known to contribute to ozone depletion. Worldwide agreements have been reached to restrict, and eventually ban, the manufacture of CFCs. These bans take place during 1994, depending on the country and the type of CFC. CFCs are typically used as coolants in air conditioners, chillers, freezers, or some processing applications. Within GrandMet Food Sector, CFCs are used primarily in chillers and in the direct contact freezing of vegetable products. Users of CFCs have been targeted by citizens and environmental groups, in order to add pressure for the early elimination of these chemicals. The public relations affects of these pronouncements has had a significant effect on some European companies.

GrandMet Food Sector originally established a program to eliminate all processing uses of CFCs by the end of 1993. In conjunction with upgrades to refrigeration systems, GrandMet replaced direct contact freezing uses of CFCs with state-of-the-art ammonia refrigeration systems in 1992, one year ahead of plan. Previous annual losses of CFCs of 300,000 pounds have been eliminated.

GrandMet also developed plans and targets for the elimination of other uses of CFCs, such as R-11, R-12, and R-502. Commercially supplied units and appliances with less than 20 pounds of CFCs were not included in this program. Also excluded were mobile units containing these CFCs. For European facilities, these other uses of CFCs will be eliminated in 1994. For US facilities, program targets require this elimination in 1995, allowing an additional year as construction lead time.

Halons, another form of CFCs, are commonly used in dedicated fire-suppression systems. These uses are also targeted for prohibition. GrandMet has identified these uses and each affected facility has developed plans for eventual halon replacement, if existing supplies are expended.

Ammonia standards

GrandMet is the third largest user of ammonia refrigeration for food processing in North America. The company also has one of the world's single largest ammonia refrigeration systems. Ammonia is considered a toxic, flammable gas, requiring special handling and safeguards. A variety of industrial standards and government requirements exist to ensure the proper design and construction of these refrigeration systems.

Prompted by some accidental releases of ammonia, GrandMet has revised its standards for ammonia refrigeration systems. Beginning in 1990, GrandMet initiated an extensive review of the standards for its ammonia refrigeration systems. This entailed investigation into a variety of industry practices, building codes and government requirements in order to develop a single standard meeting all the requirements. Requirements were researched for materials of construction, redundant systems, piping construction and labelling, pressure vessels and welding standards.

Based upon our experience, GrandMet's standards emphasize the importance of welding. Although not often, faulty welds have resulted in major accidental releases of ammonia. Standards used in the nuclear power industry were identified as state-of-the-art for this application. In addition to specifying standards for the refrigeration system itself, the newly developed requirements also consider other aspects of the facility. Emergency ventilation, explosion prevention, detector systems and storage vessel containment are also addressed in these new standards. Key features related to safety are also specified.

In 1992, GrandMet Food Sector's ammonia refrigeration standards were approved and finalized. In conjunction with development of these standards, Environmental Affairs worked closely with the mechanical engineering function in developing an ammonia refrigeration audit checklist. This checklist is designed to be used by facilities as a simple method to assess their conformance with corporate standards; it has now been translated into Spanish, French and German in order to foster its application worldwide. Significant system improvements have already resulted from the employment of this procedure.

Ammonia training

Uncontrolled releases of ammonia may have significant effects on the environment and nearby communities. As GrandMet makes such extensive use of ammonia refrigeration, Food Sector has developed extensive standards for the design and construction of its refrigeration systems. Yet, the physical parts are only a small part of the make-up of a safe refrigeration system. Therefore, attention must be given to the training and knowledge of the ammonia system operators.

Having spent considerable capital to restructure and upgrade facilities in ammonia refrigeration systems, attention now focuses on the capabilities of system operators.

During the development of the ammonia system design standards, an Ammonia Design Committee was established. This committee included representatives from major facilities, outside consultants, and internal company functions, such as mechanical engineering and Environmental Affairs. The combination of committee and consultants developed both the requirements and the content for training ammonia refrigeration system operators.

The training requirements address three levels of operator competency: Level III, Level II, and Level I. Level III training involves a general awareness of ammonia and its hazards, and a general review of refrigeration systems. For intermediate sized refrigeration systems, Level II operators are required, based on the requirements for additional expertise. Level I operators are necessary for major refrigeration facilities or for key operators.

Each level of training contains environmental and safety components. Successful candidates must currently be trained in hazardous chemical communications, emergency response, lock-out and tag-out, and respiratory protection. Upon completion of these components, as well as the ammonia-specific training, candidates can become certified for one of the three levels.

This certification program is a first of its kind within the industry. GrandMet Food Sector requires that every operator of one of its ammonia refrigeration systems must be certified at the appropriate level. Candidates who fail to complete the required training are not allowed to operate or maintain these systems.

Solid waste

Solid waste generated from food processing facilities is both an environmental issue and an economic issue. Wastes that are generated must be properly handled and disposed. These wastes were at one time purchased as a raw material, and now additional expense will be incurred in handling their disposal.

Environmental liabilities exist for all waste disposals. Assurances must exist that wastes are being properly handled and disposed. Food Sector is involved in several landfill cleanups, resulting from leachate or other contamination from these previously acceptable disposal techniques.

Having evolved from a variety of independent food businesses, it was not always clear how much or what types of waste were being generated company-wide. Amounts disposed and how they were disposed were sometimes unknown.

In 1990, GrandMet Food Sector conducted an initial survey of all facilities to determine the types and amounts of waste being generated. For each waste stream, disposal or recycle methods were requested. Based upon this survey, a solid waste tracking form was developed and distributed to all facilities. Each month facilities are required to submit details on wastes generated and disposed to Environmental Affairs. This data is in turn tabulated on a business and Sector basis.

In 1992, GrandMet established solid waste reduction goals. One goal is to reduce the amount of solid waste generated in 1990 by 15% by the year 1995. This target applies to all solid waste leaving the facility except for wastewaters. By reducing the overall waste being generated, issues of landfill disposal and costs of treating or disposing of these materials can be reduced. A second environmental goal is to reduce the amount of waste being landfilled in 1990 by 50% by 1995. This target is similar to that being faced by many municipalities based on restrictions by federal agencies. GrandMet views this goal as a way to reduce liabilities for future cleanup of landfill sites.

The monthly solid waste tracking system is in place at each facility. In Fiscal Year 1992, 126,000 tons of waste was generated and 18,000 tons of this was landfilled.

Waste reduction goals have been incorporated into business unit objectives, on which management incentives are based. The goals are divided into annual targets, seeking an annual reduction of solid waste generation of 3%, and an annual reduction in the waste landfilled by 10%.

Chemical storage and handling

GrandMet's food processing plants use a variety of chemicals to sanitize processing systems, provide utility support services, or as incidental food ingredients. Such chemicals include ammonia, chlorine, lubricating oils, acid cleaners, and laboratory supplies.

Most of these chemicals have some hazardous characteristics, mandating their careful storage and handling; worker safety can be affected by spillage or contact with them. Environmental contamination can also occur through spills. A variety of government regulations and standards have been developed to promote the safe use, handling and labelling of such chemicals. These requirements have become quite complex and confusing, making it difficult to coordinate all of them.

Food safety concerns have also resulted in requirements to protect the safety of food products as well as ensure their proper production. As a result, additional labeling and record-keeping requirements have been imposed.

Based upon findings during routine environmental assessments, Environmental Affairs developed simplified corporate standards for the storage and handling of general chemicals, flammable and combustible chemicals, chlorine gases and battery-charging units. These corporate standards are derived from a variety of industrial standards, safety requirements, fire protection standards, food safety standards, and environmental regulations. Each corporate standard contains a checklist for determining conformance to the requirements as well as a diagram showing the construction of an acceptable storage area.

To simplify chemical labels, Environmental Affairs acted jointly with Food Safety and Regulatory Affairs in the development of a combined labelling system. Plants had requested simplified labeling procedures, employing a single label in lieu of the several formats previously required. Based upon hazard-identification requirements and food safety needs, a combined label format was developed. This resulted in the in-house design of a GrandMet Food Sector labeling system.

PCBs

Polychlorinatedbiphenyls (PCBs) were used in the past, primarily as dielectric fluids in electrical transformers. Based upon their environmental persistence and toxicity, PCB's were targeted for regulation and removal as early as 1979. Shortly thereafter, a number of the company's electrical transformers were either replaced or flushed and refilled with non-PCB fluids.

Additional regulations affect food processing facilities in close proximity to PCBs. Arrangements must be made for additional fire protection and spillage control, due to the potential formation of dioxins during fires involving PCBs.

Over the years, most of GrandMet's PCB-contaminated transformers were eliminated, however, several systems remained in use.

The environmental goal for elimination of all uses of PCBs was issued in May, 1992. All electrical transformers and capacitors containing PCB dielectric fluids were to be eliminated under this program. Each facility must be able to document the PCB status of all electrical transformers at their site, whether owned by the company or by electrical utilities. All transformers with documented PCB concentrations below 50 ppm, whether owned by the company or owned by utilities but on our property, must be labelled with the standard 'non-PCB' label. Any disposal of PCB-contaminated transformers or capacitors must have the prior notice and approval of Environmental Affairs. All uses of PCBs were eliminated in 1993.

Underground storage tanks

Underground storage tanks (UST) are typically used to store petroleum products below the ground surface in order to reduce fire and explosion hazards. Being located within the ground, these systems are subject to corrosion and subsequent leakage. Due to the type and mobility of the contaminants present in the petroleum products, any leakage can potentially involve extensive and expensive environmental cleanups.

A variety of regulations have been issued at both the country and state level to force owners to safeguard these tank systems. Based on the potential cleanup costs involved, most companies opt for removal of these tanks.

GrandMet Food Sector removed all unprotected underground storage tanks in the US by the end of 1993. In Europe, all such tanks will be removed before October, 1994. Any underground storage tank being removed, or closed in place, must have third party oversight and verification, using parties approved by Environmental Affairs.

Over the last three years, GrandMet has removed approximately 41 underground tanks. Of these, about 75% had identified leaks. Removal and cleanup costs have ranged from $20,000 to $250,000 per tank.

Fryers

Several Food Sector facilities use deep-fat fryer systems in the processing of cooking of food products. The fryers may either be covered or of open top construction. Air pollution emission standards exist for most of these fryers, placing restrictions on visible emissions and odours. The more serious problem, however, is the risk of fryer fires.

Based on experience, no consistent company standards existed for fire-suppression systems on deep-fat fryers. Maintenance was not being routinely conducted, and necessary safeguards were sometimes missing or malfunctioning. GrandMet Food Sector therefore developed a fryer checklist and incorporated these requirements into corporate standards. These requirements address both fire suppression, safety interlocks and standards which indirectly minimize air contaminant emissions. The fryer checklist has been translated from English into French and German for use on a worldwide basis.

Asbestos

Friable forms of asbestos are known to cause certain forms of lung diseases and federal programs have been developed to address potential asbestos exposure in public buildings. The US Environmental Protection Agency has issued requirements covering the removal and disposal of asbestos-containing materials. The Occupational Safety and Health Administration (OSHA) has issued maximum air contaminant levels for occupational exposure to asbestos-containing materials. These issues of worker safety and potential corporate liabilities require that asbestos material be properly managed.

Food Sector does not mandate the removal of asbestos. Rather, the program ensures that friable forms are encapsulated or removed, and that all asbestos on site is properly managed. Until all asbestos is removed from our facilities, any remaining materials must be intact, protected and labeled. Accordingly, Food Sector has issued the following procedures to all facilities, applicable to all remaining asbestos-containing materials:

- Each facility is required to have an independent asbestos survey to identify and record the location, type, and condition of all asbestos-containing materials.

- A copy of this survey must be submitted to Environmental Affairs.
- All asbestos-containing materials at the facility must be labelled as asbestos.
- Records must be retained of all asbestos removed, including type and amount removed, quantity, who removed it and when, and where and when disposed, as well as documents regarding any agency notifications.
- Any asbestos-containing material at the facility must be encapsulated and protected from damage and fraying.
- Any asbestos-containing material at the facility must be inspected monthly for integrity and labelling.

An asbestos inspection form was developed and distributed to the facilities. All asbestos-containing material must be checked, the inspector identified, the date of the inspection noted and any deterioration noted and corrected.

Packaging

Packaging is used to transport and protect food products. Packaging materials need to be selected with consideration for their ultimate disposal and any potential adverse environmental affects. Initially, environmental issues of packaging centered on heavy metals or other toxic components. Upon disposal at landfills or incinerators, these trace components of packaging were identified as contributing to air pollution or groundwater contamination.

Based upon regulatory trends to limit toxic constituents in packaging, GrandMet Food Sector worked with suppliers and printers to remove these targeted toxics. These efforts were completed without compromising packaging integrity.

Emphasis has now shifted to public pressure and legislative initiatives to reduce the quantity of packaging, and to increase its recycle content or re-use potential. For food industries, meeting these developing standards is not always easy and straightforward. Packaging integrity and food safety regulations often times limit available alternatives. Functionality continues to be the primary concern.

Recycled content of cardboard or paper-based packaging has been increased. However, materials directly contacting food products must contain virgin materials to satisfy food safety standards and performance.

Transport packaging has been reduced in many cases. Cardboard is replacing plastics in binding pallets. Arrangements are being made to return or re-use pallets, and to reduce quantities for disposal.

Some packaging materials are limited in their use to increase the recycle content, as performance characteristics might suffer. Extensive research is being conducted into alternative packaging materials.

The quantity of packaging used has typically been a cost issue, not an environmental one. Steel cans have become thinner and ridged, in order to provide the necessary strength, while reducing the weight and associated material transport costs. Other packaging types, such as plastics and papers, have continually been reassessed and reduced to cut production costs.

Packaging reduction and recycle improvements will continue. The primary concern remains that of food safety. Information for discussing these issues continues to be provided to legislative and trade groups so that informed discussions can be made when regulations are being developed.

Next steps

Environmental management has now evolved from the command-and-control approach, to implementation of management systems, similar to traditional quality management concepts. GrandMet Food Sector recognizes this trend and its benefits.

As indicated in the environmental policy, line management was given the ultimate responsibility for environmental issues. At the same time, recessionary pressures have limited the opportunities to add staff and resources to administer the increasing number of regulations and programs. Government requirements reflect these new management concepts, by asking industry to prepare various prevention plans and management plans.

Those effects have prompted GrandMet Food Sector to evolve its environmental programs into management systems. Each facility has been asked to create an environmental committee, much like committees used for safety or quality. These committees will provide grassroots support for compliance programs, training and awareness, in addition to allowing for a sharing of the work load.

Affiliations are in place with outside environmental interest groups, to provide support materials, project ideas and awareness training to these plant committees.

Part IV

■

MANAGEMENT APPROACHES AND TECHNIQUES

'Environmental auditing will not be a sufficient condition of business success in the 1990s, but increasingly it will be a necessary condition.'

John Elkington, Director, Sustainability Ltd,
Business International Research Report, 1990.

'In most companies, organisational issues account for more than 90 per cent of environmental problems.'

George Moellenkamp, DRT International, quoted in *Changing Course*
(Schmidheiny, 1992).

Introduction to Part IV

■

This Part describes the management tools and their application in putting company policies on the environment and business opportunities into practice. Tools such as eco-auditing, environmental management systems and ISO 9000, target setting, and environmental impact assessment, overlap and complement each other. They grow out of well-established practices already familiar in other business areas. Total Quality Management approaches, in particular, are present in many environmental management systems.

Earlier sections have shown how the life cycle, or 'cradle-to-grave', approach to the environment has become a strong element in legislation, including corporate liability, and in corporate environmental planning. Life cycle assessment (LCA) concepts and the way they are being put into practice are described in Chapter 18. LCA as well as eco-auditing offers a framework within which to integrate the full range of environmental effects from a company's activities; it also enables corporate priorities on environmental performance investment to be set. By focusing on the objective information that eco-audits and LCA can provide, the costs and benefits of specific investment options, both from the point of view of the company, and from the wider perspective of other stakeholders, become clearer.

Chapters 19 and 20 deal with target setting and monitoring. Chapter 19 highlights the way in which the negotiation of environmental targets with managers can be harnessed to build individual commitment to overall environmental policies, and to change corporate culture. Chapter 20 deals with internal monitoring of environmental performance, and with environmental reporting. It demonstrates that monitoring not only provides accountability within a company, but can be used proactively to identify business opportunities for technology development and to plan for future environmental investment.

Chapter 21 covers the management of environmental crises, dealing especially with the need for management of public communications alongside effective technical responses. Without clear crisis management plans, confusion can build rapidly and destroy public confidence in a company's ability to control a crisis if one occurs. Openness to admitting problems, effective response systems, and rapid action from the top, are vital elements of crisis management; and rapid responses can prevent crises being as damaging to the environment, and to corporate profile, as they might otherwise be.

Executive summaries

■

17 Environmental auditing and management systems

Remarkably few organisations have as yet implemented systematic environmental programmes that include some element of auditing. One of the constraints to action has been the confusion felt by managers in many organisations about exactly what an 'audit' is, and how an environmental programme should be approached.

Nevertheless, the combined experience of industry leaders who have developed successful environmental programmes, has now identified a number of key concepts and principles which are starting to be crystallised in proposed legislation and other standards.

Ideas about auditing and environmental management have evolved over the past twenty years as the environmental pressures on organisations have changed. The term 'audit' was originally applied in the 1960s and 1970s to the narrow concept of systematically measuring compliance with pollution control legislation. During the 1980s as the pressures on organisations increased, audits were extended to cover a much wider range of activities. More recently, generic environmental management systems, designed to support a company-wide approach to continuous environmental improvement, have been developed and are likely to become established best practice during the remainder of the present decade. The 'environmental audit', in the strict sense, now refers to one stage in this process, i.e., the measurement of performance in achieving predetermined targets.

The implementation of an environmental management system is discussed, using the latest British Standard BS7750 and the EC Eco-management and audit regulation as examples of current best practice.

Further developments in environmental management practice are expected to take place during the remainder of the present decade. Already a greater emphasis is being placed on the public disclosure of environmental information, and many larger companies are now producing environmental performance reports as an integral part of their annual financial reporting cycle. In addition, environmental management is likely to become more closely integrated with other ethical issues affecting organisations.

18 Life-cycle assessment

The concept of life-cycle assessment (LCA) is a dynamic, innovative way of looking at environmental improvement. Life-cycle assessment is the evaluation of a particular material or activity from the generation of the raw materials to final disposal. Often referred to as 'cradle-to-grave' or 'earth-to-earth', life-cycle assessment has recently increased in visibility and importance due to its application to the area of environmental protection. The concept behind environmental life-cycle assessment is to address the genesis of environmental degradation rather than the standard approach of trying to remediate environmental problems after they have been created.

19 Agreeing targets for environmental improvement

The objective of this chapter is to examine some of the issues that arise when negotiating improvement targets as part of an environmental programme within a large organisation. British Standard 7750 for Environmental Management Systems is relevant to any organisation's environmental improvement programme. The very simplicity and clarity of its structure can, however, potentially give an unbalanced picture of the challenges faced at each step. In particular, great care must be taken when setting and agreeing improvement objectives and targets. Most of the important work in an environmental programme comes in the gap between quantifying the effect and defining the target. If this work has been well done, all should then be plain sailing. If it is badly done, environmental improvement will only happen by luck, if at all.

20 Reviewing and monitoring environmental performance

The public interest in the environmental performance of businesses and industries increases steadily as our economic system changes towards an ecological, social market economy. Environmental performance has become one of the key issues in the public domain and it may well become one of the decisive success factors determining a company's future.

Improvements in environmental performance are not achieved overnight. It takes responsible entrepreneurship, the framework of which is described in the second section of Chapter 20. This framework is made up of elements such as: a corporate culture enlivened and demonstrated by top management; unanimously accepted governing principles and policies; universally binding procedures; advanced and suitable technology; knowledgeable, environmentally-conscious and well-trained personnel; competent and responsible managers and supervisors; and a system of checks and balances that guarantees regular controls and corrective actions.

Once the foundations for environmentally-responsible attitude and behaviour have been laid down, one can start to monitor and measure performance. Care must be taken to balance safety, health, and environmental interests, and not to play-off one against the other. Various approaches to reviewing and monitoring environmental (in its broadest sense) performance are discussed in the third section of the chapter. They comprise: environmental auditing; reviewing investment projects; establishing eco-balances; and collecting, consolidating and reporting environmental data. The concept of SEEP (Safety, Energy and Environmental Protection) is described; this has proved to be a valuable tool to provide corporate management with a global overview of the company's environmental situation. Finally, in that section, the costs of environmental protection are elucidated.

21 Managing an environmental crisis

Public interest in environmental matters and the speed of modern communications combine to demand special management techniques for handling serious accidents or disasters.

Ever since Bhopal (1984) and Exxon Valdez (1989), environmental crises in a wide range of industries have hit the headlines. While every situation is different, certain characteristics can be identified in all these high-profile, threatening situations, and a number of operational principles have proved valuable in their management.

Since a crisis is the management of problems under the glare of public scrutiny, it is necessary to exercise control of this situation. A major component of all problems lies in lack of information, rumour and speculation, which combine to undermine management's authority. While there are many plans and procedures in place in most companies, serious environmental incidents reveal the need for special crisis management systems.

These need to centralise the information flow into and out of the crisis management group, and to ensure that the company communicates consistently through a limited number of spokespeople, trained to speak with the brevity and clarity which will assist journalists to understand and reflect the corporate viewpoint.

In addition, companies may find it helpful to use market research to get behind the headlines to understand how the company is viewed in the midst of a crisis.

17

Environmental auditing and management systems

RICHARD DALLEY

Associate Consultant, Aspinwall & Company Ltd, Consultants in Environmental Management, Shrewsbury, UK

- Generic environmental management systems (EMS) are likely to become established as best practice in the rest of the present decade, for example, using BS7750, ISO 9000, and the EC's Eco-audit regulation.
- Leading companies expect to better legislative standards wherever possible and have made commitments to apply uniform high standards worldwide.
- The EMS approach enables an organisation to manage its overall environmental position, recognising that:
  - all the activities of an organisation may have significant environmental impacts;
  - environmental management is a responsibility shared by all managers and to be integrated with other management processes;
  - strategic decisions may have environmental consequences that need to be taken into account.
- An EMS requires an organisation to adopt a programme of continuous environmental improvement following a logical sequence that is entirely consistent with established project-management practice, as routinely applied to a wide range of business problems:
  - strategic review of environmental issues affecting the business;
  - set objectives and targets to minimise these impacts;
  - devise and implement an action programme to achieve the set targets;
  - measure (i.e., audit) performance in achieving targets;
  - periodic review of the overall adequacy of the system.
- Undertaking an EMS has much in common with implementation of a Total Quality Management System (TQM); high-level commitment from senior staff and main board directors is crucial for success.

What is environmental auditing?

The word 'audit' causes more confusion than any other in the environmental lexicon. Environmental audits can mean many things to many people and in the rapidly developing field of environmental management there is as yet no consensus defining precisely what an audit is, or is not. Environmental auditing is universally held to be a worthy activity, but what exactly are its benefits to an organisation? If an audit is commissioned from one of the growing army of environmental consultants, what do you get for your money? There has been a rash of publications in recent years from trade associations, government agencies and environmental groups, describing audit practice and procedures (for a recent overview of the UK experience see Grayson, 1992 and for the USA the bibliography produced by the US Environmental Protection Agency, 1988). Nevertheless, despite all that has been written and said about auditing, remarkably few organisations have adopted a systematic approach to improving and measuring their environmental performance (Laming, 1990: Elkington and Dimmock, 1991: Touche Ross Management Consultants, 1991).

There are several reasons why the good intentions of many organisations have not been translated into action. Lack of awareness may be one cause, but in the 1990s most large organisations now recognise the environment as a significant business pressure. What most inhibits action is more, perhaps, that managers don't know where to start or what to actually do. This bar to action is exacerbated by the embryonic state of environmental management practice, and a certain mystique surrounding the 'environment' which leads some to believe, quite erroneously, that a PhD in ecology is required to understand the process.

Nevertheless, despite the confusion that continues to surround the subject, there are now many examples of organisations which have implemented successful environmental programmes (Willums and Golüke, 1992). The combined experience of these industry leaders, environment groups, consultants and legislators who have developed successful environmental programmes, has identified a number of key concepts and principles which are now starting to be crystallised in proposed legislation and other standards, such as the EC Eco-management and audit regulation (1992) and the British Standard BS7750 (1994) on Environmental Management Systems. The purpose of this chapter is to outline present ideas about auditing and management systems and to guide managers towards current best practice.

To gain an understanding of current thinking in environmental management practice, it is useful to consider some of the developments which have taken place during the last twenty years.

COMPLIANCE AUDITS

The roots of current environmental management practices can be traced back to the late 1960s, to a time when many companies, especially those with significant process emission and waste management impacts, began to experience tighter legislation. This was reinforced by new regulatory agencies with enhanced powers of enforcement, such as the US Environmental Protection Agency. Many large companies, especially in the petrochemicals sector in the USA, felt a need to systematically measure their compliance with legislation in order to control their legal liabilities. The term 'environmental audit' was introduced – perhaps unfortunately with the benefit of hindsight – to describe the process of inspecting process operations at specific plants or sites, in order to identify actual or potential breaches of legislation. Audit procedures were developed by a number of organisations, focusing on operational matters such as chemical storage, spill prevention, treatment and discharge of wastewater, emissions to atmosphere, solid and hazardous waste management and odour and noise nuisance.

PROACTIVE AUDITS

The late 1970s and 1980s witnessed a surge of popular interest in environmental issues. As a consequence, many companies became concerned not only with controlling their legal liabilities, but also with enhancing their overall environmental profiles. A number of organisations began therefore to extend the concept of auditing beyond compliance with pollution control legislation. This move stemmed from an appreciation that merely obeying the law was not sufficient to maintain an environmental leadership position. Leading companies expected to beat legislative standards wherever possible and, moreover, made commitments to apply uniform high standards worldwide, irrespective of local requirements, which might be quite low, especially in developing countries. The 3M Corporation was one of the first companies to extend the concept of auditing in this way, with its 'Pollution Prevention Pays' programme begun in 1975. This ongoing programme seeks to eliminate the causes of pollution by reformulating products, modifying manufacturing processes, redesigning production equipment and recycling byproducts (Willums and Golüke, 1992).

A bewildering range of new types of audit were proposed at this time, using a range of terminology but generally including the following:

Activity audits	To assess the environmental effects of particular activities, e.g., waste management, energy conservation, etc., that cross site and organisational boundaries.
Issue audits	Focusing on environmental issues of key concern such as the organisation's impact on ozone depletion, rainforest destruction, global warming or contaminated land.
Product audits	Examining the environmental impacts of particular products or services by taking a life cycle view from raw material sourcing through to final disposal.
Pre-acquisition and divestiture audits	To assess liabilities associated with the purchase or sale of businesses, facilities and other investments. These may be related to contaminated land and groundwater or the need for expenditure to meet pollution control standards.
Supplier audits	To measure the environmental performance of key suppliers of raw materials, components and services.
Health and safety audits	Assessing the exposure of the workforce and local populations to pollution and other nuisance such as odour and noise.

ENVIRONMENTAL AUDITS AND THE MANAGEMENT PROCESS

Current thinking – and the trend for the 1990s – recognises that piecemeal auditing of specific subject areas is not the most effective way for an organisation to manage its overall environmental position. Rather than undertaking a diverse range of specific audits in different parts of the organisation, a more strategic and comprehensive environmental management system (EMS) is required. This approach recognises that:

- All the activities of an organisation may have significant environmental impacts. Environmental management is not solely restricted to controlling pollution from manufacturing processes.
- Environmental management is a responsibility shared by all managers and is not the unique preserve of a specialised technical department. Environmental management must therefore become integrated with other management processes.

- Strategic decisions, e.g., those relating to product design, selection of plant and equipment, site acquisitions and other investments may have important environmental consequences that also need to be considered.

An environmental management system requires an organisation to adopt a programme of continuous environmental improvement, following a logical sequence of steps:

(1) undertake a strategic review to identify all the environmental issues affecting the business;

(2) set objectives and targets to minimise these impacts;

(3) devise and implement an action programme to achieve the targets set;

(4) measure (i.e., audit) performance in achieving the targets;

(5) periodically review the overall adequacy of the system.

It is at this point that the use of the term 'audit' becomes problematical and leads to confusion. Some authorities have extended the meaning of environmental auditing to apply to this new strategic approach and hence use 'environmental audit' synonymously with 'environmental management system'. The opposite view, and that advocated here, is that it is now less confusing to restrict the term 'audit' to its narrower usage, that is, the systematic measurement of performance against predetermined targets. Audits then become an integral part of the overall EMS (see point 4 above), rather than constituting the management system itself.

Thus it can be seen that the concept of environmental auditing, and the use of the term, has changed over the past twenty years as the environmental pressures on organisations have changed. A number of broad developmental phases can be recognised:

Compliance audits 1960s/1970s	A narrow concept of environmental auditing as a systematic method of measuring compliance with legislation. Applicable primarily to manufacturing and process plant.
Proactive audits 1980s	The concept widens to embrace a greater range of an organisation's activities, e.g., including products, services, suppliers, investments.

Environmental management systems 1990s	A strategic approach to continuous environmental improvement. Environmental management becomes fully integrated with other management activities throughout the organisation. A management plan is established setting targets to reduce environmental impacts. Environmental auditing in the strict sense now refers to one stage in this process, i.e., the measurement of performance in achieving the predetermined targets.

The needs of organisations

Organisations in all spheres of activity – manufacturers, utilities, retailers, transportation and other service companies and government agencies – are all experiencing a growing pressure to demonstrate high standards of environmental performance.

We have already seen that during the 1980s and 1990s a significant shift in the perception of an organisation's environmental responsibilities has taken place. Traditionally attention has focused on the control of pollution from large manufacturing and process facilities. But growing public awareness of global issues such as ozone depletion, acid rain, loss of rainforests and a host of other concerns has forced the environment up the political and social agenda. Consequently organisations are now held responsible for all actual and potential environmental impacts associated with their activities – whether or not these are regulated by specific legislation.

There is therefore a discernible shift away from legislation as the single most important force driving environmental change in companies. The pressure for change comes now from a wider range of 'stakeholders' in the business and not just from the regulators. Shareholders and other investors increasingly view environmental performance as a cipher for general management competence. Customers may discriminate against products they consider to be environmentally harmful. Employees want the reassurance of knowing that they are working as part of an environmentally-responsible organisation; and local residents, the media and pressure groups display a keen interest in companies' environmental performance.

But the rise in society's environmental awareness presents business opportunities as well as threats. This aspect is covered in more detail elsewhere in this volume, but clearly there are prizes to be won for businesses that manage their environmental issues effectively, and the environment should not be seen by managers as just another cost.

Organisations that manage their environmental affairs effectively stand to:

- reduce liabilities and insurance premiums;
- save time and resources otherwise lost to crisis management;
- save on raw materials and waste-disposal costs;
- develop more competitive products;
- meet the market needs of customers and investors;
- enhance their reputation.

An examination of companies that have established a position of environmental leadership shows that they are indeed well managed, well respected, ahead of the game and, above all, profitable.

Different organisations experience environmental pressures of varying types and degrees. For large process-orientated businesses, compliance with pollution control legislation may well remain the overriding concern, because financial liabilities in the event of a pollution incident may be very significant. For the large retailer, the concerns of the 'green' consumer may place the emphasis on the life cycle assessment of products, eco-labelling and the environmental performance of suppliers. Small, unlisted companies are unlikely to experience strong pressure from investors, but they are increasingly likely to have their environmental performance scrutinised by larger companies to whom they supply products or services.

Organisations therefore have diverse environmental management needs and no single off-the-shelf 'audit' or checklist can adequately meet the needs of all companies. Companies need to think through their own specific environmental position for themselves, identifying their environmental impacts and tailoring an appropriate response, taking account of the other business forces acting on them. The latest developments in environmental management systems are designed to provide companies with a framework around which to develop such a flexible approach.

Environmental management systems

Environmental management systems are intended to provide a systematic, comprehensive, but, above all, flexible framework around which companies can develop measurable (i.e., auditable) action programmes to minimise or remove threats to the environment. The new British Standard BS7750, 1994, provides the specification for an environmental management system and is the first such standard to be officially adopted anywhere in the world. The standard incorporates many recent concepts in this field and serves as a useful illustrative example of current good practice.

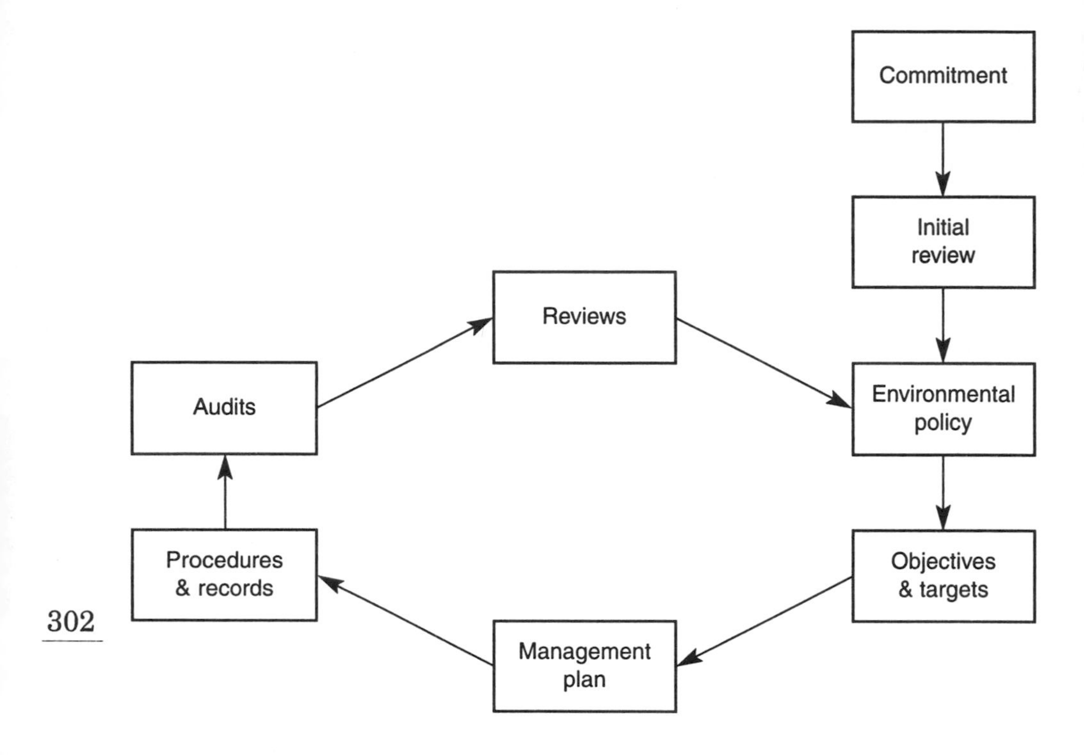

Figure 17.1 Environmental management systems – key elements

BS7750

BS7750, in common with other environmental management systems, requires that an organisation should:

- identify its environmental effects;
- state explicitly how it is going to deal with these;
- demonstrate that action has taken place.

In this respect an EMS is entirely consistent with established project-management practice as routinely applied to a wide range of business problems, i.e., define the problem; set targets which address the problem; take action to achieve the targets; measure performance in achieving the targets.

BS7750 is based around a number of key activities, undertaken sequentially, and these are common features of any good EMS:

Commitment

Undertaking an EMS has much in common with the implementation of a Total Quality System. Experience has shown time and time again that quality systems wither unless they have the strong backing and commitment from senior staff including main board directors. A similar high level of commitment is required when implementing an EMS – there is nothing more disheartening than witnessing a harassed junior manager struggling to integrate an environmental programme into existing management structures single-handed.

Initial review

The initial or preparatory environmental review is an essential prerequisite to implementation. Its purpose is to assess the broad environmental strengths, weaknesses, opportunities and threats associated with the organisation. This stage is critical to the success of the EMS; done well, much of the rest of the programme slips into place almost naturally. Without it, difficulties are bound to arise.

It is important therefore that the Initial Review should be comprehensive, examining all the actual and potential environmental effects of all the activities of the organisation, and involving as many members of the organisation as possible. BS7750 provides a helpful summary of the areas that an organisation might need to consider in the Initial Review:

- areas where environmental performance could be improved;
- views of relevant interested parties;
- environmental objectives and targets beyond regulatory requirements;
- ability to anticipate changes in regulations and legislation;
- adequacy of resources and environmental information;
- environmental records;
- environmental cost/benefit analysis and accounting methods;
- internal and external communications on environmental issues;
- environmental training plans;
- environmental aspects of products and services;
- incorporation of environmental considerations in design and marketing;
- resource consumption (energy, fuels, materials);
- waste reduction/recycling initiatives;
- use and disposal of hazardous materials, processes and products;
- transport policy;
- nature conservation;
- complaints and their recording and follow-up;
- visual impact, noise and odours;
- environmental probity of suppliers;
- environmental aspects of emergency planning;
- environmental effects of investment policies.

Environmental policy

The primary output from the Initial Review will be a comprehensive summary of the environmental issues which are collectively recognised throughout the organisation as the issues which have to be managed. The aim should also be to reach a consensus about what commitments can be made in response to these issues, in the light of the other business pressures to which the organisation is exposed. It is then helpful to prepare a formal environmental Policy Statement which crystallises these commitments in the form of policy principles relating to the key issues. (For examples of some recent corporate policy commitments see: *The Corporate Environmental Register*, 1993; and Willums and Golüke, 1992).

Objectives and targets

The next stage involves setting specific objectives and targets designed to reduce the environmental impacts identified by the Initial Review and referred to in the Policy Statement. The targets established must be documented and be as specific and measurable as possible.

The targets must also be practicable and have the support of all employees who are expected to work to their achievement. Setting targets can therefore be a challenging task, raising internal conflicts within the organisation about priorities and resourcing. The issues surrounding target setting are discussed in some detail elsewhere in this volume.

Management programme

The management programme, or 'action plans', represents the heart of the EMS. The programme describes what the organisation is actually going to do to mitigate its environmental effects. Action plans should not be over-elaborate, but it is essential that they are documented and that managerial responsibilities and time-scales are explicit.

For example, in the UK, the Environmental Protection Act 1990 places a legal Duty of Care on all waste producers to ensure that wastes are handled, stored and disposed of, following best practice and complying with all legislative requirements. Company 'X' (a medium-sized manufacturer of instrumentation equipment) may have identified in its Initial Review that it is a significant waste producer and therefore made a commitment in its Policy Statement to managing its waste-disposal activities in accordance with best practice. This broad policy commitment has then been translated into a number of specific objectives, one of which is to comply with the Duty of Care by 1 October 1993. In the case of waste management the action plan referring to waste disposal might then be documented as follows:

Waste Disposal Action Plan:

Objective Ensure full compliance with the Duty of Care by 1 October 1993.

Manager Materials Controller.

Target (1) Establish exactly what types of waste and how much of each are produced at each site.

Completion by 1 August 1993.

Target (2) Ensure all waste is properly stored and labelled on site and there are written procedures for cleaning up spills.

Completion by 25 August 1993.

Target (3) Ensure all materials are transferred only to properly authorised contractors, and disposed of at a properly authorised site, i.e., identify and record who the carriers are, what their registration number(s) is(are), who is the waste regulation authority, what is(are) the final disposal site(s).

Completion by 20 September 1993.

Target (4) Establish a system of records of all waste transfers for inspection by the Waste Regulatory Authority.

Completion by 20 September 1993.

Procedures and records

BS7750 requires that essential working procedures should be documented and a system established for maintaining key records. This extends to documenting instructions for the manner in which activities should be conducted, either by the organisation's own employees, or by contractors, and includes all monitoring and control processes.

Audits

The organisation must undertake audits to ensure that it is following the programme it has set out in the EMS and to measure performance in completing the action plans laid down.

An audit plan should be produced which identifies the areas to be audited, the audit frequency and who is responsible. This plan should also contain the protocols for conducting the audits, which may include questionnaires, checklists, interview protocols, and descriptions of observations and measurements to be made.

These operational audits must be distinguished clearly from 'certification audits', which are undertaken to assess systems compliance with BS7750 by accredited certification agencies.

Management review

At intervals probably not exceeding one year, managers responsible for the EMS should stand back and assess the overall performance of the system. If the circumstances of the organisation have changed significantly, for example in respect of its activities, products and services, the EMS may need to be modified to ensure that it continues to reflect the needs of the organisation. The management review must consider the recommendations arising from audit reports, and how to implement them if accepted. This review should also revise objectives and targets, and the action plans as appropriate.

There are also a number of additional elements which are specific to BS7750, but which may not necessarily be found in all management systems. These are:

Organisation and personnel: Appoint a management representative with overall responsibility for implementing the EMS. Define, assign and document new responsibilities where these are required, to manage, undertake and verify environmental management activities.

Register of regulations: Establish a register recording all of the organisation's legislative, regulatory and other policy responsibilities.

Register of effects: Formally document the environmental impacts of the organisation's activities, as identified in the Initial Review in a register.

Management manual: BS7750 requires the organisation to establish and maintain a manual to collate the environmental policy, targets and objectives, the management programme, and information about key personnel and their responsibilities.

Until the summer of 1993, BS7750 was the focus of an extensive pilot programme involving over 1000 organisations participating in 35 commercial or industrial-sector working groups. In addition, there were six cross-sector working groups covering topics such as training and communications. The experience gained during this period has now been incorporated into the latest version of the standard BS7750 (1994). Procedures for accreditation have however been delayed pending developments in the implementation of the EC Eco-Management and Audit Regulation, with the result that, at the time of writing, no organisation has yet formally registered under the Scheme.

THE EC ECO-MANAGEMENT AND AUDIT

The proposal for an EC Eco-Management and Audit regulation was finally agreed by 11 out of 12 member states in December 1992, having passed through several drafts and a period of consultation extending over several years. At the time of writing only Germany is withholding its consent, although this is on a technicality rather than any substantive matter of principle. The regulation is scheduled to come into force 21 months after full adoption, i.e., towards the end of 1994 at the earliest.

The development of the regulation to some extent mirrors the changes in ideas about environmental management which have taken place during recent years. The regulation was originally conceived as a procedure for auditing pollution control measures at larger manufacturing and process sites. However, as the perception of organisations' environmental management needs has grown to encompass a much wider range of activities, the regulation has broadened in scope to include the concept of the all-embracing EMS. This is reflected in the name of the regulation.

The Eco-Management and Audit regulation and BS7750 have therefore developed from different starting points. The Eco-Management and Audit regulation was originally intended to manage the environmental impacts of individual manufacturing sites, while BS7750 is a generic company-wide environmental management system. Nevertheless the two systems have grown together and the regulation now requires qualifying sites to have an EMS in place in addition to its other requirements.

The basic scheme relies on a system of site-registration and has the following requirements:

Audit	An initial baseline audit must be undertaken. This is similar to the BS7750 Initial Review, but is applied to an individual site rather than the entire organisation. In addition the regulation is more prescriptive in setting out a checklist of what must be covered in the audit. A cycle of re-auditing must be completed at intervals of not less than three years.
Environmental statement	A formal, publicly available statement must be drawn up for each site, based on the findings of the baseline audit. This must include a description of the site's activities and data on emissions and waste production etc.
Environmental management system	An environmental management system shall be established to promote continual environmental improvement at the site.

Thus, the Eco-Management and Audit regulation remains site-based, but now includes an integral requirement for a site-specific EMS. The regulation provides that companies which have been certified to recognised national or international standards for an EMS (e.g., BS7750) will be considered to have met the corresponding requirements of the regulation. It is expected that the European Commission will request the European Standards Committee (CEN) to develop a European-wide EMS and this is likely to draw strongly on the experience of BS7750.

Planning and resourcing an environmental programme

GETTING STARTED

Any organisation considering an environmental management initiative must first be clear about its overall objectives and, most importantly, decide whether or not there is adequate senior management commitment to the process. As in any other sphere of management activity, if the aims are confused, or if commitment is lacking, the exercise will fail to deliver the potential benefits.

Some organisations may be content to restrict environmental action to the minimum necessary to keep on the right side of the law. This approach may be justifiable in business terms, providing it is based on a rational appraisal of the organisations' best interests. Some companies, for example those with very tangible pollution control problems, may have clear legal responsibilities which largely dictate key actions. Most organisations, however, lie somewhere in between, and find it difficult to gauge the environmental risks and opportunities to which they are exposed, or the appropriate level of response.

Having established senior management commitment, the first step that any organisation should take is therefore to stand back and think through its environmental position. This means undertaking what, in the terms of an EMS, is the initial or preparatory environmental review. A team of staff, with members who together combine a broad knowledge of the organisation and ideally some environmental expertise, can undertake or co-ordinate the review, involving in the process as many other employees as possible. The review should examine all the actual or potential environmental issues surrounding the business. It should look beyond regulatory compliance and include past problems as well as immediate and longer-term environmental effects.

The results of the initial review will clarify the type and importance of the environmental issues confronting the organisation. These issues can then be weighed against other business imperatives, and a more rational decision reached about the appropriate level of response. Many organisations may not feel ready to commit immediately to seeking full accreditation under BS7750 or similar scheme. Nevertheless, it is recommended that the key elements of an EMS (i.e., identify key issues; set objectives and targets; implement action plans; audit progress) should be established in any event, as the best means of making measurable progress.

GENERAL PRINCIPLES

Having taken the decision to put in place the key elements of an EMS. an organisation should consider a number of general principles for implementation:

Management objectives

A primary consideration is that the EMS must be of positive benefit in running the business. If staff at all levels do not find the EMS useful there will be little point in its introduction, and the system will wither.

Management control

There is a danger that implementation may make progressively greater and greater demands on staff time and other resources as the process gathers momentum. This can be avoided through exercising tight management control, working to a defined timetable, and holding frequent reviews of progress.

Bureaucracy

Any system that imposes an onerous burden of paperwork, or unnecessarily restricts the operational freedom of staff, will not gain acceptance. The EMS should therefore be as simple and non-bureaucratic as possible.

Staff resources

The use of staff resources needs to be carefully planned and the need for employing external consultants considered. The greater the involvement of internal staff, the more ownership of the system there will be. This must be balanced against the amount of staff time that can be reasonably devoted to the project. The pros and cons of internal versus external involvement may be summarised:

External
- less ownership
- extra resources
- independent view
- additional environmental expertise
- costs more money

Internal
- better ownership, raise awareness
- strain on resources
- know the organisation
- some environmental expertise
- costs less money

Implementation

The best way of proceeding, is to take each of the key elements of an EMS in turn. Having completed the initial environmental review these are:

- define policy statement;
- set objectives and targets;
- devise and implement action plans;
- establish audit procedures and timetable;
- undertake a periodic system review.

Future developments

Environmental management has gone through a period of rapid evolution and change during the past ten years. A variety of different and competing approaches have been proposed during this time and not surprisingly this has caused some confusion among organisations considering environmental initiatives. Some consensus is now emerging, however, and it is likely that the concept of the generic EMS, designed to support a company-wide approach to continuous environmental improvement, will become firmly established as representing best practice during the remainder of the present decade. Already this approach is becoming codified in national and international standards such as BS7750 and the EC Eco-Management and Audit regulation, and these developments will help organisations frame new initiatives tailored to their own specific needs. It is also possible to identify a number of complementary developments which are likely to gain ground in the immediate future.

ENVIRONMENTAL REPORTING

Calls are increasingly being made for organisations to disclose publicly information about their environmental performance. The Eco-Management and Audit regulation recognises this, in requiring the publication of an environmental statement by qualifying facilities.

Moreover, many large organisations which have developed sound environmental programmes, now see it to be in their enlightened self-interest to produce an annual environmental report to accompany the traditional financial report. A number of good examples of corporate environmental reports produced by leading companies are now available (Macve and Carey, 1992). The best environmental reports provide clear information about the status of an organisation's environmental programme following a similar format to an EMS, i.e., including:

- a summary of the major environmental issues affecting the organisation;
- a statement of environmental policy;
- designation of responsible directors;
- statement of the environmental objectives and targets for the past year;
- performance in meeting these targets, recording failures as well as successes;
- statement of new targets for the coming year.

The accounting profession in the USA has long taken an interest in aspects of environmental reporting, particularly in relation to potential financial liabilities (American Institute of Certified Public Accountants, 1977). This interest has now been taken up in Europe and, in the UK, the Institute of Chartered Accountants and a number of other organisations have been working to develop principles of good practice in the integration of financial and environmental reporting (Macve and Carey, 1992; the Hundred Group of Finance Directors, 1992).

Integration with other ethical issues

The environment is just one of a range of ethical issues which are of concern to organisations – others include health and safety, animal testing, worker's rights, child labour, Third World exploitation, racial discrimination and community concerns such as inner city regeneration. It is not always possible to draw a clear distinction between environmental management and the management of some of these other issues. For example, some companies which place environmental requirements on suppliers of products and components have found it inappropriate to enforce environmental standards on Third World suppliers, when more pressing issues of worker health and safety, and the use of child labour, must take priority.

Moreover many stakeholders simply do not care to distinguish between all these concerns, and choose to judge the performance of the organisation on its total approach to ethical issues.

A trend is therefore developing, which is likely to see the integration of management action to address environmental issues with action on other ethical issues.

External advisers

There has been a very strong growth in the number of organisations and individuals offering their services as environmental consultants during recent years. Consultants are drawn from a number of existing fields and professions – engineers, biologists, geologists, chemists, physicists, economists and management consultants, can all be found offering environmental advice. At the present time, there are, however, no recognised professional bodies specifically representing environmental consultants. A number of trade associations have appeared in recent years and have promoted codes of practice and registration procedures for qualified environmental practitioners and auditors (AEC, 1991; EARA, 1992). In addition, as environmental management systems become codified in official standards and legislation, it is anticipated that regulatory authorities will play a greater role in approving competent practitioners. At the present time, however, the best advice that can be offered to organisations seeking help is only to use advisers with a proven track record.

References

American Institute of Certified Public Accountants, *The measurement of corporate social performance, AICPA* (1977).

Association of Environmental Consultancies, *Qualified environmental auditors: code of practice and registration procedures*, AEC (1991).

British Standards Institution, BS7750 *Specification for environmental management systems*, BSI (1994).

Elkington, J. and Dimmock, A., *The corporate environmentalists: selling sustainable development, but can they deliver?* A report on the Greenworld Survey, SustainAbility (1991).

Environmental Auditors Registration Association, *Guidelines for environmental auditing: qualification requirements for environmental auditors*, EARA, (1992).

European Commission, *Proposal for a Council Regulation establishing a Community scheme for the evaluation and improvement of environmental performance in certain activities, and the provision of relevant information to the public (Eco-management and audit),* European Commission (1992).

Grayson, L. *Environmental auditing: a guide to best practice in the UK and Europe*, The British Library (1992).

Hundred Group of Finance Directors, *Statement of good practice: environmental reporting in annual reports*, HGFD (1992).

Laming, R. Surveys of enterprises' attitudes to the environment, *UK CEED Bulletin* (1990) 30, pp. 12–13.

Macve, R. and Carey, A. (eds.) *Business, accountancy and the environment: a policy and research agenda*, The Institute of Chartered Accountants in England and Wales (1992).

The Corporate Environmental Register, *Corporate environmental register*, The Environment Press (1993).

Touche Ross Management Consultants, *The DRT international 1991 survey of manager's attitudes to the environment*, Touche Ross Management Consultants (1991).

United States Environmental Protection Agency, *Annotated bibliography on environmental auditing,* USEPA, Office of Policy, Planning and Evaluation. (1988).

Life-cycle assessment

TIMOTHY J. MOHIN

United States Senate, Committee on Environment and Public Works, Washington DC, USA

- Life-cycle Assessment (LCA) is the evaluation of a particular material or activity from raw materials through to final disposal – the 'cradle-to-grave' approach.
- The LCA concept addresses environmental problems and degradation at source, trying to prevent problems in the first place rather than creating problems for remediation at some future time.
- Through a better understanding of the fundamental systems that impact on the environment, we are better able to design new systems and new ways of doing business which are less harmful to the environment.
- LCA involves accounting for all inputs and outputs to a defined system, as well as understanding the system itself – this approach can become unwieldy if the 'system boundary' is defined too broadly, and all LCA studies contain some assumptions to maintain manageability.
- The constituent phases of LCA are generally defined as:
  - determine goals and scope (including 'system boundaries');
  - inventory analysis of all inputs and outputs;
  - impact assessment to quantify the environmental risks associated with pollution and resource-consumption identified in the LCA inventory;
  - improvement analysis to improve and optimize;
  - environmental performance along with other business elements.
- Many companies are finding that LCA approaches to planning are useful in business decisions, and as a conceptual framework for corporate policy – reinforcing the message that environmental concerns, both before and after a product leaves a plant, as well as within a plant, must become a part of everyday planning.

Introduction

The concept of life-cycle assessment (LCA) is a dynamic, innovative way of looking at environmental improvement. Life-cycle assessment is the evaluation of a particular material or activity from the generation of the raw materials to final disposal. Often referred to as 'cradle-to-grave' or 'earth-to-earth', life-cycle assessment has recently increased in visibility and importance due to its application to the area of environmental protection. The concept behind environmental life-cycle assessment is to address the genesis of environmental degradation rather than the standard approach of trying to remediate environmental problems after they have been created.

Background

For at least twenty years now the environmental movement has been gaining momentum. One of the most important lessons we can learn from its history is that even with concerted efforts directed towards abatement and mitigation of environmental problems, there remain serious environmental threats to health and habitat. Why is this? In the United States there is an impressive array of strong environmental laws. Many of these laws have been revised to keep up with changing times. For example, tail-pipe emissions from automobiles have been reduced drastically in the past twenty years, but smog continues to plague all major US cities. Clean water and hazardous waste legislation has been on the books for years but these problems are more of an issue today than ever before.

Life-cycle assessment

The emerging science of life-cycle assessment is a critical pillar of the movement toward sustainability. Environmental life-cycle assessment is an effort to better understand the *systems* that contribute to environmental degradation. This approach is fundamentally different from the 'command and control' approaches of the past that focused on the end result of environmental deterioration, not the root causes.

The promise of life-cycle assessment is that through a better understanding of the fundamental systems that impact the environment we are better able to design new systems and new ways of doing business which are less harmful to the environment. Not only can information from life-cycle assessment aid in improving the environment; it is likely that some of the same design improvements will also save money, decrease liability and increase efficiency. Pollution, after all, is simply lost product, i.e., waste.

The approach of focusing on environmental issues holistically rather than solely from an abatement standpoint has great promise. For example, a large consumer-products firm recently wanted to improve the environmental profile of a laundry detergent. They could have decreased the packaging, used less fillers, or made any of a number of other obvious improvements. Instead, the company looked at the life cycle of the product and determined that the greatest use of energy was to heat the water used to wash the clothing. By reformulating the detergent to a cold-temperature product, the company actually reduced more energy-use than was possible with other improvements.

The philosophy of life-cycle assessment is to account for all of the inputs to, and outputs from, a system as well as to understand the system itself. This approach can obviously become unwieldy if the 'system boundary' is defined too broadly. All life-cycle studies contain some assumptions to maintain manageability. For example, when assessing the environmental impacts of the energy used in a system, it is sometimes impossible to determine the source of generation – nuclear plant, coal-fired plant, solar power, etc. The source of the energy may have an important effect on the overall environmental profile. The environmental effects associated with solar power are fundamentally different than those for coal-fired power plants. Most often, assumptions are made about the average contribution of any particular generation system to the overall energy supply.

The science, practice and appropriate uses of life-cycle assessment are still emerging. Much of the pioneering work to define this science has been documented by the Society for Environmental Toxicology and Chemistry (SETAC). SETAC has defined the process of life-cycle assessment as consisting of several distinct but mutually dependent phases. Fig. 18.1 represents the constituent phases of life-cycle assessment as defined by SETAC. The following sections describe each of these phases in general terms.

Step 1: Determine the goals and scope

The first step in a life-cycle assessment is to define the goals of the study. This goal-setting exercise is used to determine the appropriate scope of the study to be undertaken. For example, less rigour would be required for a study directed at comparing generic packaging systems (such as paper versus plastic) than might be required for a comparison of individual products. Specifically, in the packaging study the goal is an environmental comparison of generic packaging materials. Such a study would rely upon generic data such as 'average' production processes, uses and disposal rates for the different kinds of packaging.

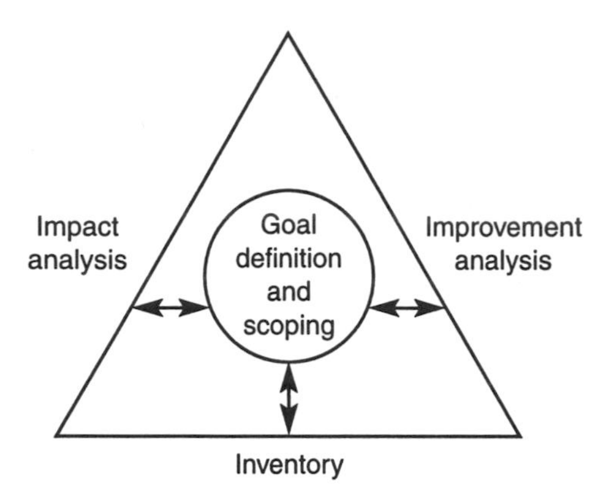

Figure 18.1 Incorporation of goal definition and scoping into the LCA technical framework

The goals of the product study, however, are to compare individual products. This study would likely require a detailed accounting of every step along the life cycle of the specific products under study. For example, this study would require data from the actual production processes of subject products rather than a generic or averaged data.

In assessing the goals of a life-cycle assessment, it is important to consider the target audience for the information. The use of life-cycle assessment for public comparison of products, or for making any environmental claims about products is still controversial. There are those who feel that the current methods for life-cycle assessment are too complicated and too easily misinterpreted so be used to support environmental claims. This criticism has gained credibility from the conflicting results of recent studies of the life-cycle environmental impacts of products such as diapers and paper cups.

In these well publicized cases, life-cycle studies of identical products resulted in very different results. To compound this negative perception of life-cycle assessment, the results of these studies invariably have favoured their sponsors.

These cases do not necessarily indict the life-cycle assessment method. More likely, they represent inappropriate or unsupportable life-cycle claims. There is a compulsion to over-interpret life-cycle results because of the wealth of data that has been collected and analysed. The essence of the goals analysis is to develop appropriate goals for the study and match the study scope and assumptions to its appropriate uses. More about the appropriate uses of life-cycle results is presented below.

The most important outcome of the goals analysis is the determination of the scope of the life-cycle study. The 'scope' of the study is defined as the system boundary. The boundary defines the breadth of the study.

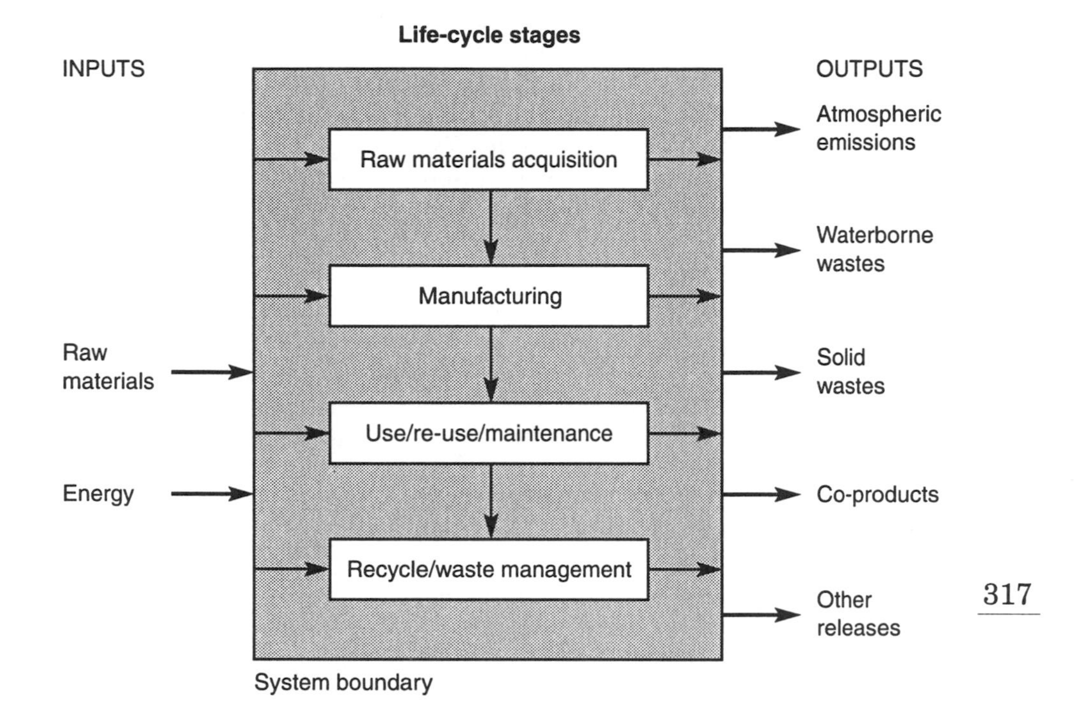

Figure 18.2 Defining system boundaries

For example, if a life-cycle study of a paper product is undertaken, the analyst must choose whether to begin the study with the trees that were used to make the paper, or go further back to include the environmental effects of the chainsaw that was uses to cut the trees, or even the truck that brought the men that cut the trees, and so on. The system boundary determines what is included in the study and what is not.

More detailed guidance on the definition of goals and the determination of the scope of life-cycle studies can be found in the SETAC's *A Conceptual Framework for Life-cycle Impact Assessment.* Fig. 18.2 illustrates the definition of the system boundary in relation to the surrounding environment.

Step 2: Inventory analysis

Once the goals and scope have been decided, the next step is to perform a detailed survey of all of the inputs (e.g., raw materials and energy) and all of the outputs (e.g., air pollution, waste) for each stage in the life cycle. Known as the 'inventory', this stage in the process is simply a tally of consumption of raw materials and energy into the system, and the generation of pollution from the system.

As part of the inventory analysis, distinct steps of life cycle are defined. Fig. 18.2 also shows the generic life-cycle stages included in an inventory analysis. Each of these life-cycle stages analysed independently to determine the total consumption of raw materials and energy, as well as the total loading from the system to the surrounding environment.

Most life-cycle studies performed to date have stopped at the inventory stage. One of the major drawbacks of these studies is that no information is provided to distinguish whether some items in the inventory constitute greater or less environmental hazard than others. For example, data collected on amounts of hazardous wastes generated or CFCs released are treated identically to data on the acreage of wetlands or rainforests destroyed. Without information on the 'impacts' of the inputs and outputs from a life cycle, the user of the study has limited information to make 'improvement' decisions. What is typically done in these cases has been called the 'less-is-best' approach. This means that the improvement option that results in the least pollution or consumption is considered the best.

In many cases, such as for decisions about energy consumption the less-is-best approach may be all that is needed. For more complex decisions, however, it is somewhat more problematic. For example, we may be far more concerned about small releases of dioxin than we might be about a fairly large release of carbon dioxide.

Most often, life-cycle investigations will reveal a complex pattern of air, water and solid waste loadings as well as the consumption of different types of natural resources. An objective system is needed to determine the relevance of the environmental impacts associated with these factors.

Step 3: Impact assessment

Life-cycle 'impact assessment' addresses this need. Similar to risk assessment, impact assessment is an effort to quantify the environmental risks associated with the pollution and resource consumption identified in the life-cycle inventory. Unlike risk assessment, however, impact assessment must also compare the relative importance of disparate risks. For example, impact assessment must be able to tell the user of LCA information if the ounce of dioxin emitted in the life cycle is more or less of importance than the consumption of 800 cubic metres of landfill space.

Clearly these are difficult judgements. There is no truly objective basis for making these trade offs, but they must be made. Albert Camus said 'not to decide is to decide'. In the practice of life-cycle assessment this is very much the case. If explicit judgements are not made within the analysis of impacts, then these judgements will be made subjectively by the uses of the study. Without impact assessment, action or inaction on the part of the consumer of life-cycle information will be based on his or her own value judgements.

CATEGORIZATION

In a recent publications, SETAC has defined the generic steps in a life-cycle impact assessment (see Fig. 18.3). The first step is to categorize all of the data collected in the inventory analysis by its anticipated effect on the environment. This is not a trivial task because many of the inputs and outputs quantified in the inventory may be associated with multiple effects. For example, the emission of sulphur oxide to the air has been associated with acid rain. Acid rain contributes to many ecological problems such as acidified lakes, leaching of metals from the soil and the loss of biodiversity in lakes and other effects.

To categorize the data from the inventory analysis, the concept of stressors has been introduced. Stressors are conditions that may lead to human health or ecological effects. These effects are called impact categories. A single stressor (in the example above, sulphur oxide emissions) could be linked to several impact categories (acid rain, acid lakes, metal leaching etc.). All of the data gathered from the inventory analysis is apportioned to an impact category (for more information see SETAC's *A Conceptual Framework for Life-cycle Impact Assessment*).

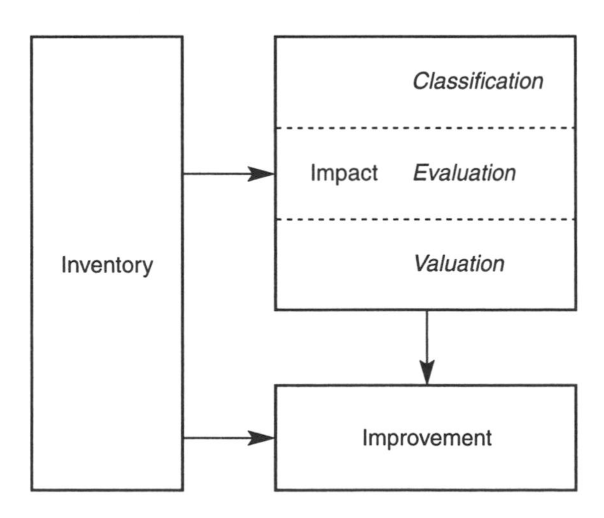

Figure 18.3 Impact analysis phases

EVALUATION

The next step is to evaluate the relative magnitude of the potential impact. This evaluation step takes into consideration both the magnitude and the potency of the loading or consumption. For example, if stressors to human health from benzene were being examined and the associated release was benzene, evaluation of the impact would require information about how much benzene is released to, and distributed through, the environment, and the levels of benzene that can cause human health effects.

The missing piece of information in this example is how much any given human is exposed to benzene. Exposure is extremely difficult to calculate in the context of life-cycle assessment because of the multitude of localities and exposure règimes that occur throughout a life cycle. But without exposure, actual impacts cannot be estimated.

An extremely important point about life-cycle impact assessment is that, because of the lack of exposure information in most life-cycle assessment studies, actual impacts are not estimated. What is estimated is the relative 'contribution' of any particular stressor to an impact category. As explained above, this contribution is generally estimated as the product of the magnitude and potency. In this way, the data collected in the inventory analysis can be grouped, evaluated and characterized by its relative contribution to an impact category.

VALUATION

The last step in life-cycle impact assessment is valuation. This is probably the most difficult and controversial aspect of life-cycle assessment. The valuation step is a subjective analysis of the importance of the various impact categories. In other words, valuation will provide the study-user with a way to distinguish between the various impact categories analysed in an impact assessment. For example, this method might compare the relative importance of acid rain against a human health impact such as cancer.

These are extremely difficult judgements to make. There are no 'right' or 'wrong' answers in estimating value. Each individual will bring his or her own value system to the analysis. For these reasons, decision models are used for this phase of the study. Models like the Analytical Hierarchy Model can encode the responses of experts to questions of value between pairs of attributes. For example, an expert may be asked to split up 10 points between two impact categories such as acid rain and human cancer. Dependent on the response of a panel of experts, this system will calculate the collective judgement of the panel on the relative importance of various impact categories.

This is method is subjective and may not be acceptable to all practitioners and users of life-cycle assessment. The advantage of using it is that the judgements of experts in the field are considered in an objective manner before any decisions are made on how to make improvements in the system.

Step 4: Improvement analysis

The final actions that are taken as a result of a life-cycle assessment fall in the domain of 'improvement' analysis. This final stage of a life-cycle assessment is probably the least studied. How the results of LCA studies should be interpreted and used is now largely dependent on the ultimate user of the study. In many cases the decision on how to improve the environmental attributes may be obvious. But to optimize these decisions, other information may need be considered, such as cost, cost-effectiveness, liability and consumer response. Improvement decisions would necessarily need to include information specific to the entity. Currently, there is no generic guidance on how to approach this problem within the context of life-cycle assessment.

Uses of life-cycle assessment

There has been a lot of interest in life-cycle assessment as a tool to support eco-labelling programs for consumer products. While this is clearly an important application for the future of LCA, this method may be too young to support these kinds of claims. Until there is a widely accepted, consistently applied and scientifically reproducible method, these kinds of claims may be unsubstantiated. This is not to say that LCA should be abandoned as a labelling tool. Rather, the message is that more research and development is needed before such uses will be credible.

There are many uses for life-cycle assessment other than eco-labelling. Many companies are finding that life-cycle planning is a good way to approach almost all business decisions. It is a conceptual framework for corporate policy. Everything from product development to plant maintainence can be viewed from a life-cycle perspective. Of course it is not practical to do a life-cycle study for every action we take. Rather, the concept is that environmental concerns, both before and after a product leaves the plant, must become part of everyday planning.

The only way to truly account for environmental concerns is by accounting for all of the known impacts throughout the life cycle. Short of this goal, there are many environmental factors that are 'upstream' or 'downstream' of the manufacturing process that could be considered in product design and development without a full LCA.

Fox example, the development of low-energy computers and other energy-saving appliances required that the manufacturers consider impacts after the product leaves the plant; the distribution system and marketing system and is in the hands of the consumer. In the case of low-energy computers, this change was not motivated by a desire to market a 'green' computer but rather to extend battery life for notebook computers. Applying this innovation to desktop computer equipment, however, could save enough energy each year to power the States of Maine, New Hampshire and Vermont. This example illustrates the kind of innovation that is possible when the product life cycle is considered. In this case, the original motivation was not environmental protection, but the end result could have a dramatic effect on energy consumption.

This kind of planning does not have to be solely motivated by good will and altruism (although those are certainly good reasons for taking this approach). Waste is simply raw material that never made it into a final product – overhead. When companies begin to analyse the life-cycle implications of their own practices, they often find that the improvements reduce waste, save money and make them more efficient and thus more competitive. Not only can life-cycle assessment help with corporate planning, it can be a useful tool for assessing the effectiveness of environmental regulations. In the United States, environmental regulations have often been based on information about a single pollutant or single media and usually the focus is on a small part, or 'snapshot,' of the life cycle. Clearly, if other aspects of the life cycle were considered, the ultimate regulatory decision could be altered.

For example, in regulatory actions that may cause industry to switch from halogenated solvents to water-based solvents the decision might be based mainly on the lower toxicity of the water-based version to humans exposed in or around the production process. If the full life-cycle of both alternatives were considered, there would be myriad effects other than human health throughout the life cycle.

A real world case involves terpenes, common water-based solvents. Terpenes have been considered as replacements for halogenated solvents because of their lower toxicity to humans. Terpenes, however, have been associated with ecological effects that have not been observed with the more common halogenated solvents. This is not to say one is better than the other, rather, that only with a full accounting of all the possible effects can a well informed regulatory decision be made. The US Environmental Protection Agency is currently investigating the use of life-cycle assessment tools is regulatory analysis and decision-making.

Conclusion

One of the main criticisms of life-cycle assessment has been the uncertainty and variability of some of the results. This criticism is based in the heavy expectations that have been ascribed to life-cycle assessment. There is an expectation that life-cycle assessment will be able to report back what is 'green' and what is not. Life-cycle assessment is a tool. It will not predict the ultimate 'truth' about what is better or worse for the environment, nor will it provide us with easy answers about the correct course of action.

This tool, however, has tremendous potential to improve our environmental decision-making abilities At all levels. It is important to guard against unreasonably high expectations and to allow the science to grow and emerge before it is judged.

The other major criticism of life-cycle assessment is its complexity. Recent attempts at life-cycle assessments have been very detailed and costly. By definition, the characterization of any life cycle will rely on a lot of data. At issue is: can the process be simplified, modified or streamlined to make it more accessible? It's probably too early to tell. But, as with the development of any new science, the early efforts will be refined by later efforts, and the process will become easier and more efficient. For example, it used to take a roomful of computers to perform the tasks of the computer on which this chapter was written.

Soon, life-cycle assessment may be more of a necessity than a luxury. The global market for environmental goods and services is growing dramatically. The Council on Environmental Quality estimates this market to be between $200 to $300 billion, and growing by as much as 10 per cent annually. With tougher environmental standards becoming more widespread, and environmental awareness increasing, this market will continue to expand and become more competitive. The company that is able to adapt to this change in the market will be profitable in the future. Now is the time to begin to analyse and improve the environmental profile of our activities, and life-cycle assessment is the tool for the job.

References

Fava, J. A. *et al,* (eds) *A Technical Framework for Life Cycle Assessment,* Society of Environmental Toxicology and Chemistry (SETAC) and SETAC Foundation for Environmental Education, Washington, DC 1991.

Fava, J. A. *et al,* (eds) *A Conceptual Framework for Life Cycle Impact Assessment,* Society of Environment Toxicology and Chemistry (SETAC) and SETAC Foundation for Environment Education, Pensacola, Florida (1993).

Boustead, L. 1992. Eco-Balance Methodology for Commodity Thermoplastics, A Report for the European Centre for Plastics in the Environment (PWMI), Brussels, Belgium.

Fecker, L. 1992. How to Calculate an Ecological Balance, Report No. 222, Eidgenössigche Materialprufungs- und Forschungranstalt (EMPA), St Gallen, Switzerland.

Franklin Associates, Ltd. unpublished 1993, 'Statistical Considerations of Data Error and Variability', Prairie Village, KS. (Obtainable from Franklin Associates).

SETAC. 1991. A Technical Framework for Life Cycle Assessment Workshop Report from the Smugglers Notch, Vermont, USA, Workshop, held in August 18–23, 1990.

SETAC. 1992. Workshop Report from the SETAC-Europe Workshop, held in Leiden, Netherlands, December 1991.

SETAC. 1993. Conceptual Framework for Impact Assessment Workshop Report from the Sandestin, Florida, USA, Workshop, held in February 1992.

SETAC. In press. A Practical Framework for LCA Data Quality Workshop Report from the Wintergreen, Virginia, USA, Workshop, held in October 4-9, 1992.

Product Life-Cycle Assessments: Principles and Methodolody. NOKD 1992:9. Nordic Council of Ministers, Copenhagen 1992.

Guinee J. B. *et al.* 1993. Quantitive Life-Cycle Assessment of Products J: Goal definition and inventory, J Cleaner Prod

Hunt R. G. *et al.* 1992. Resource and environmental profile analysis: A Life-Cycle Environmental Assessment for Products and Procedures. Environ Impact Assess Rev. 12, 245–269.

Keoleian G. A. and D. Menerey 1993. Life-Cycle Design Guidance Manual EPA600/R92/226. US Environmental Protection Agency, Cincinnati.

Rubik F. and T. Baumgartener 1992. Evaluation of Eco-balances. EUR-14737. CEC DG XII, Brussels.

Vigon B. W. e*t al.* 1993. Life-Cycle Assessment: Inventor Guidelines and Principles. EPA/600/R-92/245. US Environmental Protection Agency, Cincinnati.

19

Agreeing targets for environmental improvement

DAVID BALLARD

previously Quality and Environment Adviser, Thorn EMI Rental (UK) Ltd, now with Aspinwall Environmental Strategy Ltd, Shrewsbury, UK

- Human issues are at least as important as the technical aspects of agreeing improvement targets.
- The negotiation of targets is a superb opportunity to build commitment to environmental improvement among a critical portion of an organisation's members at all levels, including senior managers, and to empower them to demonstrate it.
- Negotiate objectives and targets wisely, so as to reinforce environmental commitment – in the early stages of work in an area it is normally better to agree process targets (e.g., identify ways to ..) than performance targets (e.g., reduce use of .. by 35%).
- Alter underlying processes where possible, for example, by amending standard work procedures and reinforcing this with training.
- Report progress regularly to demonstrate that the environmental programme is subject to normal management disciplines.
- A good Quality programme is the best environmental starting point for some companies.
- The Procurement function can be very important in making company-wide improvements – at Thorn EMI Plc the Technical Services Manager of the Stores Operation had first to approve suitable substitutes for stores items which were to be phased out on environmental grounds. Centralised procurement and standards can ensure that environmentally-friendly selected substitutes are rapidly brought into use.

Introduction

The objective of this chapter is to examine some of the issues that arise when negotiating improvement targets as part of an environmental programme within a large organisation. British Standard 7750 for Environmental Management Systems is relevant to any organisation's environmental improvement programme. The very simplicity and clarity of its structure (see Fig. 19.1) can, however, potentially give an unbalanced picture of the challenges faced at each step. In particular, great care must be taken when setting and agreeing improvement objectives and targets. Most of the important work in an environmental programme comes in the gap between quantifying the effect (Step 6) and defining the target (Step 7). If this work has been well done, all should then be plain sailing. If it is badly done, environmental improvement will only happen by luck, if at all.

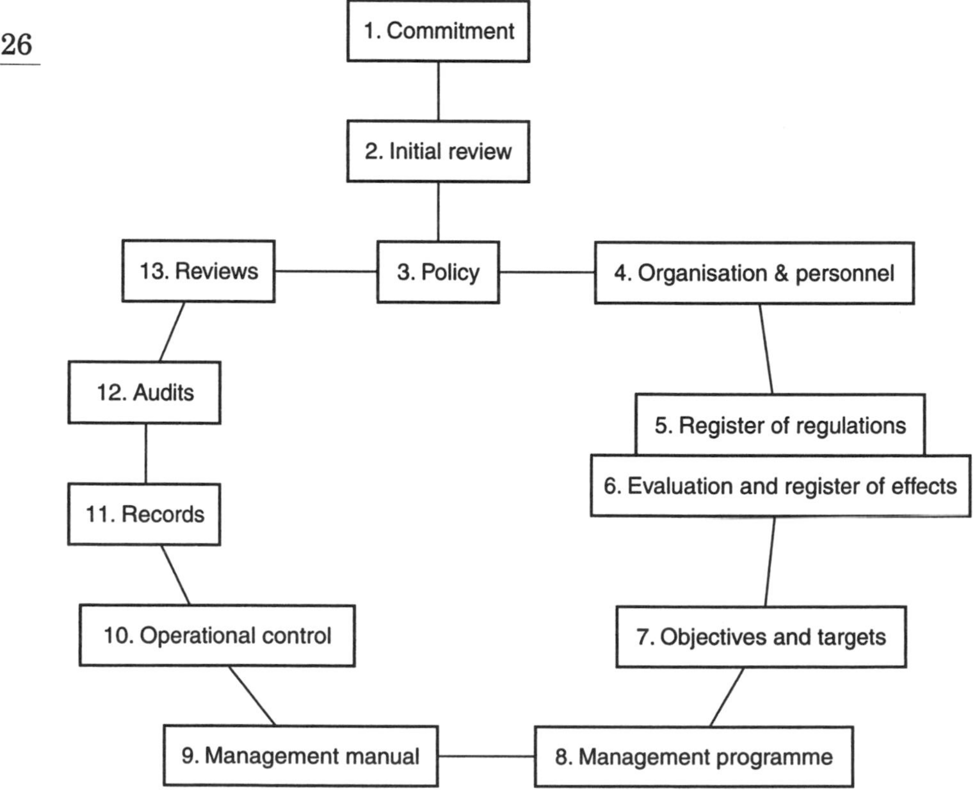

Figure 19.1 The BS7750 process

In practice, we need a good understanding of our potential freedom of movement, both in the short and the long run, before we can set a target that is meaningful. Gaining this understanding is not explicitly mentioned in the Standard, doubtless being taken for granted by its drafters. Surprisingly often, however, we hear managers going on to set reduction targets immediately after they have quantified an environmental effect (and sometimes even earlier!), without ever identifying what is feasible.

Human issues are at least as important as the technical aspects of agreeing improvement targets. Fig. 19.1 shows Commitment as the first step of the process. Indeed, it is placed outside the loop of continuous improvement, almost as if a single dose of commitment from the top is enough to last the organisation indefinitely. This is the 'Road to Damascus' approach to organisational change: all that is required after the blinding light is diligently to apply the vision to all aspects of the organisation's operations. I don't believe that this is the way that many organisations, or indeed many human beings, change. What is usually required is the long, but not unrewarding, struggle to discover and reinforce commitment to environmental improvement among a critical proportion of the organisation's members at all levels, including senior managers, and then to empower them to demonstrate it. The negotiation of targets is a superb opportunity to build commitment but, in consequence, it is also a process which runs the risk of seriously undermining management support if it is unskilfully done.

Setting environmental targets irresponsibly is therefore damaging in several ways. First, of course, there is the danger of setting a target that is undemanding. While targets need not be exclusively motivated by concern for the environment, employees will quickly become disenchanted if the environment is used to justify things that would quite obviously have happened anyway. Then there is the danger of setting an unrealistic target. This leads to failure, which is likely to undermine whatever support already exists. Third, a target may be achievable only by seriously damaging other objectives of the organisation, possibly including environmental ones. This might indicate a lack of consensus, which would be likely to bring the environmental programme into disrepute whether the targets are reached or not. A final consideration is that almost nobody likes to have targets imposed upon them, even if they are achievable. It is therefore very important to negotiate objectives and targets wisely, so as to reinforce environmental commitment rather than to threaten it.

Thorn EMI Rental (UK) Ltd's environmental programme

Thorn EMI (UK) Rental Ltd is the largest domestic consumer-durables rental operation in the world. It sells and rents brown goods (television, satellite, audio, etc.) and white goods (laundry products, dishwashers,

Thorn EMI Rental (UK) Ltd

Environment Policy (extracts)

'Environmental protection is one of the company's highest business responsibilities. We aim for continuous improvement of environmental performance and to comply with all relevant laws. We shall achieve this by working on the root causes of serious problems, rather than by just trying to reduce their effects.'

Implementation principles:

- **Awareness of environmental problems**
- **Clear targets for improvement**
- **Obeying environmental law**
- **Being organised**
- **Involving our staff and suppliers**

Figure 19.2 Company environmental policy

refrigerators, etc.) and sells small appliances throughout Great Britain and Northern Ireland. Its operations include field service, distribution, product refurbishment and shop operations, primarily through its Radio Rentals and Rumbelows brands. It has over 10,000 employees and operates from well in excess of 1,000 premises.

The early 1990s were a period of significant change for the company. The appointment of a new Managing Director, Ron Campbell (now Deputy Chief Executive, Thorn EMI Rental Ltd), in early 1991 was followed by major changes at senior management level, followed in turn by significant restructuring of Field Operations and the recentralisation of many administrative functions into 'Field Support Teams' charged with providing quality support for front-line operations. Later, Thorn EMI's retail company, Rumbelows Ltd, was integrated into Thorn EMI Rental (UK) Ltd. Following a career of six years in a variety of roles within the company, I was appointed as the company's first Quality and Environment Adviser in May 1991, right in the middle of all these changes.

The company's environmental activity had alreadv begun. An initial review of environmental issues was carried out in 1990 in response to a request from the parent company, Thorn EMI Plc. Following this, I helped draft an initial environmental policy in November 1990. This formed a basis for a revised policy that was formally agreed by the new management team in April 1992. The new policy also drew heavily on the parent company's policy, which in turn reflected Thorn EMI Plc's adoption of the ICC Charter for Sustainable Development. The new Managing Director took a personal interest in the drafting of the policy, intervening to strengthen the wording of one of its key sections. My early drafts had prioritised improvement activities towards areas where environmental improvements were consistent with the company's commercial interest. This was mainly because I was unsure of the level of support I could expect from senior management. I was bluntly told to stop pussyfooting around and to ask for what I wanted: that we should prioritise towards serious environmental problems and their underlying causes.

Interestingly, in contradiction to the view of many managers that environmental improvements cost money, the environmental improvements that have come out of the programme have in fact been very much in the company's financial interests, although this has never been a priority of the programme.

The new policy was drafted as the new BS7750 progressed and I consciously modelled it upon the requirements of the new standard. The company programme is therefore consistent with the BS7750 structure. With all the changes that were going on I did not see it as appropriate to persuade senior management that we should seek registration. The programme has been developed in such a way, however, as to leave open the option of seeking registration at some stage in the future if desired. Indeed, in my opinion, it is difficult to find any aspect of the BS7750 structure that would not be included in a meaningful environmental programme.

Environmental issues relevant to Thorn EMI Rental (UK) Ltd

Clearly, the environmental issues facing a retail service company such as Thorn EMI Rental (UK) Ltd are not in the same league as those facing a company in the petrochemical industry, or even a waste management company. Indeed, at the time of the initial review in 1990, we were not aware of any environmental legislation that affected us significantly, although it was already obvious that indirect environmental effects – those which the company can affect but does not manage directly – are fairly significant. As our programme continued we built up a much better understanding of our environmental effects.

We also identified that there is a good deal of environmental legislation that potentially affects the company. This legislation includes the 'Duty of Care for Waste Management' (1992), the Special Waste Regulations under the Control of Pollution Act, 1974, and the Air Pollution Regulations controlling paint-spray activities introduced in September 1992 under the Environmental Protection Act. Also on the horizon from the EC are a series of laws which are likely to affect our treatment of certain wastes.

Thorn EMI Rental Ltd

Some environmental effects

(a) Direct effects

Energy use	Energy used in premises, in-bound logistics and in the vehicle fleet. Total energy used in 1991/2 was around 200 million KwH.
Ozone depleters	About 6 tonnes of CFCs were used in 1990 in product refurbishment, along with about 350kg 1,1,1 Trichloroethane.
Wastes requiring care at disposal	Refurbishment depots generate around 900k batteries for disposal each year. End-of-life televisions contain small quantities of cadmium sulphide. Wastes from paint-spray operations contain solvents. There are a variety of other wastes that require careful handling.
Packaging wastes	Product packaging is largely EPS (Expanded Polystyrene). In total the company disposes of around 240 tonnes of EPS per annum.
Paper	Including press advertising, our quantified use of paper exceeds 3,300 tonnes per annum. Several areas of use have not yet been quantified. Something over 50% of paper purchases use chlorine free paper.

(b) Indirect effects

Products	Environmental impact of products is significantly greater than the direct effects above. For instance, the energy consumption of major product purchases results in emissions of around 500k tonnes carbon dioxide per annum, depending on assumed use patterns. CFCs contained in refrigerators purchased exceeds 40 tonnes per annum.

Figure 19.3 Some environmental effects of Thorn EMI Rental (UK) Ltd

Organising improvement

In line with the environmental policy, and in line with the new British Standard, the company has followed a consistent approach to organising improvement, as follows:

1 QUANTIFY IN TERMS OF ENVIRONMENTAL IMPACT

As in the British Standard, we have tried to quantify environmental impact wherever possible. This implies identifying the most significant effect associated with a particular issue (e.g., the greenhouse effect) and then quantifying appropriately.

2 WORK IN AREAS OF GREATEST IMPORTANCE

Because of this principle we have spent relatively little time trying to increase the company's use of recycled paper (for instance). Although that would, in isolation, be a good thing to do, there are bigger fish to fry first. One constraint in applying this principle is legislation: all who work on environmental issues will be aware that the law is often active on relatively small-scale local problems but rarely has anything to say on global issues such as global warming. Compliance can therefore override purely environmental priorities.

3 UNDERSTAND HOW MUCH IMPROVEMENT IS POSSIBLE

This involves respecting the manager's political constraints and the other pressures on him or her as well as understanding the technical issues. I do not mean by this that such constraints should never be challenged, only that one should try not to blunder into them unawares, and that the art of timing is very important.

4 NEGOTIATE REALISTIC TARGETS WITH MANAGERS

In the earlier stages of work in an area, it is normally better to agree process targets (e.g., 'Identify ways of reducing solvent use in paint spraying' or 'Carry out energy audit of new store design') rather than performance targets (e.g., 'Reduce solvent use in paint spraying by 35%'). This is because a meaningful performance target can only be agreed after doing preparatory work. Process targets are a way of demonstrating management commitment and of developing a culture of performing against targets during the period before performance targets can be responsibly agreed.

5 DON'T TRY TO FORCE PEOPLE TO DO WHAT THEY DON'T WANT TO

In addition to those whose freedom of movement is limited, any company will have its share of cynics. Usually it's better not to take them on, but to work around them – even if this means occasionally bypassing a significant environmental impact for the time being. If you can't succeed with people who are reasonably positive, there will be little chance with those who are not committed. If you can notch up a few successes, however, opponents will be more inclined to be receptive. Whatever you do, there's nothing more likely to get managers to dig their heels in than if they feel that you are preaching at them!

6 ALTER UNDERLYING PROCESSES WHERE POSSIBLE

There is no point in developing a special environmental procedure if the operator doesn't know about it. It is best to amend the standard work procedures and to reinforce this with training if appropriate.

7 MAKE TARGETS VISIBLE

We used the vehicle of our first annual Environmental report to the parent company to raise awareness of the area among the senior management team. Of course this is only wise if there has been sufficient preparatory work for the report to meet all normal expectations of a management report.

8 REPORT PROGRESS REGULARLY

Reporting progress regularly is essential, since it demonstrates that the programme is indeed subject to normal management disciplines. Nothing is more demotivating than agreeing a target and then have nobody take an interest in how things are turning out.

It may be worth noting in passing that it is very difficult to organise improvement in a part of the business that is out of control managerially. For this reason, the best environmental starting point for some companies might be a good Quality programme. In addition, the environmental benefits of reducing waste through improving quality are self-evident. In this respect, I was fortunate in working at a time when management within the company was being given a great deal of support in consolidating onto best practice.

At least *some* actions have been taken in most areas of environmental impact. In order to identify issues faced in our environmental programme, I will briefly discuss improvement activity to date on ozone-depleting substances, packaging wastes and energy. These examples have been chosen to reflect different levels of complexity of the issues faced, with work on ozone depleters having been simple in managerial terms, and that on energy being much more complex.

In considering these examples, please bear in mind that the company makes no claim to be particularly excellent in its environmental management programme. These examples are intended to illustrate issues encountered in environmental programmes, rather than to demonstrate model responses to these issues.

Example A – Ozone-depleting substances

In 1990, the company used 6.3 tonnes of CFCs and 0.35 tonnes 1,1,1 Trichloroethane in service operations. Both substances destroy stratospheric ozone. Following an enquiry from Friends of the Earth (FoE) in 1990, the company began to examine its use of these substances. CFCs were still used as aerosol propellants, since consumer pressure had by and large not yet affected the design of industrial aerosols. In operations, CFCs were used as a freezer in fault finding (many intermittent faults only occur at low temperatures and engineers need a safe means of cooling circuits down as they hunt for the fault). They were also used in cleaning fluids and in solvents. After the FoE enquiry, we began to eliminate these substances wherever possible.

By the end of 1991, substantial progress had been made. Many uses had been eliminated altogether. Butane was now the preferred propellant for aerosols. Since butane aerosols need special handling at the time of disposal, aerosols, wherever possible, were being replaced altogether by finger sprays. CFCs for freezing had been replaced by HCFCs. These are still ozone depleting, but to a much lesser extent than CFCs. They retain the inert characteristics of CFCs, which are important for staff safety. At a more trivial level, a replacement typing correction fluid, containing no 1,1,1 Trichloroethane, was supplied.

In order to quantify usage in terms of the environmental effect, we quantified the various substances in terms of the ozone depletion potential of CFC 11, using conversion factors supplied by FoE campaigners. Comparative usage in 1990 and in 1991 was as shown in Fig. 19.4.

Despite the progress that had already been made, when the time for environmental targeting came in the spring of 1992, there was general agreement by managers that continued improvements in this area were important. The high public profile of the CFC issue made everyone sympathetic to the need for continued progress.

How much improvement was possible? In order to delist a product, the Technical Services Manager of our Stores Operation had first to approve a suitable substitute. He advised that he needed to do substantial work in a couple of the remaining areas, but that he expected to have this finished by the end of the financial year (31 March 1993). On the other hand, he forecast that he would be able to make little progress on an HCFC replacement for use as a freezer in the planning period and

Substance	*1990 (tonnes)*	*1991 (tonnes)*
CFC	6.3	1.4
HCFC	0	1.9
1,1,1 Trichloroethane	0.35	0.39
Total (converted to CFC 11 equivalent)	6.3	1.6

Figure 19.4 Thorn EMI Rental (UK) Ltd's use of ozone-depleting substances

that this would take much longer. We therefore agreed, as a target for the planning year 1992/93, that we would have stopped buying any ozone depleters by 31 March 1993, with the single exception of HCFC used as a freezer, which was trivial in relative ozone-depletion impact, and where a longer timescale was necessary.

There was little opposition and the objectives were met. A flurry of activity in February and March resulted in the last CFCs being delisted just in time to meet the deadline we had set ourselves, thereby showing how the certainty of reporting against clear targets can help in getting commitments carried through. Usage for the year was 1.2 tonnes CFC 11 equivalent – a reduction of 25% on the previous year. More importantly, no more significant purchases would be made, thereby virtually eliminating this area as an environmental issue. The fact that the responsible manager was also the co-ordinator for the Stores' Quality Assurance procedures was helpful in ensuring that underlying processes were indeed amended where necessary.

In this case, it proved fairly easy to agree targets with managers. Having and publishing agreed targets for environmental improvement certainly helped expedite implementation.

Example B – Expanded polystyrene packaging wastes

All white and brown goods are delivered encased in expanded polystyrene (EPS). EPS is extremely light at around 9kg per cubic metre when carefully packed. When thrown loose into a skip its effective density will be much lower than this. A video recorder is packed in around 150g EPS. A colour television uses almost 500g. Disposing of EPS causes a number of significant problems to managers. For example:

- Its low density makes collection and disposal very expensive. We estimate the current cost of disposal at over £1 per kg and this cost may be expected to rise. On-site compaction is not feasible, since compaction equipment is expensive and our EPS waste is spread over approximately 30 sites. Many sites were having daily skip collections – almost entirely of EPS.
- Several sites have suffered from skips being broken into and the EPS inside them being set on fire by vandals. Under the Duty of Care for Waste Management failure to provide protection against vandals can lead to prosecution. At some sites, however, securely locked skips were still being broken into regularly.
- Avoidance of EPS in packaging is very difficult, since its very low density makes it very effective in protecting sensitive products in transportation.

Environmental problems are the direct result of the substance's low density. The transportation of EPS is very energy-inefficient, whether in a skip or, uncompacted, in any other dedicated container. Final disposal takes up a significant volume of landfill, with all the associated problems of heavy vehicles carrying waste through quiet country roads on the way to final disposal.

Quantifying environmental effect is therefore best done in volume terms. To give an idea of the significant volumes involved, we calculated that our product purchases each year generate around 46,000 cubic metres of waste EPS. This is a volume equivalent to an olympic running track piled almost 24 feet high with carefully packed EPS! Looking at it in more parochial terms, this would fill our main headquarters building 3½ times each year. Of this volume, we dispose of about 60% directly from our own sites, costing a quarter of a million pounds each year. The rest is either kept with the product by our customers or is sent for disposal by them through their local authority collection service. We should note, however, that European legislation makes it increasingly likely that industry will assume direct responsibility for 100% of such wastes.

This area was important for the reasons above, but also because it is a visible area of environmental impact for our staff and doing something very visible helps to demonstrate management commitment. How much improvement was possible?

Technically a number of solutions were possible. Compacting-equipment capable of dealing with large volumes is available, although not at a cost that would allow it to be installed at the places where the waste is generated. Thus the first issue was whether it would be possible to concentrate the waste at a single location. In principle we established that there was sufficient space to site such a machine at our main Winsford Distribution Depot and that backloading EPS appeared, in principle, to be feasible.

Next we had to choose a compacting device. After considering and rejecting various thermal compaction devices, we settled on a compactor to be provided and operated by a Midlands-based packaging manufacturer, who uses high quality compacted EPS to make new packaging and other styrene products. This company quoted for the supply of labour and plant on a sliding scale, depending on the volume of EPS generated. At projected volumes, we would receive a modest net income from this operation.

The main constraint was managerial rather than technical. At this time (mid-1992) the company's distribution operations had been substantially restructured and a bedding-down period was required. Understandably, the responsible Director, despite supporting the plan in principle, was not prepared to divert any management time away from making the new structure work.

By the time of the annual planning round in late 1992 operational pressures were less extreme. At the same time financial pressures for budgetary savings were rising. The prospect of making net savings of £$\frac{1}{4}$ million became very attractive. The responsible Director gave the go-ahead for a simple trial of whether backloading would work operationally and to confirm the economics of the scheme. This trial was carried out with the support of the Regional Distribution Controllers at three separate depots. We established that there would be ample space on the return wagons and that incremental costs would not be significant. Eventually the necessary expenditure approvals were raised and the savings were included in the Logistics Department's annual financial budget.

The environmental target was, of course, consistent with the financial budget, but less stretching. Overall, I believe that the long-term interests of the programme are better served by targets being realistically achievable than by them being very stretching. We agreed that 90% of packaging that we dispose of would be being recycled by the end of the next financial year, allowing more room for errors than the financial budget, which assumed an equal rate of waste avoidance throughout the year. At the time of writing the recycling plant has been installed and is operating successfully and there is every prospect of the target being reached.

Recycling EPS is an example of a technically simple environmental issue where progress was slowed down by other valid pressures on a management team that was under significant short-term stress. Rather than add to these temporary pressures, I felt it better to wait until the management team as a whole was keen to proceed. In due course management implemented wholeheartedly. I believe that in the long run more sustained progress is normally possible by accepting a slight delay than by trying to press on regardless against the management's wishes.

This story shows how much work is necessary before a target can be set with any prospect of success. Only after work over several months by the department's internal champion, and by myself, was the department as a whole ready to 'own' the initiative. *In the end*, target setting was very easy. But in this case it would have been totally inappropriate to set a target immediately after quantifying the environmental effect.

Example C – Energy in operations

In common with much of the rest of the developed world, energy use is responsible for many of the company's most significant environmental impacts. Energy is used in virtually all areas of the company's operations. In operations, prime areas of energy use include buildings, logistics, and vehicles. In none of these is there the option of doing without energy altogether, in the way that we can do without CFCs. For each of these areas, the factors affecting the level of use are many and complex. For instance, taking a simple example, fuel use in logistics is affected by the choice of vehicles, by their frequency of servicing, by the effectiveness of their routing, by the company's siting of distribution depots, shops and refurbishment facilities, by the skill with which vehicles are driven and by a host of other factors. Each of these areas would merit a study in itself to determine potential to reduce energy. Some areas will be adequately managed by the company's budgetary and other control systems. Resources are limited. How can we prioritise? How can we set targets responsibly?

The first step is to quantify in terms of environmental impact. The impacts of energy use include carbon dioxide and other emissions during the extraction, transportation, processing and burning of the fuels. What matters therefore is not how many Kilowatt Hours come through the electric socket but how many units of energy in total were consumed in getting them to the point of use. Unfortunately, energy sources vary significantly in this respect. For example, electricity might be produced by burning gas at a power station to evaporate water to drive a turbine which produces electricity. This uses approximately three units of heat energy to produce one unit of electrical energy, the rest going to heat up water or up the chimney. If this electrical energy is then used to heat a house, a task which could have been almost as well performed by burning the gas in the house, it will have consumed much more energy in total in doing so. If we compare our bills, however, we will see that we have been billed for delivered energy only. In this case, therefore, *environmental impact can only be quantified by calculating the total energy consumed, including energy used at the power station, in transportation, in refining, etc.* – it is seriously misleading to look at billed energy units by themselves.

We used conversion factors to give the lifetime energy consumed in delivering units of electrical energy, gas, etc., cribbed from a workshop that I had attended in 1991. The results were as follows and relate to the year 1991/92, prior to the integration of Radio Rentals' operations with Rumbelows:

Category of use	*Delivered energy (million KwH)*	*Total energy (million KwH)*
Premises (mostly electricity)	54 (28%)	150 (48%)
Vehicles	133 (68%)	152 (48%)
Inbound logistics (includes estimates of energy in shipments from manufacturers' depots)	9 (5%)	12 (4%)
Total operations energy	197	314

Figure 19.5 Thorn EMI Rental (UK) Ltd's use of energy in operations

This analysis shows clearly that despite the much lower delivered units of energy, the environmental impact of energy use in premises is approximately equal in impact to that in vehicles, due to the high use of electricity. It also shows the relatively low impact of logistics. (We also did outline calculations on our customers' use of energy in visiting our shops to pay bills or to upgrade their equipment, etc., and found that this was insignificant compared to the two main categories of use.) We shall concentrate on the issues associated with energy use in premises.

Having identified energy use in premises as a significant area of environmental impact, we still had no idea how much improvement was possible. Top-level data couldn't tell us this by themselves. Was 150 million KwH per annum good or bad compared to best practice? A set of benchmarks was needed.

The Normalised Performance Indicator (NPI) system set up by the Energy Efficiency Office provided a starting point. This classifies energy consumption in KwH per square metre, showing the levels at which performance might be regarded as good (NPI category 1 covers the top 25% of the market compared with similar outlets) and bad (level 3 covers the bottom 25%). We did a quick review of 21 stores which showed that there appears to be significant potential for savings even after laboriously adjusting the NPI figures to take account of display-product energy consumption (which does not vary directly with the selling area of the outlet). We also did a formal NPI calculation of two management offices, showing that one had good energy efficiency but that the other was very poor. Overall, there seemed to be potential for significant energy improvements in the long run.

Unfortunately, it is a long way from identifying that *some* improvement is possible to being able to say that (say) 20% reduction (or, more meaningfully, achieving a 50% reduction in the number of sites with an NPI rating of 3) over a 5-year period ought to be possible. Our financial system provides little or no usage data for energy – not even the difference between fixed and variable costs. Tariffs vary significantly between premises. Tracking back data from past invoices would be a Herculean task, even had there not been a recent merging of two accounting departments. How could we quantify potential to save without meaningful data on site energy usage?

In my view, there was no alternative to building a site-based energy management system. This involved capturing unit energy data at the invoice-approval stage for further analysis in a database. It would be possible to construct a history using meter readings from the Regional Electricity Companies. Constructing such a database would not be a small job, but there were other benefits. It would be possible to identify unusually large bills prior to payment (this is a particular issue with estimated bills, when it is helpful to be prompted as to when the meter estimate should be checked locally). Since privatisation of the electricity supply, industry tariffs have become more complex and the complexity is set to increase. There were potentially large savings to be made from better tariff management and with good software this could be done in-house much more cheaply than if it were put out to contract.

Recognising this, we decided to build an energy database in 1992. The construction of the database was a target, albeit a *process* rather than a quantified target, in our 1992/93 environmental programme. We started to build the database and indeed entered a good deal of data into it. Unfortunately, the tariff management side in particular was rather more complex than we had realised. We decided to move to a competitively priced off-the-shelf package that seems to be capable of supplying all our needs much more robustly. The expenditure for this package was approved in early 1993. While we did not achieve our target of completing construction by the end of March 1993, the earlier commitment to do so was certainly instrumental in getting the project restarted and the capital expenditure approved after the initial delay.

In the 1993/94 environmental planning round, we agreed a number of process targets. With the Finance Department, we agreed that the system would be installed with one year's history backloaded. We also agreed that they would use the system to provide a list of shops by NPI rating. With the Property Department, we agreed that they would discuss a work programme to improve the NPI position once this list became available. We also agreed to devise a simple energy-efficiency checklist for refurbishment of shops and other premises.

One year into the exercise we have set process targets, which have been useful, but are not yet ready to set a quantified performance target. Had we actually set a performance target, we would have been a step further on in the BSI structure but no more meaningful work would have been done – probably less. Once the process targets now agreed have in fact been implemented, it will become easy to track energy performance year on year, and indeed to see the effect of various management actions on a location's energy-usage pattern. Once again, doing the preparatory work providing a basis for the target is the difficult job!

Conclusion

These three examples highlight a number of important points about the negotiation of environmental targets.

First, visible targets that have been agreed by the managers involved are very helpful in ensuring that progress on environmental issues is maintained. Nobody likes to fail against a target that they have agreed. Negotiating realistic targets and then making it very clear that progress against these will be followed up is straightforward common sense.

Second, environmental issues vary tremendously in the amount of work that is necessary before a quantified target can be set. Sometimes this is relatively easy, as in the case of ozone-depleting substances. Sometimes it is relatively difficult, as with energy, where little or no data may exist in a useable form. The issues that need to be addressed before a target can be agreed include the non-technical managerial constraints on a manager's freedom of movement. It is essential that these issues are not ignored since, if they are, the target would either not be achievable or would risk the opposition of the manager, and either result would harm the programme in the longer run. On the other hand, if this preparatory work is not skimped, implementation will be easy.

Third, there is an obvious danger that momentum will be lost during this period of discovery. Setting process targets can help by giving some milestones against which to report without shortcutting the issues described above. If these are carefully chosen, the setting of process targets can help to build ownership of the environmental programme among a wider constituency of managers within an organisation and, by being achieved, can help to build the overall momentum of the programme.

Setting targets in an environmental programme may be relatively straightforward and quick. Often, however, moving from a reasonable quantification of a top-level environmental effect to the setting of detailed improvement targets will take time – time to collect data in a form relevant to an improvement programme, to find out what is technically and managerially feasible and to allow managers' commitment to environmental improvement to ripen. It is important not to rush this process. The overall programme *must* allow time for differences in speed of progress in different areas of environmental impact.

The personal challenge facing the co-ordinator of the environmental programme is to maintain focus on improvement, while having confidence that by investing time where necessary, eventual implementation will be both quicker and more reliable. It is easy to become despondent when progress seems to be slow. This is a critical stage of the programme, however, and delays are normal (and to some extent desirable) as the organisation commits to actual changes. Overall, progress is likely to be quickest if the co-ordinator can find the confidence to relax, while at the same time being very attentive to what is practically achievable in the here-and-now, and being committed to eventual improvement.

20

Reviewing and monitoring environmental performance

WERNER B. ROTHWEILER

Head of Liaison and Audits, Corporate Safety and Environment, Ciba-Geigy, Basel, Switzerland

- Reviewing and monitoring environmental performance is one of the many elements of Total Quality Management (TQM) that constitute the framework of responsible entrepreneurship.
- Ciba's Directives, Guidelines and Standard Operating Procedures, provide detailed and explicit instructions to put Ciba's Vision and Governing Principles, which include a commitment to economic, social and environmental responsibility, into practical action in all areas and situations.
- Standards are elaborated by a Corporate Safety and Environmental Commission that represents a cross-section of divisions, manufacturing plants, development and engineering departments.
- Research and development work on safety, health, and the environment has resulted in an array of technical standards that are widely employed throughout the industry, and which are used as a yardstick for reviews and internal audits.
- An internal Unit undertakes independent safety and environmental audits of all larger manufacturing sites; these audits are based on internal standards, which often go beyond regulatory requirements.
- Ciba's safety and environmental audits also include an assessment of the technical standards and future needs of the facility, using a comparison with best available practices and techniques.
- As stakeholders and the public make use of their right-to-know, and begin to show a growing interest in industry's environmental behaviour, companies must prepare to disclose their ecological performance, just as they put their balance sheet in the public domain.

Introduction

When we discuss environmental performance we use a language and a terminology that would have puzzled our forefathers. They would not have understood what we are talking about. Performance is something that can be measured. When we talk about environmental performance today, there is probably no doubt in anyone's mind that we refer to our efforts to preserve the environment and to minimize environmental pollution, notions that did not exist for earlier generations. It is interesting and intriguing to note that in encyclopaedias and dictionaries, which we used as students some 30 years ago, neither the term environmental pollution nor the term environmental protection can be found.[1] Thus we are not surprised that in those days students of chemistry used the sink for disposal of all waste from their chemical experiments, a practice that was widely accepted at the time. An environmental consciousness did not yet exist.

The environment was, until very recently, perceived as man's surroundings which influence and affect him, and on which he depends for a living. But it was hardly perceived that, conversely, man influences his environment, possibly to the extent of inflicting irreversible damage. For thousands of years man was living in relative harmony with mother Earth. Relative, because science does attribute the fall and ruin of some archaic cultures to the exploitation of the environment, particularly to the overuse of wood used as building material and as a source of energy. However, if there was relative harmony between man and nature, it definitely ended with the onset of the industrial revolution some two hundred years ago.

Some of the negative effects of industrialization soon became visible and called for countermeasures. The first concern was about workers' health and resulted in the factory laws. Later on concerns were raised about forests, plants and wildlife, which led to the enactment of respective laws and ordinances. However, the need for the protection of water, air, and soil, the bases of life, was recognized very late. For a long time these fundamental elements were looked upon as being abundant and therefore not needing any particular care. Only about twenty years ago, when the Club of Rome[2] and other concerned world citizens, such as US President Jimmy Carter[3] and environmental activists like Greenpeace, rang the alarm bell, did a worldwide movement of stricter environmental legislation set in. In 1990, on the eve of the United Nations' Conference on Environment and Development (UNCED), some 50 reputable leaders of transnational businesses formed the Business Council for Sustainable Development (BCSD) and made a commitment to promoting *eco-effciency* and to changing course towards a *sustainable development* of our common future.[4] But even today, after UNCED's Earth Summit Conference in Rio, and the publication of the 'New Limits to Growth'[5] there are numerous environmental issues which are still unresolved and lend themselves to political disputes instead. Few things have been more difficult than reaching a consensus about what man must do to preserve this planet for future generations.

Under these circumstances voluntary approaches by industry become all the more important. Ralph Saemann, Vice-Chairman of the Swiss Academy of Engineering Sciences, and board member of the Swiss Association of Environmentally Conscious Management, made the following pledge for redefining industrial responsibility:

> *A modification of our economic system towards an ecological, social market economy, complete with new objectives, responsibilities, practices and modes of leadership in industry and technology, as well as a renewed emphasis on personal ethics for personal decisionmaking, will help us to cope successfully with the new environmental challenge. Finding a solution to the global environmental threat will require revolutionary restructuring of both the philosophic and technological underpinnings of industrialized society. Rather than meeting only the traditional objectives of technical elegance and economic success, technology will have to fulfill two additional objectives: societal acceptance and environmental compatibility.*[6]

It is against this background that environmental performance should be discussed. Measuring performance implies that we have a yardstick, that judgemental criteria exist, and that there are standards for comparison. If these standards have not been proclaimed as part of regulatory requirements, we have to develop and implement standards of our own. This presupposes an *entrepreneurial culture* that fosters *responsible care* for our environment, and it requires leaders who accept the criterion of *sustainability* for their businesses' development and growth.

> *The key to sustainable development . . . is to pursue a path that permits the people alive today to meet their needs without jeopardizing those of future generations.*
>
> Dr Alex Krauer, President Ciba-Geigy, *Ciba Journal*, February 1992

> *It is a primary responsibility of the management of the chemical industry in both the short and long term to protect the environment as integral part of good business practice.*
>
> CIA Environmental Objectives

> *There is nothing to stop industry itself setting higher standards. Indeed that is exactly what the Responsible Care programme, established by the chemical industry, sets out to do.*
>
> The Rt Hon. Michael Heseltine, May 1991

> *Change in response to environmental pressure is inevitable. The sooner companies and investors incorporate this fact into their strategic planning and future R&D investment decisions, the sooner they will be able to reap positive commercial advantages.*
>
> *Greenpeace Business Newsletter*, 1992

Thus, *reviewing and monitoring environmental performance* is but one of many elements of *total quality management* that constitute the framework of *responsible entrepreneurship*.

The elements of environmentally-responsible entrepreneurship

Corporate culture

The framework of environmentally-responsible entrepreneurship must be firmly anchored in a solid foundation of an unequivocal *corporate culture* that defines *beliefs* and *values* with which all employees can identify themselves. This is depicted in Fig. 20.1.

As a transnational chemical enterprise operating on all continents, Ciba has adopted a *Vision* and *Governing Principles* wherein management expressly acknowledges the acceptance and the importance of certain *ethical standards*, and a scale of values by which all entrepreneurial activity is to be judged. This is all the more important in an increasingly decentralized organization where the individual business units have authority and responsibility. The Vision and the Governing Principles, just like the pole star or a compass, serve as means of orientation, and indicate the right direction to be followed by all employees.

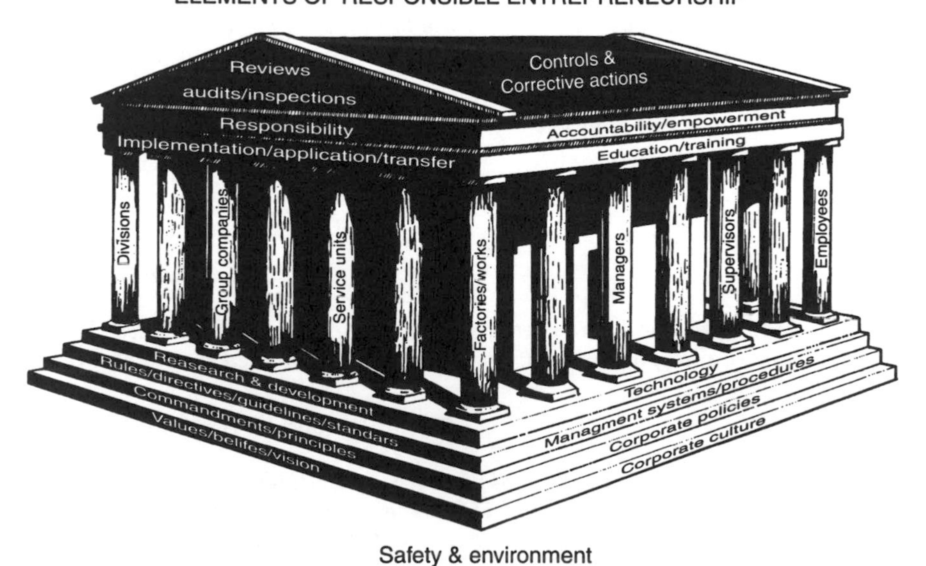

Figure 20.1 An 'environmentally responsible entrepreneurship' framework

Ciba's *Vision* is expressed in one sentence:

> *By striking a balance between our economic, social, and environmental responsibilities we want to ensure the prosperity of the enterprise beyond the year 2000*

This deceptively simple statement is expanded by an elucidation of the three responsibilities.

VISION 2000

- **Responsibility for long-term economic success**
  We aim to generate appropriate financial results through sustainable growth and constant renewal of a balanced business structure, so that we justify the confidence of all those who rely on our company – stockholders, employees, business partners and the public.

- **Social responsibility**
  We want to be open and trustworthy towards society. Through our business activities we wish to make a worthwhile contribution to the solution of global issues and to the progress of mankind.
  We recognize our responsibility when turning new discoveries in science and technology into commercial reality; we carefully evaluate benefits and risks in all our activities, processes, and products.

- **Environmental responsibility**
  Respect for the environment must be a part of everything we do. We design products and processes to fulfill their purpose safely and with as little environmental impact as possible.
  We use natural resources and energy in the best possible way and reduce waste in all forms.
  We dispose safely of all unavoidable wastes using the best available technology.

The commitment to responsible behaviour and sustainable practices is further documented in the following *Governing Principles*:

- In our activities we take account of the fact that raw materials, soil, water and air are finite resources that we must use carefully and responsibly.

- We want to make an active contribution to environmental protection and, even without official regulations, take measures required to put our concept of environmental protection into practice.
- We see to it that our products will not have untoward effects on the environment, either during manufacture or when properly applied or disposed of.
- We shall not compromise health, safety, or the environment for economic or productivity gains.
- Safety, health, and environmental problems shall be resolved by using the same scientific approach as is used for the research and development of new products.
- Wastes shall be handled, treated, and disposed of with the same care as is devoted to sales products.

CORPORATE POLICIES

What is captured in the *Vision* statement and in the *Governing Principles* is expanded in the *Safety, Health, and Environmental Policies*. These *commandments* provide guidance to insure the health and safety of all our stakeholders and the protection of our environment. They comprise policies on:

- Safety and environmental management systems
- Safety and environmental protection in production
- Safety and environmental protection in contract manufacturing
- Safety and environmental compatibility of products (product stewardship)
- Energy conservation
- Project management

These serve as a code of conduct in research, development, production, and marketing.

The term 'safety' is henceforth used in a comprehensive sense including occupational health and industrial hygiene.

PROCEDURES

Governing principles, commandments, and codes of conduct, however, are not sufficient. For all practical purposes and situations, detailed and explicit instructions are required. These are embodied in *Directives, Guidelines*, and *Standard Operating Procedures*, covering

- Chemical and physical processes and operations
- Transport and handling of chemicals
- Use of energy and utilities
- Avoidance, minimization, re-use, and disposal of wastes
- Hazard analysis, risk assessment, environmental compatibility assessment

These documents comprise the company's entire knowledge pertaining to safety and environmental protection which has been acquired over the years by generations of chemists and engineers, sometimes as a result of painful experiences. The know-how has been documented in codified form to be used and applied by all employees world-wide.

These *standards* are elaborated by a *Corporate Safety and Environmental Commission* that represents a cross-section of divisions, manufacturing plants, development and engineering departments. The practical experience of the commission's members guarantees the realistic and practicable nature of the standards. The representation of all major business units and the practice of refining all regulations and standards until they are *unanimously* accepted by the commission members, guarantees the appropriateness of the measures to be enforced and helps them to find acceptance throughout the corporation world-wide.

R & D IN ENVIRONMENTAL AND SAFETY TECHNOLOGY

Signing an environmental charter and proclaiming good intentions is not enough. Our colleagues at the production front want to be provided with the necessary tools that enable them to comply with the ever-increasing *regulatory requirements* or *self-imposed standards*. When standard solutions are not readily available from the shelf, new methods and techniques need to be researched, developed and engineered.

Two Governing Principles pertaining to research and development require that:

- Safety, health, and environmental problems shall be resolved by using the same scientific approach as is used for the research and development of new products.
- Wastes shall be handled, treated, and disposed of with the same care as is devoted to sales products.

Therefore Ciba maintain a number of dedicated service units, where novel methods and technologies are developed, or where existing techniques are improved and refined in fields, such as:

Environmental technology
- Chemical and biological effluent treatment
- Biodegradation of special wastes
- Wet air oxidation of non-biodegradable wastes
- Incineration of liquid and solid special wastes
- Biofiltration for waste air purification/deodorization
- Off-gas purification by absorption, catalytic oxidation, thermo-reaction, incineration, adsorption on carbon fibre
- Flue-gas purification, including removal of NO_X
- Immobilisation and stabilisation of slags and ashes
- Site remediation
- Groundwater decontamination
- Ecotoxicology
- Environmental trace analysis
- Biospheric monitoring
- Noise abatement

Safety technology
- Material and product safety testing
- Thermal process safety testing
- Explosion protection technology
- Electrostatic hazards research
- Occupational toxicology
- Biological monitoring methodology

R and D work in these disciplines has resulted in an array of *technical standards* that are widely employed throughout industry and also used as a yardstick for reviews and internal audits.

EDUCATION AND TRAINING

Policies, directives, and guidelines are of no avail if they are not followed. New technologies and novel solutions to safety and environmental problems are of little use if they are not applied and put into practice. However, if rules and regulations are followed merely as a result of subservient obedience instead of conviction, we have gained little. If novel techniques are not properly sponsored and propagated they will never get beyond the R & D stage.

To ensure the implementation of safety and environmental procedures and codes of good practice, as well as the introduction and the application of new technologies, it is important that they are well understood and accepted. This can only be achieved by adequate education and intensive training at all levels.

Ciba have therefore developed a respectable array of courses and workshops for senior staff and technical personnel in which master foremen, chemists, and engineers are provided with the necessary knowledge and know-how based on well-defined training objectives. The programmes range from a safety and environmental introductory course for newly engaged scientific and technical staff, over basic and advanced courses in process safety, risk analysis, and environmental protection for chemists and engineers, to special courses for environmental officers and safety advisers.

To reach a uniform safety and environmental standard throughout the worldwide group, distributing documents does not suffice. The necessary technical expertise and the understanding of the basic principles involved must also be systematically disseminated. On the one hand we try to achieve this by application of the 'snowball principle'. Managers from Ciba's Group Companies are invited to participate in six-day workshops on either environmental protection or process safety. They will apply the methods learned in their own plants and thus contribute to an effective promulgation of the company's safety and environmental knowledge. On the other hand, safety and environmental experts from the Parent Company visit the Group Companies' Works regularly to render advice and support.

ACCOUNTABILITY

In our complex world-wide organization, the number of management layers has been reduced, modern forms of teamwork have been introduced, and efforts have been made to encourage individual initiative and risk taking. Therefore it is all the more important that economic, social, and environmental performance goals are clearly defined and that the responsibility for meeting these goals is unequivocally assigned.

The **President** and the **Board of Directors** have a *duty of care* and are obliged to oversee all activities to ensure that the company does not leave the path of economically, socially, and environmentally responsible entrepreneurship.

The **CEO** and the **Executive Committee**, as the highest executive body, bear the overall responsibility, not only for economic, but for social and environmental performance as well. They achieve this by:

- providing binding corporate policies and governing principles;
- installing an adequate organization and clearly defining responsibilities;
- appointing ethically and intellectually qualified managers;
- exercising regular supervision and control.

The **Heads of Divisions** and their deputies who set their own targets are, among others, responsible for the safety and environmental compatibility of their products and manufacturing processes, and for procuring and disseminating all relevant information.

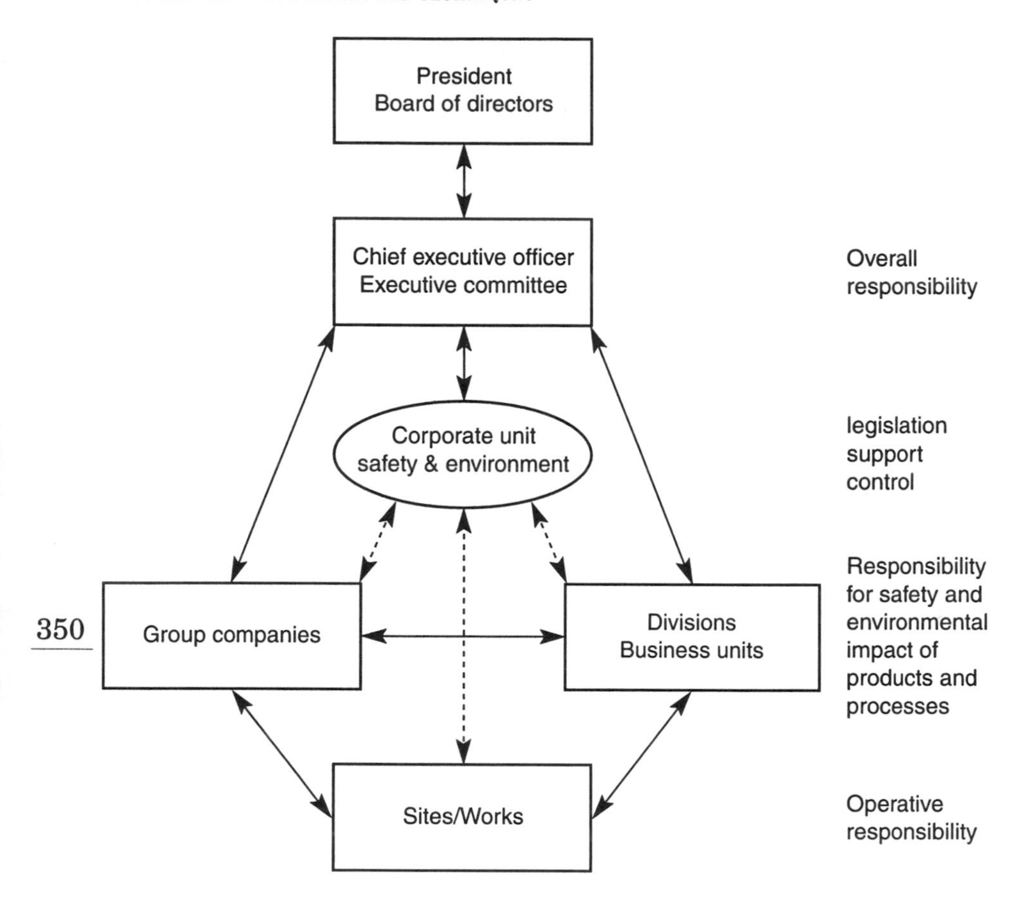

Figure 20.2 Accountability

The **Heads of the Group Companies**, as representatives of the Executive Committee in their respective countries, have overall responsibility for the compliance of their group's activities with local safety and environmental legislation, with company policies, directives, and guidelines in these fields.

The **Site or Works Managers** are responsible for all operational matters at their site, including safety, health, and environmental performance. They have the right to stop production or any other business activity if there are well-founded misgivings about safety, health or environmental protection.

The **Corporate Unit Safety and Environment** is a staff function of the Executive Committee. On their behalf it oversees and strives to continually improve the state of safety and environmental protection through:

- procurement and documentation of safety and environmental data;
- elaboration of safety and environmental standards;
- promulgation of scientific knowledge and engineering know-how;
- support and assistance from safety and environmental specialists;
- supervision and assessment of safety, environmental compatibility, and regulatory compliance.

CONTROLS AND CORRECTIVE ACTIONS

Policies, rules and codes of conduct remain empty phrases as long as they are not put into practice; directives and guidelines are useless if they are not adhered to. In the Ciba Group of Companies adequate controls and follow-up are therefore provided, and care is taken that problems and deficiencies are corrected or repaired.

Certain control functions are inherent in the responsibilities assigned to the Heads of Divisions, Business Units, Group Companies, and plant sites. These managers are supported by experienced staff and technical experts as well as safety and environmental specialists. The Heads of larger Group Companies have their own technical department with safety, health, and environmental specialists, who maintain close contact and share experiences with their Parent Company counterparts. The Site Managers are supported by their own deputies of safety and environmental protection. The Divisions and Business Units ensure that relevant information, which is technically and factually correct, is disseminated, and that a comparable standard of safety, health, and environmental protection is achieved world-wide.

The Corporate Unit Safety and Environment, a staff function of the Executive Committee, undertakes independent safety and environmental *audits* of all larger manufacturing sites. With these corporate audits Ciba strives to achieve far more than merely assessing the *environmental management systems*, as is intended in the European Community's much-talked about Eco-audit Directive. Not only do we audit against our *internal standards,* which often go beyond *regulatory requirements*, but our experts also make an *assessment of the technical standard* and of *future needs* of the facility. This includes a comparison with *best available practices and techniques*, some of which have been developed within the company by its own safety and environmental specialists. Any deficiencies and shortcomings are corrected expeditiously, and the auditors' advice and recommendations must be followed and implemented within reasonable time.

To summarize: Ciba's culture is largely determined by the 'Vision 2000' and by common values which have been adopted by all employees. These values are expressed by governing principles for safety and environmental protection, which in turn are embodied in rules of conduct for various situations. The responsibilities for setting safety and environmental goals and objectives, and for attaining them, have been clearly defined. In uncertain cases, specialists can be relied upon for expert advice and guidance. Unsolved problems are researched using scientific methods. Experience and newly acquired know-how are codified and documented in an easily comprehensible form, so that they can be readily applied to practice and translated into action. The adherence to the rules of conduct and the acceptance and implementation of new findings are periodically checked, and in case of deviations the necessary corrections are initiated.

Reviewing and monitoring environmental performance

THE ENVIRONMENT *VERSUS* SAFETY

Safety of employees and plant was for many years, until the late 1960s and early 1970s, before the clean water laws were enacted, of primary importance in the chemical industry. Consequently, aqueous processes and highly chlorinated solvents were favored because their use eliminated the much feared fire and explosion hazard. The environmental impact of these safety measures was hardly considered. Only in recent times has the pendulum swung back completely, so that one of the fastest growing concerns has become the protection of the environment. A company's environmental performance is coming under increasingly close scrutiny from government bodies, pressure groups and the public alike. While all the attention and resources currently being lavished on the environment are to be welcomed, accidents make no distinction between damage to the environment, people or plant. Improvements to protect one of these three areas should not have a detrimental impact on the other two. In the frantic clamour to assure everybody of environmental credibility, care must be taken not to jeopardize safety and occupational health.

Nothing is to be gained by playing one against the other. Remember that any safety incident of a certain magnitude may inadvertently result in pollution of the environment. Seveso, Chernobyl, Schweizerhalle, Exxon Valdez, Braer and many other accidents testify that a lack of safety can become the cause of an environmental disaster. On the other hand, factory inspectors and industrial hygienists are increasingly concerned about an overemphasis of environmental protection at the expense of occupational health and safety at the workplace. This conflict between safety and environmental protection should simply not exist.

They are not mutually exclusive, but closely linked, and require careful balancing if the underlying cause of the problem cannot be eliminated first.

In the context of reviewing performance, the term 'environmental' will therefore be used in a very broad sense. It will include process safety, occupational safety and health, as well as energy consumption.

APPROACHES TO REVIEWING AND MONITORING

There are essentially two different approaches to reviewing performance, i.e., *direct* or *active,* and *indirect* or *passive*.

Choosing the direct or active approach requires one to go out to see with one's own eyes, to verify and to physically check the object of the review. This is normally referred to as inspecting or auditing. What is the difference between an *inspection* and an *audit* then? Unfortunately the term audit is misused, more often than not, in that it is applied to ordinary controls, which are an integral part of any manager's or supervisor's *duty of care*. In this case one should distinguish rather between *routine checks* and *formal inspections*. Routine checks are a natural supervisory function, whereas inspections are of a more formal nature and performed by a line manager of one of the next higher hierarchical levels or by a specialist from a staff function serving the line management, e.g., the safety or environmental officer of a manufacturing site. The term 'audit' should be reserved for systematic and objective evaluations that are carried out and documented by well-qualified and independent experts who have been given their mandate by top management (i.e., corporate, divisional, business unit, or group company management). The so-called 'self-audit' is a contradiction in terms and the expression is therefore to be rejected.

The indirect or passive approach is to let the units under review report periodically. Reporting key economic and financial data has a long tradition. So why not use this method for environmental data? While auditing is more effective, but also much more time consuming, reporting is more efficient and provides management with a worldwide view and a *status report* of the corporation's environmental situation. In an audit, Ciba review the software, physically check the hardware, and initiate an expeditious remediation of any shortcomings. Reports consist of a compilation of key data that allows management to follow trends and to identify problem areas. From this, management can derive general as well as individual, site-specific performance goals.

ENVIRONMENTAL AUDITING

Environmental auditing is a comparatively young discipline. It is gaining rapidly in importance, as legislation is getting stricter and ever more demanding, and as the hazards of environmental *liability* – not to mention the indirect costs of losing *credibility* with the authorities, the local community, and the customers – increase. Principles, methodology, and techniques of environmental auditing are treated elsewhere and can be looked up in the rapidly expanding body of literature.[7] Indeed, ecological balancing, using life-cycle assessment methodologies to recognize and evaluate environmental impacts of products and processes, forms part of Ciba-Geigy's approach to monitoring environmental performance, but is not dealt with in this chapter. The following exposition is therefore limited to a few Ciba-specific features and personal experiences.

Corporate auditing, as we conduct it is a tool of *strategic management*. It does not only help to ensure compliance with today's legal requirements, but it helps to minimize future risks and liabilities, as it takes a proactive approach and tries to anticipate future developments. Furthermore, it helps to increase operating efficiency, effectiveness, and quality, as it not only looks at management and control systems but also at technical, organizational, and personnel measures. Moreover, it helps to improve the company's credibility and its relations with the stakeholders and the sites' neighbours. Finally, auditing can result in gaining distinct competitive advantages and thus in ensuring the company's survival.

Ciba has been carrying out corporate audits since 1980. Sites are audited every 4–5 years. We have no intention of increasing this frequency rate for the following reasons:

- recommendations derived from a technical assessment can be so demanding that their implementation may take one to three years;
- shorter intervals would undermine the credibility and jeopardize the high prestige that the corporate audits now enjoy;
- divisions and business units, group companies, and sites have to be given a chance to live up to their proper responsibilities and to do their own internal audits and inspections.

If corporate audits are too frequent, e.g., every year, as has been proposed, there is a tendency for the divisional and local management to wait for the corporate auditors to come round and tell them what to do, instead of feeling responsible themselves and thus becoming proactive.

The *objectives of corporate audits* are to provide:

- an independent assessment
  - (a) of compliance with company policies, directives, guidelines and procedures, as well as with national or regional environmental legislation and local consents;
  - (b) of the effectiveness of measures taken to protect personnel, plant, equipment and the environment (technical, organizational, and

personnel measures, as well as appropriate management and control systems);

(c) of measures required in the future;

- assurance to management that operations are consistent with good practices;
- assistance and support to the divisions and group companies in the setting up of their own audit and control functions.

Additional benefits that accrue from corporate audits are:

- increased safety and environmental awareness at the audited site;
- improved understanding and universally uniform interpretation of company policies;
- improved flow of information on safety and environmental matters throughout the company;
- promotion of effective solutions and management systems;
- reduction of potential liabilities through the identification of problem areas and potential risks;
- evidence of the commitment to safety and environmental protection as a Responsible Care company.

'Responsible Care' is an umbrella programme of the chemical industry designed to improve performance in the fields of health, safety, environment, product safety, distribution, emergency response, and relations with the public; it also enables companies to demonstrate that these improvements are in fact taking place. Responsible Care is about performance; it is not a public relations programme.

The corporate audits are supplemented by audits conducted by the group companies and the divisions respectively. This makes sense and is almost a necessity, taking into account that the regulatory requirements vary so widely among the different countries and that it is hardly possible for a corporate auditor to know them all. Additionally, in a diversified company, each division has its specific problems and technical solutions which are better known to a divisional than a corporate auditor.

The *scope of corporate audits* of plant sites can be summarized as follows:

Functional scope

- risk management
- operational and process safety

Locational scope

- production facilities*
- warehouses and tankfarms*

- occupational safety and health
- impact on air
- impact on surface/groundwater
- waste management (avoid, minimize, recycle, dispose)
- noise emissions
- education and training
- engineering, maintenance
- emergency planning/response
- workshops
- utilities
- effluent treatment plants*
- waste treatment and storage facilities*
- incinerators*
- recycling units*
- landfills*

* (including contractor's facilities)

The *corporate audit of divisions and business units* aims at assessing how well they manage safety and environmental matters. It reviews the systems and procedures as well as the organizational and personnel measures that the management of a division or business unit has adopted to ensure that they live up to their social and environmental responsibilities as defined in our Vision statement, in the Governing Principles, and as specified in corporate policies, directives, and guidelines.

Of a special category are *pre-acquisition / -joint venture / -disinvestment audits*. These are carried out as part of the *due diligence* procedure either by the division or the group company involved. The objective is to identify potential environmental problems and assess the possible consequences (e.g., liabilities) and the cost of correction or remediation before a decision is taken to buy, merge with, or sell a facility.

What about *validation or certifcation* of audits by *accredited verifiers* and *public disclosure* of the audit results?

While we strongly believe in auditing as a strategic management tool, we can see no benefit derived from having audits validated and verified by an independent third party, as has been proposed. How can anyone verify the results of an audit unless he has taken part in the audit or carried out one himself? It simply cannot be done. Believing otherwise would be a delusion. In addition, duplicating an audit would not make much sense.

Moreover, *auditing* as we perceive it, is a *management function* that covers not only control of regulatory compliance but also is a tool to ferret out weaknesses in a site's environmental performance. A co-operative relationship based on mutual trust is required between the auditor and the auditee. Fear of adverse publicity or personal sanctions resulting from the public disclosure of audit reports would be counterproductive, as it would jeopardize the delicate relationship between auditor and auditee. Furthermore, it is in the Corporate Management's highest interest that auditors be as independent and objective as possible and that they uncover all shortcomings and weaknesses with scrutiny. Consequently we believe that audit reports need not be verified by third parties and should not be disclosed to the public. We do, however, support the certification or validation of the *audit process* to accepted standards.

With environmental impact or emission statements the situation is quite different. The public has a *right-to-know* about the environmental performance of the industry in its neighbourhood. This right can be satisfied by a *public disclosure of impact or emission statements*. It is the local authorities' responsibility to verify such statements.

REVIEWING INVESTMENT PROJECTS

When planning *capital investment projects*, safety and environmental aspects must be given the same conscientious and thorough consideration as production technology, marketing, economic, and other aspects. Project management must document that all relevant safety and environmental problems related to handling, manufacturing, processing, transportation, storage, and disposal of chemicals and wastes have been identified; that these problems are properly resolved by applying up-to-date technology; and that all *internal safety and environmental objectives and standards*, as well as any *regulatory compliance requirements*, are met.

All projects must be reviewed by safety and environmental specialists who must give their formal consent before the project can be approved. Normally this does not pose a problem as the respective specialists are involved already at an early planning stage and are often part of the project team.

Important aspects that must be reviewed, particularly with respect to manufacturing processes, are *integrated waste/pollution prevention* and *energy conservation*. The time for end-of-pipe solutions is over, as they are too unspecific and inefficient, often not effective enough, and in most cases too costly. The costs of waste management increase in the order of declining acceptability of the method applied, i.e., elimination, reduction at the source, recovery/re-use/recycling, treatment, incineration and disposal. Additionally some of the end-of-pipe solutions, such as effluent and off-gas treatment, are heavy energy consumers, and only transfer environmental problems from one compartment to another.

The project review thus ascertains that integrated pollution prevention and energy conservation techniques are being applied.

ENVIRONMENTAL REPORTING

On the one hand we in Ciba were faced with the impossibility of auditing all of our manufacturing sites every year. On the other hand there was a strong desire to have a world-wide survey of the environmental state of the corporation, and to be able to monitor safety, energy consumption, and environmental pollution. Only with the relevant information can one correctly set priorities and initiate the necessary corrective actions. As an example, it makes little sense to call for a 10% reduction of VOC emissions across the board, when three sites alone account for over half of the corporation's total world-wide emissions. However, knowing this enables one to make a well-directed concerted effort that, in the end, is more cost-efficient and benefits the environment even better. Ciba therefore decided to introduce annual reports on safety, energy, and environmental protection (so-called 'SEEP' reports) with the objective:

- to obtain an accountability of safety, energy conservation and environmental protection performance;
- to determine performance criteria for the sites;
- to serve as a management tool;
- to provide a database for communication of corporate performance.

When we drafted the reporting scheme three years ago and came up with a very comprehensive questionnaire, we were somewhat reluctant. We were not sure whether we had gone too far, and we were concerned about how well this 30-page questionnaire would be received by the Group Companies and those sites who had to deliver a report every year.

We had, apart from the parent company, approached group companies in 15 other countries. The SEEP report was declared mandatory for 27 larger manufacturing plants. For the smaller sites and joint ventures it was voluntary. To our surprise, a total of 63 sites, including some joint-venture partners, submitted reports. The reactions were quite positive although the task of preparing the report has proved to be an additional burden. There was a common understanding and a general consent among the sites that they needed this information, as a basis of sound environmental management, and to be able to answer queries of the authorities and the public as a Responsible Care company. Obviously the sites were satisfied. Not only were they provided with a useful format for collecting environmental data, but SEEP proved to be a useful instrument for managing environmental matters and for communicating environmental performance to the employees and the public.

The SEEP report consists of the following sections (status 1992):

1. General site data

• Production data	– production volumes by product group – processes discontinued for safety or environmental reasons

• Water balance	– quantity of input, capacity of source(s) – use as process, cooling, or sanitary water
• Effluent discharge	– type and quantity of discharge – discharge points, receiving water body (average and minimum flow)
• Noise	– cases of non-compliance with legal limits
• Soil, groundwater	– cases of contamination
• Audits, inspections	– by authorities, division or group company

2. Water pollution control/effluent discharges

• Sampling points	– at production outfalls – site total before treatment – site total after treatment – any direct discharges (e.g., storm water)
• Effluent volumes	– at the above sampling points
• Effluent quality COD/TOC AOX/FOCL Inorganic salts	 – loads (tons per annum) – regulatory/consent limit values – number of samples analysed – number of excursions

• Reportable substances:

reportable critical substances

aromatic amines	arsenic	cyanides
phenols	cadmium	sulphides
dinitrotoluenes	chromium	fluorides
ethyl benzene	cobalt	phosphates
toluene	copper	organic phosphorous
xylene	lead	total nitrogen
naphtalene	mercury	ammonium nitrogen
halogenated compounds	nickel	
organic tin compounds	zinc	BOD

• Biodegradability: — in case of external biological treatment

3. Air pollution control/emissions

• Summary of Emissions (tons/year):
- non-halogenated organics
- halogenated organics
- carcinogens

- organic dusts
- inorganic dusts
- inorganic gases

- Emission of individual critical substances. (Class 1 substances of the *Swiss Air Pollution Control Ordinance* (LRV 1992) have been used for guidance.)

Inorganic dusts	*Inorganic gases / vapours*
cadmium and its compounds	arsine
mercury and its compounds	cyanogen chloride
thallium and its compounds	phosgene
	phosphine

Organic gases / vapours, dusts

acetaldehyde	dioxane	phenol
acrylic acid	diphenyl	propenal
aniline	ethylacrylate	pyridine
chloroacetaldehyde	ethylamine	tetrachloroethane
chloroacetic acid	formaldehyde	tetrachloroethylene
chloroethane	formic acid	tetrachloromethane
chloromethane	furaldehyde	thio-alcohols
chlorotoluene	lead alkyls	thio-ethers
cresols	maleic anhydride	toluene-diisocyanate
dichlorobenzene	methylacrylate	trichloroethane
dichlorodifluoromethane	methylamine	trichloroethylene
dichloroethane	nitrobenzene	trichlorofluoromethane
dichloroethylene	nitrocresols	trichloromethane
dichloromethane	nitrophenols	trichlorophenols
dichlorophenol	nitrotoluenes	triethylamine
diethylamine	o-toluidine	xylenols
dimethylamine		

- Chlorofluorocarbons:
  - quantities purchased, purpose
- Emissions from boilers and incinerators:
  - SO_2 (tons/year)
  - NO_X (tons/year)
  - HCI (tons/year)
  - CO_2 (tons/year)

4. Solvent use/losses

- Type and quantity of solvent:
  - halogenated solvents
  - non-halogenated solvents

• Solvent balance	– total input of solvents – input of new solvent – solvent recycled – waste solvent incinerated – solvent losses

5. Waste management

• Nature/quantity of waste:	– general wastes – sludges from effluent treatment – wastes for recycling – special wastes
• Waste treatment technique:	– recycling – incineration – landfilling – special treatment
• Waste treatment facility:	– type – internal – external

6. Safety and occupational health

• Accident/Incident Statistics	– type of incident: occupational accident occupational illness material damage environmental pollution – causes: technical human error organizational other – statistical data: number of incidents number of lost days frequency/severity loss
• Exposure Monitoring:	– critical substances – monitoring method – number of samples – reference value – number of excursions

- Status of Risk Analysis:
  - processes requiring risk analysis
  - existing risk analyses
  - pending risk analyses
  - re-validations required

7. Energy management

- Primary energy sources/use
- Energy conversion/co-generation/losses
- Energy consumption breakdown by category
- Fuel quality, sulphur content, low NO_X technology

8. Expenditure and personnel for process safety, product safety and environmental protection

- Capital investments
- Period costs
- Personnel costs
- Number of personnel

9. Cases (reports on 'SEEP' achievements)

- Prevention/reduction of water/air pollution
- Avoidance/reduction of waste
- Recovery, re-use, recycling of wastes
- Integrated pollution control measures
- Progress in environmental technology
- Reduction of process risks and major hazard potentials
- Energy conservation

10. SEEP – Highlights (report on changes, problems, developments)

- Major problems and challenges
- Legislation
- Objectives (site, division, group company)
- Organization and responsibilities
- Technology/processes/products
- Monitoring/controls
- Waste reduction programmes
- Personnel awareness and instruction
- Accidents/incidents
- Communication with public and authorities
- Recognition awards

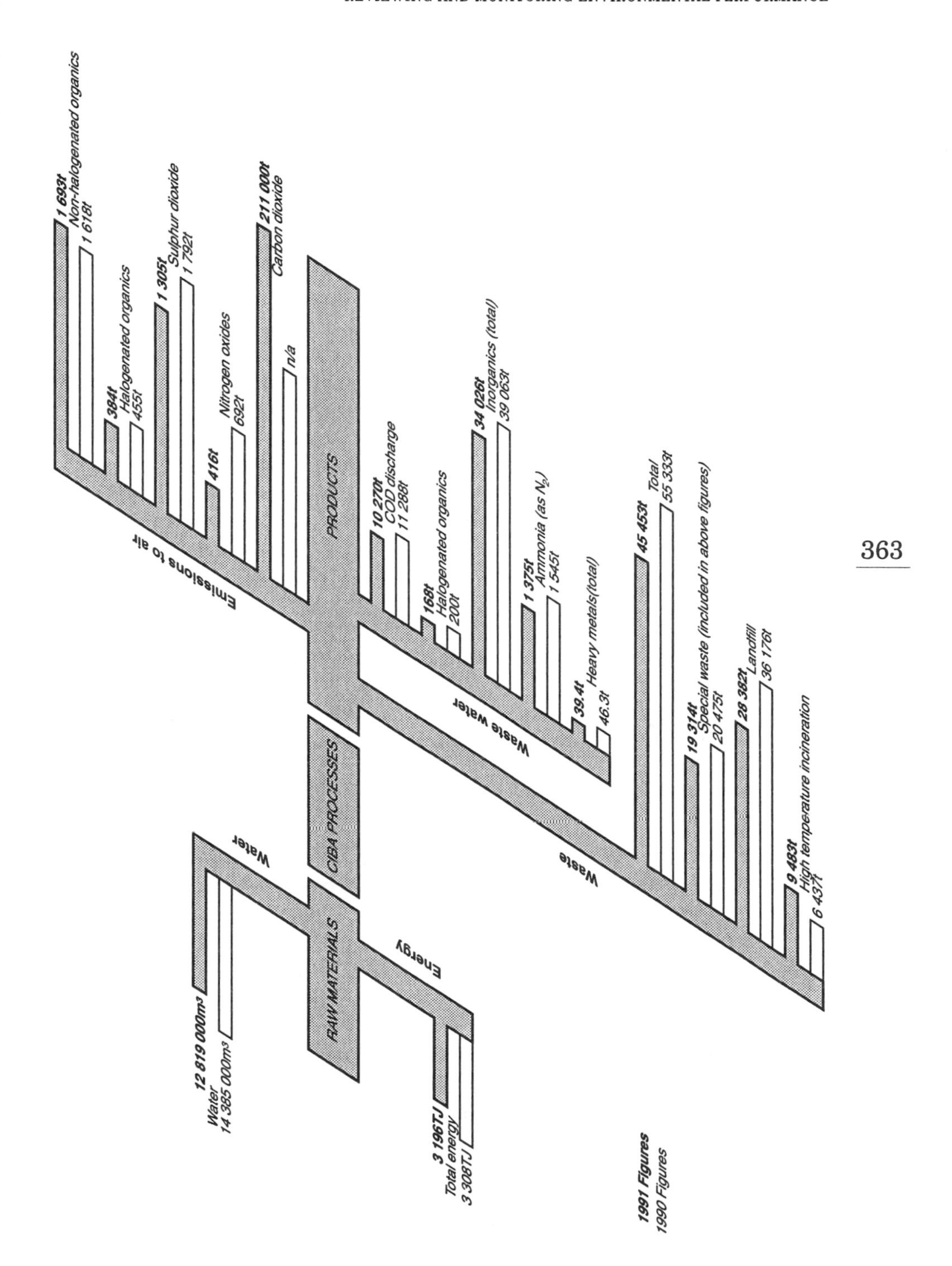

Figure 20.3 1991 Emission statement for the CIBA (UK) Group

The SEEP report has proved to be a valuable tool for corporate environmental data collection in a universally uniform format. Corporate management is provided with a global overview of a company's environmental situation, and with the information necessary to set *strategic environmental goals*. At a group company and site level the awareness of the local environmental situation is enhanced. The chances that problems are tackled and opportunities exploited increase. Reference data which has been established allows Ciba to monitor the *future development toward a sustainable environmental performance*. Finally, local management is provided with information for an *emission statement* that can be disclosed to the public.

An example of such an emission statement is given in the graph, Fig. 20.3. The information provided summarizes releases to the environment for the entire Ciba (UK) Group, covering discharges to water courses and to air, and the amounts of waste being landfilled or disposed of by high-temperature incineration. The data-collection system is a dynamic one, which is why certain information is only available for 1991. As the system evolves, it will give an ever clearer picture, so that we can set – and continue to achieve – ever higher targets.

COSTS OF ENVIRONMENTAL PROTECTION

One can distinguish between the following basic categories of environmental costs:

- end-of-pipe treatment costs;
- integrated pollution prevention costs;
- remediation costs;
- environmental taxes.

Before environmental legislation was enacted some 30 years ago, man had used waterbodies, land, and the atmosphere as sinks for liquid, solid, and gaseous wastes respectively, free of any charge.

The environmental laws brought about the installation of *end-of-pipe treatment* facilities such as effluent treatment plants, secure landfills, and – some time later – special waste incinerators. These end-of-pipe treatments incurred significant costs that had to be recovered. As a result, environmental expenses were included in the cost accounting systems of the industry, and the products were charged with their share of environmental costs. End-of-pipe solutions, although they made a significant contribution to pollution control when they were first introduced, will not suffice any longer as many of them are unspecific and therefore have a low efficiency. Biologically non-degradable substances pass a conventional effluent treatment plant unaffected. Landfilled waste, often persistent and toxic, may pose a threat to groundwater and biosphere for many generations if the deposit is not properly secured and monitored.

As these weaknesses became apparent, legislation tightened and the requirements became increasingly stricter. As a result, ever more sophisticated end-of-pipe solutions were developed, such as wet-air-oxidation for non-biodegradable effluents, multi-stage flue-gas purification, thermal waste-gas treatment, catalytic oxidation, etc. All of these new methods require equipment of highest technology and consequently have a high price.

Costs of waste treatment

Waste-water treatment	*Swiss francs per m^3*
biological effluent treatment	3–4
wet air oxidation	150–200
incineration of effluent	200–500
Off-gas treatment	*Swiss francs per 1000m^3*
biofiltration	1.25–3.75
adsorption on active carbon	2.50–12.00
catalytic oxidation	4–10
incineration (thermoreactor)	10–24
absorption in high boiling solvent	16–22
Waste incineration	*Swiss francs per ton*
liquid organic waste	100–500
solid ordinary industrial waste	160–800
solid hazardous chemical waste	2400–4000
Landfilling	*Swiss francs per ton*
ordinary industrial waste	260–370
hazardous chemical waste (inorganic)	500–1000

Each treatment method has its specific field of application, depending on the nature of waste, the concentration of waste products in water or air, and the volume of the waste streams. The costs for one particular treatment method depend, among others, upon the size of the plant and its capacity utilization. In special cases, such as the disposal of a mixture of laboratory chemicals, third parties may charge as much as SFr 20,000–40,000 per ton.

For Ciba the total costs for environmental protection including product safety and process safety amounted to 1,420 million Swiss francs or 6.4% of turnover (1992).

environmental protection:	investment costs	1.1%
	current costs	2.4%
product safety:		1.1%
process safety:	investment costs	0.6%
	current costs	1.2%
		6.4%

These are the costs which can unequivocally be attributed to environmental protection. They do not include the costs of integrated pollution prevention and product or process inherent safety. Those are difficult to determine because they are an integral part of any product's research, development, and manufacturing costs. As integrated pollution prevention and inherent safety increase, the specific environmental costs may stabilize and possibly decrease, since it is impossible to account separately for costs of integrated pollution prevention.

As end-of-pipe solutions prove to be of questionable value, and as costs become exorbitant, chemists and engineers are getting more interested in *integrated pollution, prevention, recovery*, and *recycling*. This results in *re-engineering* old processes, and in developing and implementing novel techniques. Often these redesigned processes are not only ecologically, but also economically, superior to the old ones. The following example illustrates the point.

After 20 years the process for the manufacture of an important intermediate was redesigned. In the original process the crude reaction mass was dissolved in an organic solvent and the product was purified and recovered by *crystallization*, centrifuging, and drying. The solvent was recovered, but it had to be rectified before its re-use. In the re-engineered process the product was purified and recovered by *distillation*. The advantages are obvious: omission of centrifuge, dryer, solvent recovery and rectification system, and of solvent tanks; no solvent and therefore no solvent emissions, and a much smaller fire hazard; higher yield, less waste, and a dust-free product.

As plausible as the solution may seem, the development of the re-engineered process was no easy task and took several years. The idea as such was not new, but when it was first tried out 20 years ago it failed because the distillation technology was not mature enough to render a product of satisfactory quality. Although the costs for re-engineering and the investment for a new plant were substantial, the effort paid off as the manufacturing costs were lowered by 15% and the environmental burden was reduced manifold. This example illustrates that environmental protection does not necessarily mean additional costs, but that it can be paralleled by added value.

Thus, an integrated approach to pollution prevention, i.e., tackling the problems at source, must be inherent to any technology deserving the attribute 'sustainable'.

As the sins of the past manifest themselves, and as the need to correct the detrimental effects of inadequate technology rises, so do the costs of remediation. Soil and groundwater decontamination, excavating, closing and securing old landfills cost dearly. Capital expenditures for closing and securing old landfills are in the order of 30–100 million Swiss francs per site. Current costs of the treatment of leachate and of groundwater decontamination range from 0.3 to 1 million Swiss francs per annum and per site. Decontamination of soil and gravel by a washing process runs in the order of 100 Swiss francs per cubic meter. We can only hope that mankind has learned the lessons from the past and now realizes that correcting mistakes costs much more than doing things right the first time, and that any waste avoided is money saved.

The last category of environmental costs, i.e., taxes, is just appearing on the horizon. CO_2 contributes approximately 50% to the greenhouse effect that may cause an estimated temperature rise of 3–4°C within the next three decades, if the increase in the emission of greenhouse gases cannot be stopped. The Earth Summit Conference (Rio 1992) debated the necessity and the means of reducing the quantity of man-made CO_2 in order to reach conditions of sustainable development, unfortunately without reaching a consensus. Discussions in Europe circle around a CO_2-tax (levy, 'Lenkungsabgabe') as a regulating instrument with incentive character to induce a change in energy consumption, and to curb the use of fossil fuels. Estimates based on one Swiss scenario suggest that the goal of a 20% reduction of CO_2-emission by the year 2005 can be achieved (in Switzerland) by introducing a tax in the order of SFr. 70.0/ton CO_2. Considering the specific CO_2 contribution of the various fuels such a tax would result in price increases of +50% for natural gas and light oil, +100% for gasoline and heavy oil, +200% for coal. Such a levy could be a feasible instrument provided that fiscal neutrality – meaning redistribution of the CO_2 - tax revenues to consumers and businesses (e.g., direct refunding ('Oekobonus'), reduction of other taxes, etc.) — and a co-ordination among the western European nations were guaranteed.

Conclusion

In the early days of modern environmentalism, proponents (over)emphasized singular problem areas to sound the cry of warning and to catch public attention. Opponents, in turn, tended to view environmentalism simply as a negative movement, opposed to progress, economic development, and the advance of civilization. We are now moving past such confrontation. It is becoming more widely accepted that a concern with the environment is simply good economics and planning, good management practice, and an element of total quality management.

As stakeholders and the public make use of their right-to-know and begin to show a growing interest in industry's environmental behavior, companies must prepare to disclose their ecological performance, just as they put their financial situation in the public domain. This presupposes responsible entrepreneurship and a management that regularly reviews and monitors the company's activities concerning the environmental burdens caused by them, as has been described in this chapter. The toolbox for environmental monitoring will undoubtedly be filled with more powerful and sharper tools in the future, and it may be that eventually a standardization of methods is achieved among the industry world wide.

References

1 Dictionaries and reference works, such as those of Cassell (1953), Duden (1961), Larousse (1962), Brockhaus (1964) and Webster (1968).
2 Meadows, D. L. *The Limits to Growth*, Universe Books, New York (1972).
3 United States Department of State. *Global Future: Time To Act*, Report to the President on Global Resources, Environment and Pollution (1981).
4 Schmidheiny, Stephan, with the Business Council for Sustainable Development. *Changing Course: A Global Perspective on Development and the Environment*, MIT Press, Cambridge, Massachusetts (1992).
5 Meadows, D. H., Meadows, D. L. and Randers, J. *Beyond the Limits: Global Collapse or a Sustainable Future?* Earthscan Publications, London (1992).
6 Saemann R. 'The Environment and the Need for New Technology, Empowerment and Ethical Values'. *The Columbia Journal of World Business*, Fall and Winter 1992, pp. 186–93.
7 The Literature includes such publications as the following five:

Greeno, J. L., Hedstrom, G. S., and DiBerto, M. *Environmental Auditing, Fundamentals and Techniques*, Arthur D. Little Inc., Massachusetts (1989).
ICC Guide to Effective Environmental Auditing, ICC Publication no. 483, ICC Publishing S. A., Paris (1991).
Guidance on Safety, Occupational Health and Environmental Protection Auditing, Chemical Industries Association, London (1991).
Environmental Self-Assessment Program, Global Environmental Management Initiative, GEMI, Washington DC (1992).
Grayson, L. *Environmental Auditing: A Guide to Best Practice in the UK and Europe*, The British Library, London (1992).

Recommended reading

Botkin, D. B., Caswell, M. F., Estes, J. E. and Orio, A. A. *Changing the Globall Environment: Perspectives on Human Involvement*, Academic Press, Boston, Massachusetts (1989).

Capra, F. *The Turning Point: Science, Society and the Rising Culture*, Fontana Paperbacks (1983).

Capra, F. *Wendezeit: Bausteine für ein neues Weltbild*, Knaur, Munich (1988).

Meadows, D. H., Meadows, D. L. and Randers, J. *Beyond the Limits: Global Collapse or a Sustainable Future?* Earthscan Publications, London (1992).

Schmidheiny, Stephan, with the Business Council for Sustainable Development. *Changing Course: A Global Perspective on Development and the Environment*, MIT Press, Cambridge, Massachusetts (1992).

Saemann, R. 'The Environment and the Need for New Technology, Empowerment and Ethical Values', *The Columbia Journal of World Business*, Fall and Winter 1992, pp. 186–93.

Vester, F. *Leitmotiv vernetztes Denken: Für ein neues Weltbild*, Heyne, Munich (1989).

Weizsäcker, E. U. von. *Erdpolitik: Oekologische Realpolitik an der Schwelle zum Jahrhundert der Umwelt*, Wissenschaftliche Buchgesellschaft, Darmstadt (1990).

Willums, J.-O. and Golüke, U. *From Ideas to Action: Business and Sustainable Development*, ICC report on the Greening of Enterprise. ICC Publishing, Oslo (1992).

21

Managing an environmental crisis

MIKE SEYMOUR

Executive Director, Issues and Crisis Management, Burson-Marsteller, London

- Crisis management is defined as the solving of issues and problems catapulted onto the public agenda.
- Characteristics of all high-profile crisis situations are:
  - events unfold in an unexpected manner at unusual times;
  - intense pressure on manaqement for answers to a wide ranqe of questions and concerns;
  - lack of accurate information and technical data, particularly in the early critical hours;
  - rumour and speculation compound the problems for management;
  - management is inclined to adopt a 'bunker' attitude, seeking to shut out their critics and detractors.
- Valuable operational principles for management of environmental crises are:
  - define the real problems and focus on solving them;
  - isolate crisis managers from other business responsibilities;
  - base decision-making on 'worst case' assumptions.
- Control can be exercised through communications by centralising the flow of information into and out of the crisis management group, using market research to assess perception of the company in the midst of crisis, and ensuring that company spokespeople are clear and brief.
- Advance preparation of special crisis management systems and procedures is valuable, and would cover a management audit and risk assessment as well as communications planning and training.

News and the environment

The environment and the protection of the planet have produced headlines which have achieved impact for journalists or news programmes over the last ten to fifteen years. But increasing public interest has coincided with a series of changes in the communications networks which overlays and intrudes into every aspect of modern life.

Access to instant information is accepted today as a normal way of life. News on the other side of the world is routinely played out through radio and TV newscasts – often as it happens. Such easy availability has bred an intense public interest in every issue of the day, where the environment is high on political and social agendas around the globe.

What is a crisis?

Businesses are complex organisations striving to operate profitably within a series of financial and regulatory constraints. All companies, be they large or small, face a range of risks and problems in the daily conduct of their business. Under normal operating conditions, managers solve these problems quickly and efficiently; that i why they are selected, trained and paid to run their organisations.

The rules of normal operations change when routine situations are escalated or thrust onto the public agenda. Circumstances are suddenly altered when the problem is picked up by citizen or community groups, highlighted by activists and discussed by politicians. Suddenly situations escalate out of control, when the media are at the office door or besieging the gate to the plant.

Any group of managers will list environmental risks as high in the hierarchy of potential problems facing their business. Recent events have proved that environment issues, accidents and disasters engage the attention of the public and the media quickly and frequently.

What is crisis management?

Against this background of inherent risks, corporate decision-makers have had to respond to the speed of modern communications. Reacting to the ebb and flow of news in the global village, managements have been forced to handle their problems under the glare of public and media scrutiny. They have had to master new management techniques.

Today crisis management can be defined as 'solving issues, problems and accidents which have been catapulted onto the public agenda'. Since environmental aspects of any business are under constant review and observation, in a time of a crisis, corporate decision-makers are forced to manage in a 'fishbowl' environment, in the full focus of public scrutiny.

A series of accidents in Germany occurred in 1993, initially at a BASF plant, which were followed by problems in other parts of the Hoechst AG organisation near Frankfurt. Yellow-coloured rain emissions were followed by several fatal accidents in different plants. Management had to struggle to solve this series of problems against a blitz of media discussion. The public had their concerns fanned by Government officials, Greenpeace, BUND and other activist groups.

Managing environmental crises

This chapter is concerned with crisis situations which relate to such environmental disasters. However, the characteristics and principles discussed relate to situations experienced in other aspects of operational and corporate management. In any crisis, whatever the root cause, the one major common factor is the potential risk of adverse or negative impacts on the business, which is seen as inherent in all these situations as soon as they are brought into the public domain.

Environmental crises demand solutions to complex problems which are accelerated and escalated by the communications that define and affect businesses and private lives. Pressures are caused by harsh, ill-informed public judgements about the actions and decisions of management. Initially critical discussions need solutions in hours, before deadlines are reduced to minutes by individuals and groups acting outside the company.

An accident at a chemical dump at Holmestrand in Norway in 1992, stemming from a simple error, led to the release of a cloud of poisonous gas. Poor procedures resulted in a long delay in informing the Norwegian Pollution Control Board. By the time management attempted to control the situation, the media and the local community were calling for immediate action and evacuation, before it was announced that the chemical cloud was not dangerous.

The history of crisis management

Environmental disasters and modern crisis management can be traced back to Bhopal and Union Carbide's handling of the catastrophic gas leaks in 1984. Five years later, the general public watched with horror when the Exxon Valdez ran aground in Prince William Sound in Alaska; the subsequent management of the devastating oil spill caused widespread damage, which could not be forgiven by the public watching through the eyes of the media.

Two of the many recent cases serve to recall a series of accidents and incidents which have suddenly dragged companies, plants and managers under the glare of public attention.

- In 1992 a toxic cloud hanging over an Italian city was traced to a small chemical plant. The management denied that their operation had caused smells and emissions. Thirty minutes later the plant exploded killing four people. There were immediate demands for manaement action and a review of procedures, led by media and activist groups.
- A chemical fire in Cologne in 1993 was followed by dioxene rain which spread over the plant and the environs. The plant was closed and soil clearance of the surrounding area was necessary. Greenpeace and local groups pressed for tighter controls, including the closure of the plant.

The constituent parts of any crisis

All environmental crises will be different. However several common factors can be identified within each of these individual, high-pressure situations. Management needs to recognise the constituent parts of any crisis and take these factors into account when planning special handling techniques.

THE ELEMENT OF SURPRISE

The focus of an environmental issue or the initial problem will always come as a surprise. Any plans that do exist will not have anticipated the exact set of circumstances facing the decision-makers. Furthermore the intensity of interventions from outside will always catch management off guard.

Timing – or mis-timing – plays its part in the surprise. Accidents always seem to happen on a Sunday, during a public holiday or in the middle of the night!

PRESSURE FOR ANSWERS

Pressures on managers increase quickly as demands for answers flow in. Calls come from the media and other interested parties, blocking vital telephone links. Worried relatives, the local community, customers and consumers start calling for information because they have heard the news.

Local newspapers, radio and TV channels pick up the story, but their attentions are quickly overshadowed by regional, national and even international media. Regulatory authorities, environmental and safety officers, plus government departments, demand facts, figures and statistics. Soon public officials are seeking to brief Ministers and politicians.

Against this upward spiral of escalating events, management seeks to handle swelling interest, while sensing an increasing lack of control.

Since environmental issues and accidents always increase public concern, citizen and activist groups will be upon the scene quickly. Greenpeace, Friends of the Earth, Legambiente in Italy, Vereniging Milieudefensie in the Netherlands, Bellona in Norway, plus a range of conservation groups, all stand ready to seize upon any incident where industry and business could be to blame. Shrewd and skilled use of the media result in speedy vocal demands for action and answers to queries. Management will quickly find themselves facing a series of groups of knowledgeable and persistent critics.

INSUFFICIENT INFORMATION

At this time, managers need facts, figures and details to allow them to assess the situation and to answer the tidal wave of questions from within and outside. But initially, in any environmental crisis, information will be sparse; any facts that are available are often inaccurate or based on a limited assessment. Often vital technical data will be unobtainable.

RUMOUR AND DISTORTION

Rumour and speculation will compound all these problems. While management is struggling to define the extent of the true problem, outside 'experts' will be seeking publicly to interpret each twist in the escalating situation. Quickly, facts merge with perceptions, while all media discussion appears to distort and exaggerate every aspect of the deteriorating situation. The basic story is quickly extended by the media, who will call on a wide range of sources to add depth and colour to the few facts that are available.

UNDER JUDGEMENT

At this stage, corporate management is forced to react to the rapid build-up of activity with full realisation of the degree of scrutiny from outside the organisation. Each decision appears to be observed minutely, suffering judgement by internal and external audiences. Pressures are increased by the apparent lack of support and understanding against which these judgements are being made.

SHUTTING OUT THE OUTSIDE

Next the management team reacts typically by attempting to shut out the world outside, as all efforts become focused on 'solving the problem'.

At this time a bunker attitude breeds an unwillingness – or inability – to communicate internally or externally.

The final twist in any crisis is seen when the headlines and news reports become the facts of the situation. Management focuses on these headlines as their primary problem – rather than face the requirements to communicate with employees, the local community, customers, regulatory authorities, environmental officials, etc.

Operational principles for managing environmental crises

The constituent parts of any crisis place a severe pressure on management, as they struggle to solve complex, technical problems. Experience of working alongside corporate managers has revealed a series of principles which have come to the aid of decision-makers, working under intense and demanding circumstances. Over and over again, these guidelines have been proved in practice.

PROBLEM DEFINITION

Earlier in this chapter, it was seen how the media and other groups unwittingly contribute to management difficulties by forcing inaccurate, fragmented information into the public domain. In any environmental crisis these actions attract other interest groups and activists to the problem. Together they seek to interpret data, so compounding the situation. Thus in the middle of a crisis, when facts are distorted by rumour and speculation, it is essential to *define the real issues*. Only then can management start to prioritise the options for solving the problems.

ISOLATION OF KEY MANAGEMENT

In the maelstrom of a public crisis, key managers need to be able to concentrate exclusively on the complex, demanding aspects of crisis management. In order to achieve the necessary cushioning from distracting day-to-day problems, all key management team members must, *delegate other responsibilities* for the duration of the crisis.

ASSUMPTION OF WORSE-CASE SCENARIOS

Management focusing on the situation face a rapidly accelerating escalation of events. Since so many of the facts are missing, and information is being distorted, *a worst-case assumption must be made.*

In order to ensure that plans anticipate the most difficult problems, management must assume that the toxic cloud will spread over the local community as the wind changes. The first statistics should not be trusted – plan on the number of casualties and deaths increasing. In spite of any evidence to the contrary, assume that the company is to blame.

But beyond unfolding events, managers must anticipate the hidden factors that will have lain undiscovered until this moment. Over and over again, '*a smokin gun*' is revealed such as:

- A disaffected employee, maybe recently dismissed, who is ready to tell the media that similar situations have happened before – but top management did not reveal these incidents.
- A memo lies forgotten in an old file which rejects a proposal for the very safety measure or device which could have prevented the accident.
- A whistle-blower claims that management were briefed on the risks but chose not to take any action.

Exercising control through communications

In June 1993, an accident in a chemical plant in South Italy caused an explosion which killed seven workers. While the cause of the accident was still being established, the plant owners stood accused on a reputation which had been created by the media and pressure groups. Over a period of several months the company had stayed silent while a series of allegations were discussed publicly. Pressures caused the Ministry for the Environment to classify the plant as a 'high risk' facility. The company faced a communication battle where their opposition had decided that management were 'guilty', even before the facts of the accident had been established.

Information, facts and perceptions will always be central to the shaping of any environmental crisis. Emotional responses to misunderstood risks and dangers will increase the need to communicate quickly and widely.

But communications, particularly in the public domain, appear unfamiliar and menacing to the conventional, operationally oriented manager. Threats appear to grow, when external aspects seem to be conspiring to reduce the control exercised by any crisis management group. Experience of crisis handling has permitted communications and management consultants to define some guidelines for seizing and maintaining control. These points that follow all focus on communications aspects as essential elements in successful management techniques. Each point provides essential support to the operational management of a crisis.

CENTRALISE INFORMATION FLOW

The first requirement is to see that the key crisis managers are taking decisions based on the best available information. This process will require a concerted effort to draw all sources towards the *crisis management centre*, where careful collation and assessment must take place.

In any environmental accident, casualty and damage statistics become an early focus of attention. However, numbers will fluctuate and alter, especially during the first critical hours. The police and fire services will release information often without checking with the company, which takes the initiative away from management. Thus every effort must be made to establish *accurate facts and figures* before exercising tight control over release of this critical data. It is particularly important to apply rigid rules over the release of names of any dead or injured persons.

SPEAK OUT CONSISTENTLY

A most effective way to demonstrate control to the outside world is to speak consistently. In early 1993 during the first twenty-four hours following the grounding of the 'Braer' tanker off the Shetland Isles, twenty-three different 'experts' appeared on TV news reports to discuss and speculate on the accident. During that time the tanker owners said little and responded timidly to media's queries. The company only began to still the discussion and reduce criticism when they spoke out clearly, using one spokesperson.

During an environmental crisis, external communications should be consistently presented through *a limited number of authorised and trained spokespeople*. At the same time, clear instructions need to be given to all employees on how outside enquiries are to be handled and where media calls should be directed.

Once the press and other external audiences come to recognise that the outflow of information is under control, speculation levels are inclined to drop, thus permitting crisis management to begin to slow the pace of events.

UNDERSTAND THE MEDIA

As is seen in the aftermath of any major accident, and was demonstrated graphically during the 'Braer' disaster in early 1993, consultants are called in with the sole remit of controlling the media and reducing press interest. However, this approach ignores the fact that the media will continue to pursue a story just as long as editors believe that the situation is newsworthy.

It is better to adopt a worst-case approach when planning how to respond to demands from the media. After years of treating the media as an unwelcome intrusion into management problems at the scene of any disaster, the Central Office of Information (COI) in the UK has come to *accept journalists and camera crews as part of any major incident scenario*. Special media-handling courses are run for Emergency Planning Officers; a team of COI Press Officers is on permanent standby for deployment to the scene of any serious environmental accident, explosion or major transport crash.

It is regrettable that many companies still view the media with suspicion, preferring to stay silent in the hope that all the journalists will go away. To offset this attitude, it is necessary to appreciate the following facts about the media:

- The media are interested in finding out what happened and who is to blame, and to convey that information to the public. Reporters need to find out enough to make the facts clear and easy to understand for a typical reader, viewer or listener. Without exception, they will seek to simplify the scientific and technical aspects of any problems. Spokespeople can gain control if they learn to *speak with clarity, in a simple style*, avoiding jargon, always remembering that many journalists have little knowledge of the complex issues that surround any serious environmental problem.

- An editor is always short of time and usually short of space. This search for brevity often causes the media to reduce its questioning to three main categories:

  (a) *How did the accident happen?* What will be the likely environmental effects and damage?
  (b) *What went wrong?* Who is to blame?
  (c) *What steps are you taking to clear up the mess?* How much compensation will you be paying out?

Long, complicated press releases may have their place as background documents, but will not answer the needs of a journalist under the pressures of time and space. Management should remember that:

- Deadlines are a key factor in preparing radio and TV programmes or newspaper production. They should be respected; answers should be given by the time promised or, if the facts are not known by the deadline, the reporter should be called.

- If the company does not supply information, the journalist will be forced to go elsewhere.

- Since the definition of news requires an emphasis on things that go wrong, a company's bad news is good news for the news editor. When anticipating how the media will respond to environmental crises, key managers have to accept this paradox. It must be realised that the very problem or issue that a company will hope not to see featured is often the story a journalist hopes to uncover.

RESEARCH BEHIND THE HEADLINES

When a crisis is receiving full media and public attention, the pressure of events may cause management to become transfixed by newscasts and newspaper headlines. Just at the time when information is confused and solid facts are unavailable within the company, everyone outside appears to know everything. At this time it is important to *discover what critical audiences really feel* about the problem and the company reputation.

Crisis management experts advise the early setting up of market research against relevant target audiences. In an environmental crisis, (e.g., following a toxic spill or accidental leak to a river), market research would be carried out among parents, teachers, local community leaders and citizen groups. Whenever possible, research should be conducted by telephone to get results as quickly as possible. It should be feasible to set such research up within a half day and deliver results the following day.

The results will show what are the real attitudes towards the crisis, the responses by plant management, and clean-up activities, while indicating how responsibly the company is seen to be acting in the minds of key audiences. Repeating the research at regular intervals allows the company to see how attitudes are changing – for better or worse.

Then, based on a solid, foundation of knowledge. it is possible for the company to prioritise audiences and decide what messages should be directed to whom, thus answering the needs of all those who have an interest in solving the crisis and protecting the environment, without sending out conflicting messages to diverse audiences.

COMMUNICATE TO ALL AUDIENCES

Many different groups and individuals become involved in any company crisis; each will have a different view of the situation and comprehension of the problems. If corporate management is to be perceived as proactive and in control, it will need to act and communicate in terms clearly understood and appreciated by all these diverse groupings.

Thus the communications challenge is extended beyond a vain attempt to hold off the media and switches to *the development of a programme*, which seeks to reassure, show concern and demonstrate an understanding of the various concerns being felt.

The prime communications task is to address a wide range of audiences, which must include:

- Employees and their families, since they are the company ambassadors.
- Supervisors and plant managers, divisional and corporate management, because they can ensure internal understanding.

- Trade unions, because they will be questioned and will feel the need to comment.
- Customers, because they will be concerned about interruption to supplies or services.
- Contractors, distributors and suppliers, because their business could be at risk.
- Technical experts and financial analysts, because the media will ask for their views.
- Activist and citizens groups, because they will be called upon to comment and need the balance of the company viewpoint – whether they want it or not.
- Local and national elected representatives, because they will want to be seen to be in the know.
- Leaders of communities in which employees live and work.
- The media, who will pursue different storylines, depending on whether they come from the local, regional or national media and specialist publications.

Since these tasks are so demanding, companies often call upon the assistance of public relations and public affairs experts to bring their expertise to bear upon the planning and execution of integrated communications.

Preparing ahead of the crisis

All business organisations have plans and procedures for handling day-to-day problems. Environmental standards, regulatory requirements and workplace legislation set out the parameters within which all industries must operate. However, over and over again, environmental crises show up the inadequacy of normal contingency plans. Special steps are needed to prepare for handling these high-pressure, high-profile management challenges.

Since operational and environmental problems can strike any business organisation at any time, companies, both large and small, are calling upon consultants to help them plan and execute programmes for getting ready to handle crisis situations. A typical preparedness programme would cover:

- Management audit and risk assessment.
- Plan preparation and manual production.
- The formation of an effective crisis communications team.
- Training, testing and validation.

MANAGEMENT AUDIT AND RISK ASSESSMENT

Since the concept of environmental crisis management is concerned with the handling of complex problems while managing outside perceptions of the company, the first step to getting prepared should be an objective assessment of potential environmental risks and vulnerabilities.

In Europe today, senior management is realising the value of bringing in independent experts to evaluate management, financial, operational and communications procedures. As business becomes more complex, these consultants are finding it necessary to work closer together to provide joint assessment teams which include environmental and technical expertise. Typically these teams conduct a comprehensive assessment and planning process which would include:

- Assessment of risk areas and business vulnerabilities.
- Objective analysis of current operational and management procedures, plus preparation of contingency plans.
- Development of tailored crisis-management plans.

PLAN PREPARATION AND MANUAL PRODUCTION

One of the major recommendations, to come in the aftermath of many environmental accidents and crises, is the need for action-oriented plans and back-up materials. This second stage of preparation covers the building of plans to create crisis management teams and the production of a procedures manual.

Again, this specialist area is being undertaken by external consultants although prudent companies rightly insist on the work being undertaken in close consultation with internal project teams and technical experts.

Since plans must be quickly understood and adapted to fast-moving, high-pressure situations, it is essential that management be confident and comfortable with them. Final formats for procedures will have to be adapted to mesh with the unique culture, organisation and management style of each company; top management should be suspicious of template solutions which are unlikely to suit their particular needs. A single plan cannot apply universally in light of the differences in regulatory aspects which demand variations in procedures for different markets.

Once the plans have been agreed, they should be contained in a special manual, for issue on a limited distribution to crisis managers in the company. A typical manual would cover:

- Crisis management teams and support staff, showing nominees and deputies with their roles and responsibilities.

- Crisis management procedures, to include alerting and triggering systems, plus detailed plans for handling areas of high operational and environmental risk.
- Standby statements for use in press releases or by spokespeople in discussion with the media.
- Lists of telephone, fax and mobile numbers for key internal and external contacts.

While the manual will be compiled to answer crisis management needs, plans must be constructed to match with specialist support groups which will be in existence in most business organisations, e.g.:

- regulatory and legal teams;
- quality control and technical teams;
- environmental, health and safety teams.

While these teams will require specialist support plans, it will be important to ensure that these dovetail into procedures being used by the crisis management team.

TRAINING, TESTING AND VALIDATION

The last stage in preparation for crisis training, which should take place as soon as plans have been completed, should cover:

- Awareness training.
- Testing and validation through simulated environmental crises.
- Media and interview training for spokespeople.

AWARENESS TRAINING

Comprehensive plans will have little worth if all employees are not alert to problems and know how to report early warning signs. It will be necessary to brief operational and office staff on preparedness plans, and on their roles in making them work. Since different levels of understanding will be needed, it may be necessary to conduct separate briefings or training sessions for each of the following groups:

- Crisis management and specialist support teams.
- Other corporate managers.
- Middle managers and supervisory grades.

TESTING AND VALIDATING THROUGH SIMULATION

Companies are making increased use of simulation to test and validate crisis procedures and teams. This flexible training medium is best run by training experts with experience of facilitating these demanding sessions. If carefully planned and imaginatively developed, simulation can answer a range of requirements, which could include:

- Testing decision-making procedures.
- Building effective crisis management teams.
- Validating the composition of the crisis teams, management procedures and technical support.
- Testing preparedness levels and the handling of communications through realistic media training.

MEDIA INTERVIEW TRAINING FOR SPOKESPEOPLE

Executives selected to speak to the media on behalf of a company in the middle of a crisis, face a daunting task for which they require special preparation. Media interviews, under these stressful conditions need to be conducted with the help of communication specialists, with direct experience of handling radio, TV and newspaper reporters when in pursuit of crisis stories.

A typical training session would cover:

- understanding how the media works;
- message development for issues and risk areas;
- learning or refreshing interview skills;
- practical experience of radio, TV and down-the-line telephone interviews.

Subsequently it will be necessary to update and refresh interview techniques, especially if a serious accident occurs and a crisis looms.

Handling a crisis – the ultimate test

A manufacturing company in London, whose plant was bounded by residential housing, a road and a canal, had reason to test plans and procedures – for real.

At lunchtime during a normal working day a load of packaging materials was being unloaded by a forklift truck. While picking up the first pallet, a 45-gallon drum of di-isocynate was knocked accidentally; the drum fell off the truck and split on hitting the ground.

In the emergency that followed, two employees were trapped inside the warehouse area, their only exit blocked by the spilt chemical which was toxic if inhaled. Because the accident happened very quickly, it was not known if toxic materials had gone into the drainage system. The wind direction threatened local houses, and neighbouring industrial premises. The Corporate management and operational safety teams swung into action, initiating the plans and procedures which had been completed and tested four months previously. Close liaison with the Emergency Services, and a comprehensive communications plan kept employees and outside agencies fully informed. Local residents were warned and evacuation plans worked efficiently.

The damaged drum and residual chemical was sealed and recovered from the site. The spillage was cleared and the drainage system was flushed and cleared out using a 'sludge-gulper' .

After five hours of intense activity the emergency was contained and the hazard cleared. As a result of the speedy, responsible actions by the management at all levels, local and media reports focused on the successful and safe handling of a potentially dangerous situation.

This, and other well handled incidents, demonstrate that successful management of environmental crises comes from being prepared and ready in advance. No company can afford to ignore this important aspect of corporate responsibility.

Recommended reading

Berge, Dieudonnée ten, *The First Twenty-Four Hours: A Comprehensive Guide to Successful Crisis Management*, Basil Blackwell, Oxford (1980).

Bland, Michael, *The Crisis Checklist*, Public Relations Consultants Association (1992).

Dealing with Disaster, HMSO, London (1992).

Gillions, Paul, *Crisis Management: The Communications Imperative*, Disaster Management (1988).

Lawson, Dominic, 'The Master of Disaster', *Daily Telegraph Weekend Magazine*, London, March 1989.

Le Risk Management: Pour une meilleure maîtrise des risques de l'enterprise, Enterprise Moderne d'Edition (1982).

Lindheim, James B. and Wodin, Frederick, *Communicating in the New Environmental Age*, Burson Marsteller

Lindheim, James B., European President Burson-Marsteller, 'Environmental Communication: Why Bother?' Address to The Association of Petrochemical Producers in Europe, May 1992.

'Management Brief: When the Bubble Burst', *The Economist*, London, August 1991.

Meyers, Gerald C. with Holusha, J. *When it Hits the Fan: Managing the Nine Crises of Business*, Houghton Mifflin, Boston, Massachusetts (1986).

Regester, Michael, *Crisis Management: How to Turn Crisis into an Opportunity*, Hutchinson, London (1987).

Seymour, Mike, 'Craft a Crisis Communications Plan', *Directors and Boards*, Summer 1991.

'Tales of the Unexpected', *Director* magazine, London, March 1990.

Part V

■

ORGANISING THE CHANGE

'The Stationary State would make fewer demands on our environmental resources, but much greater demands on our moral resources.'

Herman Daly, Steady-State Economics, 1991.

'I have never seen a stronger force for coalescing the organisation about a common purpose than the environment.'

Edward Woolard, CEO Du Pont, quotes in 'Changing Course' (Schmidheiny, 1992).

Introduction to Part V

■

The four contributors to Part V discuss some of the options available to business as it endeavours to incorporate sustainable development into its working practices.

In Chapter 22, David Wheeler focuses upon the interactive nature of environmental management within The Body Shop. He describes the structure in place, both for internal and external audiences, to encourage maximum participation and ownership of The Body Shop's environmental policies.

In contrast, Axel Wenblad, in Chapter 23, concentrates more upon the challenge of implementing change in a large, decentralised organisation. He explains how Volvo's environmental policy was established and how it has been implemented.

John Lawrence, in Chapter 24, describes the role of an environmental adviser and how environmental advisers can play a key part in 'cascading' environmental policies. He draws upon his experience as an environmental adviser to ICI.

The fourth contributor, Tony Hill, makes the important point, in Chapter 25, that Procter & Gamble are only as environmentally excellent as their suppliers. He emphasises the need to work with suppliers to achieve continuous environmental improvement. The chapter contains detailed, practical advice on how to achieve this to the benefit of all concerned (suppliers, their customers, their customers' customers and so on).

In discussing how to organise their businesses for long-term, 'quality' profit, the contributors are drawing upon the concepts and techniques discussed in the preceding sections such as sustainable development, life cycle analysis, continuous improvement, environmental auditing, environmental management systems, environmental policy and the pressures for change. To fully appreciate these concepts and techniques it is necessary to refer back to their relevant sections.

A number of key learning points emerge from Part V, many of which will not surprise managers experienced in organising change. They include the importance of: top-level support for improved environmental performance; a clear, regular communications strategy; education and training; rewarding improvements at individual and collective levels; allocating the necessary human, financial and technological resources, and monitoring progress.

An important issue that emerges is the creation of a corporate culture which encourages all individuals to contribute to the improvement of the organisation's environmental performance. To create such a culture, the organisation's environmental policy must be integrated with its overall business strategy and then monitored accordingly.

Executive summaries

■

22 Using effective communications to improve environmental performance

The Body Shop is committed to the highest level of environmental management which is incorporated into every aspect of its operations. This chapter examines the communications philosophy and practices of a company which combines a creative and innovative approach to the production and sale of skin and hair-care preparations, with a deep commitment to social change and ecological responsibility.

The Body Shop International does not make extravagant claims for its achievements; the Company recognises that only by continued vigilance, good organisation and a high level of commitment to communication and education can progress continue. For The Body Shop, care for people, care for the environment and concern for animal rights are essential components in its business strategy. This requires a dynamic relationship between the Company, its staff and its customers,

This chapter touches on several aspects of environmental management which are relevant to The Body Shop, focusing especially on communications and education. It also includes information about the development of environmental management systems in retail outlets and worldwide operations via comprehensive systems of environmental reviews, audits and awareness-raising.

23 Defining the vision and implementing the change

Volvo was prompted to initiate its environmental protection activities by the growing attention focused on pollution problems towards the end of the 1980s, a decision which required not only that the vision of a pollution-free environment be defined, but that the manner of its achievement be described. The environmental policy developed by the company at that time was based on an overall view of the environmental impact of its products 'from cradle-to-grave'.

Since policy implementation procedures are based on the Group's highly decentralized organization, training has been an essential element in implementing the necessary changes. Environmental auditing – a procedure designed primarily to monitor policy implementation and to provide support as part of a dedicated environmental programme – is another major element in this context.

In keeping with its overall view of the environmental impact of its products, Volvo has taken the lead in developing a tool for environmentally sound product development. Known as EPS (Environmental Priority Strategies), the system enables environmental impact to be calculated in a manner transparent to all interested parties.

Whereas the policy was formulated originally during a boom period, its implementation has coincided largely with a recession in the world economy. This change in the economic climate has shifted the emphasis from capital investment to monitoring, optimization and the development of solutions for the future.

Although the programme is carried out at local level within a decentralized organization, a strategy common to the entire Group is essential as a common reference datum. A strategy document, in which the environmental policy is defined in concrete terms, has been prepared for the period 1993–95.

24 The Environmental Manager as a change agent

The task of the Environmental Manager is to act as the environmental conscience of the company and to ensure that, in various ways, changes occur in response to the wider expectations of society. By keeping everybody informed and by exerting his/her influence and authority he has to ensure that good environmental policies and targets are set at the highest level and that they are understood and implemented right through to the shop floor. All this must be done with an awareness of the economic realities. Environmental regulation world wide is extensive and complex and an important task is to establish processes by which the company is kept informed of what is required and of future trends.

The concepts are illustrated extensively by reference to the practices of Imperial Chemical Industries PLC which has a very large number of products and operating sites throughout the world. ICI has set itself a clear environmental policy, also targets to reduce waste and to improve its environmental performance in other ways. The results are reported annually to employees, shareholders and the public in an annual environmental review that accompanies the Company's Annual Report. Mandatory standards are set with respect to a wide range of issues, and to the management structures necessary to conform to quality systems such as ISO 9000. In addition, a great deal of non-mandatory guidance is provided to help operating units to achieve high standards in relation to specific local circumstances.

Environmental expectations and requirements have increased astronomically in recent years. It is the Environmental Manager's job to ensure that the company responds by changing in a way that ensures the continuing respect of the community at large and at the same time remains successful.

25 Working with suppliers to reduce waste and environmental impact

Businesses that plan for long-term growth recognise they must reduce the impact on the environment of their products over their total life cycle. Working with suppliers to reduce waste and environmental pollution is discussed in this chapter, as part of this. The programme is strategic in motivation – ensuring long-term growth by doing what is right and valued by customers. However, examples are also given that illustrate how working with suppliers can reduce waste now, and bring immediate cost savings.

A policy statement, which needs support from the top of the company, forms the basis for setting up a programme. Four environmental purchasing programmes are examined to illustrate how some leading companies have set about this. The chapter ends with practical ideas and a list of recommendations to bring continuous improvement into an environmental purchasing programme.

22

Using effective communications to improve environmental performance

DAVID WHEELER

General Manager, Ethical Audit,
The Body Shop International, UK

- The Company does not believe that environmental commitment is a form of beauty contest and, unlike some companies, does not wish to play on the environmental fears of its customers simply to increase the sales of products.
- Environmental management systems and auditing procedures may identify problems or opportunities for improvement, but both depend entirely on effective environmental communication in order to facilitate and achieve progress.
- Environmental advisers act as the main conduit for environmental information within the Company. Examples of good environmental practice are rapidly circulated via networks of advisers at retail, head office and international levels.
- The Body Shop has put in place a formal system of product stewardship, including Life Cycle Analysis (LCA) which will ultimately establish the environmental credentials of every one of the company's product ingredients and packaging types.

The need for a clear message

In recent years the environmental debate has become a battleground, increasingly dominated by technical experts and commentators while policy-makers appear unable or unwilling to intervene. Sadly, the rhetoric of concern, exemplified by the Rio Earth Summit declarations, and Agenda 21, has yet to be turned into real action by governments.

For The Body Shop, all the technical and scientific uncertainties surrounding environmental issues simply mean that the Company has an even greater responsibility to provide factual information to its customers. Like everyone else, The Body Shop's customers are not sure about the likely consequences of global warming, ozone depletion and deforestation. They are sure, however, that in using cosmetics and toiletries they do not wish to contribute to the worsening of these problems.

Thus, The Body Shop has no problem whatsoever in accepting the 'precautionary principle', and is absolutely committed to minimising the environmental impacts of its own operations. The Company's customers and staff would accept nothing less. But the immediate consequence of this stance is that the Company also has a responsibility to educate, communicate and campaign for change in order that environmental degradation can be arrested and potential catastrophes avoided. But understanding, and to a certain extent leading, public opinion on issues as politically controversial as tropical deforestation, or the rights of indigenous peoples, is not the conventional role of a retailer. Why does The Body Shop seek to play this role?

> When my grandchildren ask me what I did in the war against environmental destruction, I want to be able to say that I tried. Can I look them in the eye and say anything else?
>
> Robet Galvin, Campaigner

First, it should be recognised that even if customers did not expect such a high level of environmental responsibility from The Body Shop, it would not affect the Company's position. In this respect, The Body Shop differs a great deal from those commercial enterprises which seek to exploit thc environmental sympathies of the consumer. From its very beginnings in 1976, the Company adopted an ecological stance, developing products based on natural ingredients, avoiding cruelty to animals, maximising opportunities for the re-use and recycling of its packaging, and striving to minimise waste in every aspect of its operations. The Company was committed to high standards of environmental performance well before this became fashionable.

Second, The Body Shop has never sought and never will seek commercial advantage because of the environmental principles of the Company or its customers. The Body Shop rejects 'eco-opportunism'. The Company welcomes the rise of environmental concerns and hopes it is doing its best to respond to environmental challenges. But it does not advertise its products, and it does not make extravagant claims about its green credentials.

For The Body Shop, the environment is too important to be used simply as a marketing weapon. The Company does not believe that environmental commitment is a form of beauty contest and, unlike some companies, does not wish to play on the environmental fears of its customers simply to increase the sales of products. Sadly, there are many examples of this phenomenon in the UK. One of the most obvious is the tendency of retailers to claim 'environmental friendliness' (a meaningless term), 'recyclable packaging' (even when no infrastructure exists to recycle) and 'ozone friendliness' (even when CFCs or other ozone-depleting chemicals have never been used with the product in question).

Cutting through the hype on the high street

It has become fashionable for commentators to assert that ordinary people have lost their enthusiasm for green consumerism since its peak in the late 1980s. At that time, a string of publications emerged, for example *The Green Consumer Guide*[1] and *1001 Ways to Save the Planet*,[2] to provide guidance on how consumer power could be turned into positive pressure for environmental improvements. But the original optimism was soured by a combination of economic recession and a failure of the retail and manufacturing sectors to provide consumers with honest information and products that worked.

The resulting demoralisation and confusion was enough to prevent many people embracing the green consumer ethic. Thus one of the most powerful signals which can be sent to commerce and industry – increased (or decreased) market share – never really materialised. Consumers were faced not so much with a choice between greener or less green products, but between differing noise levels and competing claims.

Sadly, the upsurge in product-related green hype coincided with an antipathy on the part of governments to intervene at a regulatory level. In Britain, The Trade Descriptions Act of 1968 was of only marginal value in stopping the worst excesses of consumer deception. The recommendation of the House of Commons Environment Committee that the 1968 Act should be amended in order to put beyond any doubt the fact that it may cover environmental claims, remains only a recommendation.[3] Instead, throughout Europe increasing attention was devoted to market-led approaches (notably official eco-labelling schemes) and voluntary codes on advertising. Neither show any sign either of increasing the authority of genuine green claims, or of eliminating those claims which range from the meaningless to the downright deceptive.[4]

The European Community eco-label was heavily criticised even before it started because criteria were deemed unacceptably low by environmentalists.[5] Non-governmental organisations were especially concerned that too much importance was attached to the views of industrialists and not enough to those groups representing the environment, animal rights and human rights – all of which were considered relevant by ecologists.

Many leading retailers and supermarket chains who have made public noises about their ecological concern, and who placed appropriate literature on their shelves, still stock brands (and in some cases their own brands) which claim ozone friendliness, environmental friendliness, recyclability, biodegradability, etc., on wholly inappropriate products. Little wonder then that a number of these chains started removing 'deeper green' brands from their shelves: they discovered that green claims were sufficient to satisfy all but the best-informed consumers.

This is the background against which The Body Shop has to compete for its own market share: retail chains that apply a thin green veneer to their operations (and an even thinner veneer to their products) and a policy framework which cannot even define criteria for official eco-labels without giving overriding influence to retail and industrial lobby groups. A disadvantage The Body Shop has suffered in comparison with its competition is that the Company does not pay to advertise its products; neither does The Body Shop make any green claims on its product labels.

However, The Body Shop does maintain three crucial advantages which count for far more than the hype of its competitors, and which mean that the Company continues to perform well in the market-place while maintaining its principles intact.

The first and probably most important advantage is that The Body Shop produces high quality and popular products that work. A second advantage is that there is a close relationship between The Body Shop and its customers; this relationship is one of trust and mutual respect. Finally the dialogue which The Body Shop has with its staff is vibrant. Consequently, the dialogue between shop staff and customers is animated, engaging and meaningful. The dialogue is not just related to product performance and sales information, important though these undoubtedly are. The dialogue also includes stories behind particular ingredients or products which reflect The Body Shop's values: social and environmental concern; animal rights; and a commitment to fair trade with less-developed countries. Thus, when there is a need for direct mobilisation of The Body Shop's customers on campaigns these customers are already sympathetic and willing to participate.

The rest of this chapter deals with the way in which communication on environmental issues works via networks of committed staff at head office and retail level around the world and how campaigning provides the motivating force for much of their activity.

Communication with staff and customers

One of the most important components of any organisation's programme of environmental management is its ability to raise awareness of environmental issues among its staff and customers. Environmental management systems and auditing procedures may identify problems or opportunities for improvement, but both depend entirely on effective environmental communication in order to facilitate and achieve progress.

With respect to internal education and communication, The Body Shop maintains a devolved system of environmental management which ensures that highly motivated staff adopt responsibility for the environmental performance of their department or shop. In nearly all cases these individuals do not have specific expertise and they discharge their environmental duties on a part-time basis. These 'environmental advisers' act as the primary focus for communications in their area of the company, and together with the Corporate, Ethical Audit Department, collect information on specific 'performance indicators' for their area. They also help identify opportunities for improvement when these arise.

The environmental advisers act as the main conduit for environmental information within the Company. Examples of good environmental practice are rapidly circulated via networks of advisers at retail, head office and international levels. Information dissemination is co-ordinated by the Ethical Audit Department but its delivery within different areas and departments of the company is heavily dependent on the environmental advisers.

In early 1992, The Body Shop formally extended its environmental management system from its corporate headquarters and manufacturing site to its UK retail outlets. This involved the development of a Shop Environmental Guide and the training of nearly two hundred Shop Environmental Advisers in one-day sessions around the country.

Before this training took place, an eight-page booklet was distributed throughout the shops to explain how they should select thier environmental advisers, and what these advisers would be expected to do. The booklet was carefully designed to:

- explain why environmental advisers are needed;
- explain why The Body Shop is determined to improve its environmental performance;
- give pragmatic and motivating advice about choosing an environmental adviser;
- explain the role of an environmental adviser in relation to other parts of The Body Shop.

Figs. 22.1, 22.2, 22.3 and 22.4 show extracts from the booklet.

- **The Environmental Adviser – a person who can take responsibility for establishing the shop environmental management system.**
- **The Environmental Guide – gives all the necessary basic information to set up your system.**
- **The Environmental Audit – carried out once a year by the Environmental Adviser. This will help to establish where greatest improvement has been made and importantly will show the public exactly how good The Body Shop Environmental Management really is!!**

Figure 22.1 Extract from The Body Shop's booklet on selecting an environmental adviser

Retail-level communications and campaigns

The UK retail programme was developed in-house by a member of the Company's Environmental Department with support from Training and Communications staff. The Programme was not designed specifically to raise awareness of global environmental issues although this has undoubtedly been a welcome side-effect. It was envisaged that as Shop Environmental Advisers gained experience they would inevitably develop a deeper interest in environmentalism. Instead, the Programme focused on practical actions which staff could initiate in order to improve the environmental performance of their shops, thereby making a tangible contribution to the reduction of the Company's overall environmental impact.

During training sessions, Shop Environmental Advisers were given

The Environmental Adviser should be:

- **Highly motivated**
- **Committed**
- **Interested in Environmental Issues**
- **Non-Management member**
- **Not committed to a large number of other shop responsibilities**

Figure 22.2 Extract from The Body Shop's booklet on selecting an environmental adviser

The Responsibilities of the Environmental Adviser

1) Raise staff awareness and understanding of environmental management.
2) Enthuse and motivate staff about environmental issues.
3) Ensure that all environmental guidelines are implemented through the team as a whole.

To dispel any myths about what you might think this may involve, here are some things that you will and will not be expected to do.

You will *not* be expected to:	**You *will* be expected to:**
Become a vegetarian	Know about the reasons why some people are vegetarians on environmental grounds.
Grow your hair and never wash	Tell customers about the environmental practices of the Company.
Give up your car	Think of ways that all the shop as a whole can reduce its impact through transport. Think about car sharing.
Operate the refill bar single-handed. Do anything without support	Think of ways to get everyone to promote the refill service. Tell the Environmental Department if you have any good ideas so that we can share them with every shop. If you have any problems, we are here to help as well.
Become a crew member on the Greenpeace ship Rainbow Warrior	Make sure that everyone knows how the environmental management system works, whether through notices, an update session regularly or chats with individuals.
Religiously follow all the guidelines	Some of the guidelines won't work in your shop, don't worry. Set up a system that suits you. All we are aiming for is improvement.

Figure 22.3 Extract from The Body Shop's booklet on selecting an environmental adviser

How the network of support works:

Area Managers, Business Development Managers
- ensure that EA has management support
- keep environmental issue a priority
- support and guide

Education Department
- information/advice re. community talks etc.

Shop Managers
- allocate time
- listen to ideas
- encourage team participation
- support and guide

Environmental Adviser
- motivate team
- communicate and use all relevant information
- encourage team effort
- liaise with the rest of the support network
- audit the shop

Environmental Department
- back-up info
- updates (pack etc.)
- collation of info received from shops
- advice
- dissemination of results and good ideas

Training School
- instore training notes
- updates (training notes etc)
- field trainers/product trainers
- advice on training etc.
- conduct relevant training

Staff Members
- participate
- exchange of ideas etc.
- ACTION

Figure 22.4 The network of support – from The Body Shop's guideline booklet

Figure 22.5 Guide for shop environmental advisers

three practical exercises: energy efficiency (calculating the theoretical cost of a shop's annual electricity bill); life cycle assessment (constructing a comparative matrix of environmental criteria relevant to washing-up liquids); and communications (designing a campaign for promoting the re-use and recycling of post consumer packaging). In each case, the exercises were linked into issues of wider environmental concern, but these provided the context not the focus. Thus the programme was built on the assumption that the majority of The Body Shop staff are already convinced of the urgency of environmental problems – the priority now is for action, not academic debate.

After training, the Shop Environmental Advisers' first task was the completion of a mini-audit of their shops. Then, following their Environmental Guide, they and their colleagues were equipped to make progress in shopkeeping, energy efficiency, waste management and water use – the four key areas of environmental impact under their control. The mini-audits are repeated annually and results circulated to the shops mid-year. A summary of information from the 1992 audit was included in the Company's formal environmental audit statement for 1991/92, and more detailed reports were produced in subsequent years for internal circulation.

External education and communication activities are organised centrally and the main outlets for environmental messages are the shops. The Shop Environmental Advisers have a special role to play in these campaigns. The Company has produced educational material for its customers for several years. This material has at various times covered a range of environmental topics. In many cases it has been related to specific campaigns, for example endangered species, rainforests, acid rain and ozone depletion. In addition a constant educational effort is required on issues of daily operational importance to the Company. These include promotion of the refilling of bottles, re-use and recycling of primary and secondary packaging and reduction of general waste. Leaflets and broadsheets available to customers are supplemented by promotional posters, shelf cards and educational messages on till receipts and carrier bags. These messages explain The Body Shop's services in refilling and recycling bottles and the environmental reasons for reducing packaging waste.

In 1992, a three-week campaign on refilling and recycling occupied all shop windows in the United Kingdom. The campaign was designed in consultation with shop staff who implemented it. The campaign included the usual incentives for customers to return their bottles for refilling or recycling but each shop had specific targets to meet in order to qualify for prizes. The aim of the UK campaign was to increase the rate of refilling and the amount of post-consumer plastic returned for recycling – already running at approximately 1 tonne per week. The campaign was a great success with refills increased nationally by 22% and recycling by 10%.

Another mechanism for delivering The Body Shop's environmental mes-

Figure 22.6 The Body Shop campaign material

sage is the local press and news media. Although The Body Shop has a policy of not advertising its products, it is quite prepared to disseminate its views on environmental issues with a high degree of energy and creativity. The UK Retail Company's Education Department has a vigorous programme of promoting talks to schools and colleges. Typically, these are delivered by shop staff designated as Public Relations Officers or PROs. Even the delivery lorries play their part, carrying social and environmental messages on their sides.

Corporate and international communications and campaigns

As noted earlier, the network of Shop Environmental Advisers is paralleled at Head Office and International levels. In every department of The Body Shop International and its principal subsidiaries there is a Departmental Environmental Adviser. And in every one of the international markets there is an environmental co-ordinator or manager. Training and awareness-raising for these staff is of paramount importance.

Head Office DEAs receive formal annual training of two days duration. They also meet monthly to discuss issues of common interest. These include everything from environmental auditing to maximising the efficiency of waste paper re-use and recycling. Every quarter, DEAs are trained on a special topic and provided with training packs which enable them to make presentations to their own departments. In each case the training packs address three elements: (i) the issue itself; (ii) its rele-

Day 1

9.30–10.30	The Philosophy and Practice of Environmental Auditing at BSI: The Green Book (Talk with slides and discussion)
10.30–11.00	International Environmental Reviews: The Reasons for and Development of the Checklists. Aims for the future
11.00–11.30	Break – Tea/Coffee
11.30–12.30	The Checklists Using the checklists and guidance notes Section by section approach (Hard copies of checklists & guidance notes given out)
12.30–14.00	Lunch
14.00–14.45	Energy Exercise (Group Exercise)
14.45–15.30	Energy Checklists Role play demonstration with discussion
15.30–16.00	Break: Tea
16.00–16.15	Waste Video
16.15–17.00	Waste Checklists Group role play exercise with discussion

Day 2

9.30–10.00	LCA/Eco-Label Exercise
10.00–11.00	Environmental Review of the Shops (Shop Checklists & Shop Eco-Audit Form)
11.00–11.30	Break: Tea/Coffee
11.30–12.00	The State of Refilling/Recycling Individual reports on level of activity, campaigning and legislation.
12.00–12.30	Packaging – The Latest on Global Developments
12.30–14.00	Lunch
14.00–15.00	Completion of 2 sections of review checklists in pairs – Role of Environmental Co-ordinator – Purchasing/Transport – cars
15.00–15.30	Completing the Review Checklists (Twinning)
15.30–16.00	Break: Tea
16.00–16.30	Summary

Figure 22.7 International environmental review training workshop timetable

vance to The Body Shop; and (iii) the actions which individuals and departments can take to reduce their own environmental impacts. DEA training packs are produced in-house by a full-time Environment, Health and Safety Trainer.

Presenting environmental information in a decentralised way maximises the speed of information transfer, it enhances the profile of the DEA; and it ensures the accessibility and relevance of information to staff at all levels.

International environmental co-ordinators also receive at least two days' formal environmental training annually. These events occur regionally in order to maximise travel efficiency. In 1993, regional environmental training events were held in Holland, Italy, Bahrain, Hong Kong, the US and the UK. Fig. 22.7 shows the timetable for such an event.

The first regional training events focused on environmental auditing and environmental management procedures. This was primarily to assist in the achievement of a corporate goal which was to conduct environmental reviews in all international markets during 1993/94.

In addition to the annual training, international co-ordinators receive a monthly bulletin which looks at the world's environmental news from the perspective of The Body Shop. This helps ensure consistency of policy and campaigning messages on issues as diverse as whaling and PVC.

Following its success in the UK, international markets are now developing head office and retail-level networks of environmental advisers, thus replicating UK systems of decentralised communication and education.

Another vital component of corporate and international communications is The Body Shop TV. Every week in the UK and every month internationally, The Body Shop's TV company Jacaranda produces a 15–20 minute news video which covers all the major items of relevance to staff around the world. BSTV features new product information, campaigns, merchandising tips, and coverage of a plethora of general interest items, including the environment. Whether it is publication of an environmental audit report, progress on The Body Shop's windfarm investment, or campaigns on packaging, BSTV is there and the information is relayed direct to staff in 44 countries.

Environmental auditing – communicating with peers

Previous sections have dealt with mechanisms for educating and communicating with staff and the general public. However, there are other important groups with whom it is important to communicate effectively. These include regulatory agencies, environmental organisations and the

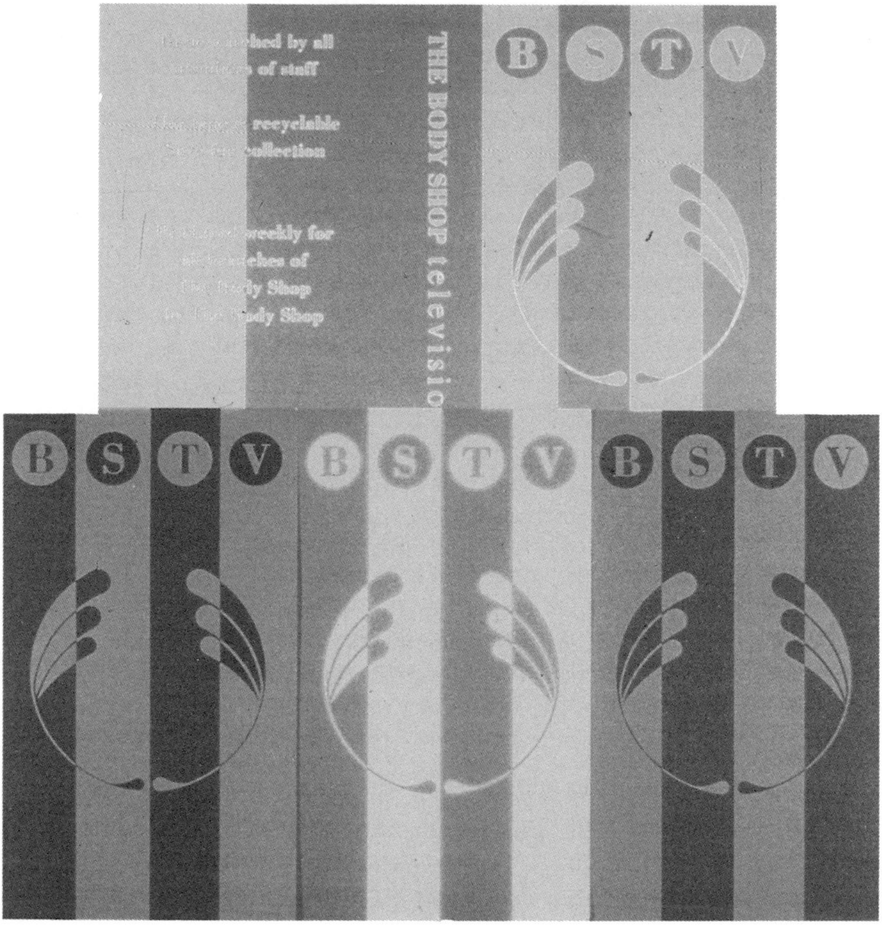

Figure 22.8 BSTV relays environmental information

rest of industry. Conferences provide one forum for information exchange, but increasingly, one of the most important vehicles for communications with these groups is the publication of The Body Shop's annual environmental audit statement.

When the draft EC regulation on environmental auditing was circulated during 1991, The Body Shop elected to follow its provisions in their entirety. Even though the cosmetics industry is not a high priority for eco-auditing and even though the EC regulation will not become mandatory for years, The Body Shop was sufficiently committed to the principle of full public disclosure of environmental information to publish its 1991/92 audit in line with draft EC regulation.

In doing this, the Company hoped that it might set an example to other retailers and other industries which (at present) seem to fear open disclosure. The Body Shop's 1991/92 audit report was not a publicity vehicle. It contained information about all of the environmental achievements of The Body Shop. But it also contained information on those aspects of the Company's operation which are not as good as The Body Shop would like them to be. In these cases the Company committed itself to new targets and a programme of innovation and research to help the Company continue on the path to environmental excellence. The targets covered subjects as diverse as renewable energy, energy efficiency, waste minimisation, purchasing standards and product stewardship.

An important aspect of the EC Eco-audit process is the independent validation of the audit by external assessors.[6] The Body Shop appointed such assessors and their comments were included in the audit report. In this way – by a combination of full disclosure and independent validation – The Body Shop ensures that honesty and integrity characterise every aspect of environmental communications with its customers, its staff, environmental interest groups, regulatory bodies and the rest of industry.

Since publication of the first environmental statement in 1992 (*The Green Book*), The Body Shop has repeated the process, publishing *The Green Book 2 and 3* alongside the annual report and accounts for the third year running. This pattern is now established and will be a constant feature of the Company's annual cycle of performance reporting.

Figure 22.9 ***The Green Book 1***

The Body Shop believes this to be consistent not only with the EC Eco-management and Audit Regulation (as it was re-named in early 1993), but also with the European Community 5th Environmental Action Programme, Agenda 21 and the CERES (Valdez) principles – all of which endorse the concept of regular disclosure of environmental information by commerce and industry.[7]

Product stewardship – communicating with suppliers

The Environmental Policy of The Body Shop commits the Company 'to use renewable resources wherever possible' and to 'conserve natural resources where renewable options are not available'. These public commitments ensure that The Body Shop takes great care in the procurement of product ingredients and items of packaging. All product ingredients must not have been tested on animals for the cosmetics and toiletries industry in the previous five years and suppliers must provide regular reassurance on this point. In addition, natural ingredients from renewable sources are favoured by the Research and Development department whenever new product formulations are tested. Where ingredients and packaging types come from non-renewable resources, e.g., from synthetic chemicals, great care is taken to minimise quantities and maximise characteristics of biodegradability (in the case of ingredients) and recyclability (in the case of packaging).

This policy has tended to focus The Body Shop's product range around natural ingredients with useful properties for skin and hair care but with low allergenic potential. The range also includes some ingredients which are sourced in less developed countries under non-exploitive trading arrangements. These trading links provide income for otherwise culturally and economically threatened communities. One example is a hair conditioner which incorporates Brazil nut oil from the Kayapo Indians of the Amazonas.

However, as the Company expands and the list of ingredients and suppliers grows, it becomes increasingly important to adopt a systematic approach to the sourcing and procurement of new materials. The Body Shop is certainly capable of exercising good environmental management in its own operations, but what of those of its suppliers?

The Body Shop has put in place a formal system of product stewardship, including life cycle analysis (LCA) which will ultimately establish the environmental credentials of every one of the Company's product ingredients and packaging types. The LCA scheme provides an overall numerical score for each item, based on approximately 100 environmental criteria. This enables the Company to track the environmental performance of its suppliers and the resulting database influences day-to-day decisions in product formulation and purchasing. This

Figure 22.10 The Kayapo project

approach requires a large amount of communication and effort both on the part of the Company and its suppliers. But the Company also recognises that there is no alternative if it is to minimise the environmental impacts of products in future. Fig. 22.11 shows the five levels of The Body Shop's 'supplier accreditation' scheme.

In addition to the development of a database of life cycle assessments, The Body Shop's purchasing department has devised a system of supplier accreditation. Accreditation provides recognition to suppliers who assist with the LCA scheme, who adopt formal policies and systems of environmental management and who demonstrate a practical commitment to reducing their own impact on the environment. Both the LCA scheme and the accreditation system were discussed with suppliers in special seminars run in 1992 and 1993.

The Body Shop also hopes that the LCA system it is adopting for its own internal decision-making may also help improve customer awareness regardless of whether an acceptable European-wide scheme for eco-labelling is eventually adopted. It is difficult to imagine how the European Community can introduce credible eco-labelling schemes without detailed and comprehensive life cycle analysis. And yet the temptation to make awards on the basis of simple assessments is very great. Consumer understanding will not be assisted if eco-labels are awarded on the basis of superficial analysis.[8]

SUPPLIER ACCREDITATION –

5 STAR SUPPLIER AWARD SCHEME

1 STAR CRITERIA	– To achieve 1 star status, suppliers will need to demonstrate clear evidence of a willingness to improve environmental performance beyond simple compliance with regulatory obligations. In addition, there should be no unacceptable or intractable environmental impacts associated with processes or raw materials in The Body Shop's supply chain. – A minimum of 11 points needed on answers to checklist 1
2 STAR CRITERIA	– **As 1 star, plus** – A minimum of 17 points needed onanswers to checklist 1 (a supplementry checklist will be issued at this stage to cover emissions and raw materials)
3 STAR CRITERIA	**As 2 Star, plus** – Provision of summary emissions and waste information (checklist 2) – submission of satisfactory Life Cycle Analysis information
4 STAR CRITERIA	**As 3 Star, plus** – Published Environmental Audits show that all identified impacts are being addressed to the satisfaction of an independent external authority – Auditing is done periodically (in line with national or international regulations)
5 STAR CRITERIA	**As 4 Star, plus** – All significant environmental problems overcome – Systems to evaluate supplier performance in place, with evidence that BSI's supplier chain is environmentally secure – Will be registeres for EC Eco-audit regulation/or registered for equivalent environmental management systems standard (BS7750 or ISO equivalent)

Figure 22.11 Supplier accreditation rating

Conclusion

One of the main principles which guides The Body Shop's development of new products is concern for the environment. This concern has resulted in the Company becoming a successful business selling high quality skin and hair-care preparations containing a range of natural ingredients and sold with the minimum of packaging.

The Company does not advertise its products and it does not seek to make commercial gain from the confusion or environmental fears of its customers. On the contrary, The Body Shop hopes to achieve the highest standards of environmental performance for its operations simply because it believes this is the only ethical position to adopt. Dissemination of knowledge and understanding of environmental issues are vital components in raising environmental standards and establishing credibility, and considerable efforts are devoted both to internal and external awareness-raising.

The Company does not pretend to be perfect or to have all of the answers. Neither does it claim that in buying products from The Body Shop, customers are going to avert global environmental disasters. However, the Company does believe that through its staff and customers, positive choices can be exercised which help minimise environmental impacts in the production and use of skin and hair-care products. Together with the educational and campaigning activities which the company promotes, The Body Shop hopes to provide an example of how a practical and honest commitment to the environment can co-exist with successful retailing and customer support.

References

1 Elkington, J. and Hailes, J. *The Green Consumer Guide*, Victor Gollancz, London (1988).
2 Vallely, B. *1001 Ways to Save the Planet*, Penguin, London (1990).
3 Environment Committee, 1991: Eco-labelling. Eighth Report of the House of Commons Environment Committee, HMSO, London (1991).
4 Wheeler, D. 'What future for product life cycle assessment?' *Integrated Environmental Management*, June 1993, pp. 15–19.
5 Porritt, J. *Green Magazine*, London, April 1993.
6 Commission for the European Communities, 1993. Proposal for a Council Regulation (EEC) allowing voluntary participation by companies in the industrial sector in a community Eco-Management and Audit scheme. S218/98 Env 64 Com (91) 459. Final.
7 Wheeler, D. 'Environmental auditing: philosophy and practice of The Body Shop International', *Eco-Management and Auditing (1) 1*, 1993.
8 Wheeler, D. 'Eco-labels or eco-alibis?' *Chemistry and Industry*, 5 April 1993, pp. 260.

Recommended reading

Deloitte Touche Tohmatsu International. *Coming Clean: Corporate Environmental Reporting*. DTTI, London (1993).

International Institute for Sustainable Development. *Business Strategy for Sustainable Development*. IISD, Winnipeg (1992).

House of Lords Select Committee on the European Communities. A Community Eco-audit Scheme. 12th Report of Sub Committee F, House of Lords Paper 42. HMSO, London (1993).

23

Defining the vision and implementing the change

AXEL WENBLAD

Corporate Manager, Environmental Auditing, AB Volvo, Gothenburg, Sweden

- Since policy implementation procedures are based on the Group's highly decentralized organization, training has been an essential element in implementing the necessary changes.
- Each company is obliged to develop a three- to five-year action programme, which must be updated annually. These progammes are used as guidelines for developing annual 'minimum standards', in other words, specific action plans incorporating the measures to be taken.
- The fact that several disciplines are involved in formulating and implementing the plans means that environmental thinking is strongly rooted with the organization.
- The environmental co-ordinators at the company's plants have attended a two-day training course based on the ICC's Business Charter for Sustainable Development; Volvo was one of the first companies to subscribe to this sixteen-point programme.
- Environmental responsibility is coincident with line responsibilities; in other words, the president or plant manager is legally responsible for environmental matters.
- The strong support of the management has contributed significantly to the speed with which the programme has been developed and launched.
- With the Federation of Swedish Industries and the Swedish Environmental Research Institute (IVL), Volvo has developed a method of calculating the environmental impact of a product over its complete life cycle, known as EPS (Environmental Priority Strategies).

Background

Environmental pollution became a focus of interest in Sweden at an early stage and debate on the environment became fairly intensive during the 1960s, leading to the establishment of a national environmental protection agency in 1967 and to the enactment of environmental protection legislation in 1969. The Act represented what was then a novel view of the environmental impact of industry by imposing controls on all emissions and nuisances, such as water and air pollution, waste, noise, etc. This was later to become a key topic of discussion within the EC under the title of 'Integrated Pollution Prevention and Control'. Environmental protection activities were already under way at Volvo when the United Nations' conference on the environment was held in Stockholm in 1972. In this context, it was appropriate that the company's first policy in the area should be articulated by Volvo's CEO Pehr G. Gyllenhammar, who declared:

> Volvo has no wish to defend the car at any cost or in every situation. It is in our interest that the car is used in a manner which does not damage the environment. Volvo recognizes that its responsibility is not confined to ensuring that its products are practical for transport purposes; they must also be functional in the widest sense. We are firmly convinced that an urban environment which is both vital and favourable to its human inhabitants can be combined with effective transport resources. Society needs both.

In the mid-1970s, Volvo developed the three-way catalytic converter with an oxygen sensor (or Lambda sond) in collaboration with Robert Bosch, making it possible to purify exhaust emissions of carbon monoxide, hydrocarbons and nitrogen oxides simultaneously. The technology has since become standard and is undergoing continuous refinement.

On the production plant side, interest during this period focused largely on waste water emissions and on certain types of environmentally hazardous waste; a large number of purification plants were installed. Treatment facilities of basically the same standard as those in Sweden were installed in the company's plants abroad, whether or not required by local legislation. As an example, the Volvo plant at Ghent in Belgium was equipped, in the mid-1970s, to treat all sanitary waste waters, a facility still lacked by the city of Brussels.

The 1980s brought a growing realization that the scale of the environmental problems was much greater than had been thought, and that the situation had deteriorated despite major efforts to reduce industrial emissions. As environmental awareness grew, the problems became increasingly politicized and, according to opinion polls carried out in 1988, the environment was the main issue in the Swedish general election held that year.

Under these circumstances, it was both logical and important that the Volvo Group should review its environmental policy and that it should launch an environmental programme.

In autumn 1988, Pehr G. Gyllenhammar appointed a steering group (whose members included the heads of Volvo's product companies) to draft a new environmental policy to replace the 1983 version, and to suggest mechanisms for reviewing and monitoring its implementation on an ongoing basis.

Translating vision into reality

At the time, not many companies had formulated comprehensive environmental policies and programmes, and few models were available. Thus, the project was novel to the extent that it required the basic thinking to be translated into concrete terms to define the thrust of the programme, while it was also clear that the policy itself would have to be based on an overall view of the environmental impact of the products.

In the course of its work, the steering group suggested that a management conference be held to discuss the problems involved. This took place in January 1989, when 400 of Volvo's top managers in Sweden participated in a one-day seminar on environmental issues.

Opened by P.G. Gyllenhammar, the seminar introduced the participants to the fundamentals of ecology and the main environmental problems. A number of leading Swedish researchers and environmental experts dealt with topics such as acidification, eutrophication, ozone depletion, climatic changes and health effects. The seminar ended with product company executives giving their views on the problems involved and on how the various companies should implement Group policy.

The event contributed significantly to creating awareness of the environmental problems among company managers, as well as the challenge facing them. In effect, acknowledgement of the problems provided a basis for formulating corrective action.

The steering group submitted its report to CEO Pehr G Gyllenhammar in mid-1989 and the new policy was finalized that August.

Volvo's environmental policy

The environmental policy of the Volvo Group is expressed in the form of six statements based on an overall view of the environmental impact of the company's products 'from cradle-to-grave'.

It should be noted, at the outset, that the Group's involvement in the transport sector means that production operations and products alike demand environmental protection resources. This is a particularly important premise in that open recognition of a problem makes it easier to identify solutions and implement protective measures – a fact demonstrated by the title of our first environmental policy brochure: *Our products create pollution, noise and waste.*

The relationship between profitability and environmental activities forms the second premise on which the policy is founded. High profitability provides scope for rational, environment-oriented investment in products and production facilities, while effective environmental protection lays the foundation for long-term profitability and satisfactory economic growth.

In its environmental policy, Volvo undertakes to minimize the environmental impact of its operations by:

- developing and marketing products with superior environmental properties, meeting the highest efficiency requirements;
- opting for manufacturing processes that have the least possible impact on the environment;
- actively participating in and conducting its own research and development in the environmental field;
- selecting environmentally compatible and recyclable material in connection with the development and manufacture of its products, and when purchasing components from its suppliers;
- applying a total view regarding the adverse impact of its products on the environment;
- striving to attain a uniform, worldwide environmental standard for its processes and products.

These statements give practical expression to Volvo's perception of the problems involved, and to the main thrust of its action programme. However, declarations alone are not sufficient; systems are required to translate theory into practice.

Policy implementation

The systems devised to implement the policy were, of necessity, designed to suit the existing organization. Although the temptation is (and was) to adopt ready-made environmental management systems to conserve resources and save time, the results frequently fall short of expectations, in addition to which the systems available when the policy was formulated were not sufficiently comprehensive.

In the highly decentralized Volvo organization, responsibility for all activities from product development to marketing resides with the five product companies:

- Volvo Car Corporation
- Volvo Truck Corporation including Volvo Bus Corporation
- Volvo Penta (marine and industrial engines)
- Volvo Flygmotor (aircraft and aerospace engines)

Since head office is responsible mainly for defining the thrust and monitoring the implementation of the policy, its precise application has become the responsibility of the product companies, as suggested in the steering group report. Each company is obliged to develop a three- to five-year action programme, which must be updated annually. These programmes are used as guidelines for developing annual 'minimum standards'; in other words, specific action plans incorporating the measures to be taken.

In this rolling planning process, the minimum standards for a particular year are evaluated, and those for the coming year defined, at year's end, the relevant decisions being taken by the presidents of the individual companies.

In the larger companies, this calls for a structure in which activities take place on several different levels. This is best illustrated by the example for Volvo Car Corporation (VCC) shown in Fig. 1, in which the Group's overall environmental policy is shown at the top of the pyramid, with VCC's environmental strategy and the three- to five-year action programmes underneath. The latter are then broken down by division and, in appropriate cases, by production unit. Similar structures are employed by the other product companies.

The minimum standards applicable to an individual plant are usually quite specific and quantifiable. A typical formulation may state that the 'the use of solvents in underbody treatment shall be reduced by 20% in 1993' or that 'all supervisors must undergo basic training in environmental matters and in the company's environmental policy'.

Overall, this means that environmental action plans are formulated at different levels within each company, calling for co-ordination within the structure shown in Fig. 23.1. The fact that several disciplines are involved in formulating and implementing the plans also means that environmental thinking is strongly rooted within the organization, and that there is a high demand for the training required to develop the necessary skills.

Training is a priority aspect of the Group's environmental policy. In this context, the environmental seminar attended by 400 managers in January 1989 may be regarded as the start of systematic training activities.

The purpose of the training programmes is (a) to demonstrate that Volvo is fully committed to its environmental policy, (b) to disseminate information concerning major environmental problems and Volvo's contribution to their solution, and (c) to provide information on Volvo's environmental programmes and the contributions which can be made by company personnel. Several general training courses have been based on the findings of the first environmental seminar, experts from official agencies and environmental organizations being engaged to speak about the problems.

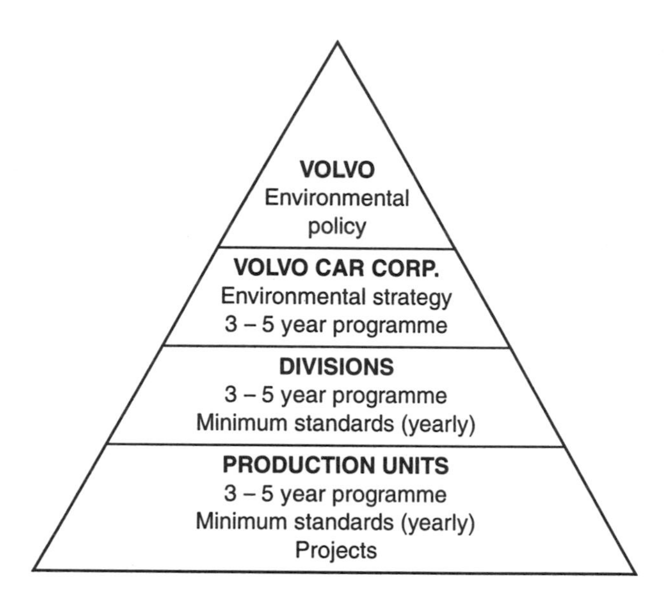

Figure 23.1 Environmental action plans – formulation at different levels

No common training programmes have been developed because of the highly decentralized structure of the Volvo organization. The training courses have become more specific and customized over time, and are also being designed increasingly for specific categories of personnel. For example, designers may attend courses on the environmental impact of different materials, while purchasers may undergo training in the hazards involved in handling chemicals.

As a further example, the environmental co-ordinators at the company's plants have attended a two-day training course based on the International Chamber of Commerce (ICC) Business Charter for Sustainable Development. Volvo was one of the first companies to subscribe to this sixteen-point programme.

In all, over 10,000 of the Group's 60,000 or so employees have undergone some form of environmental training.

Environmental responsibility at Group level is exercised by a member of the Group Executive Committee, assisted by a small specialist function and by an environmental committee consisting of experts from the different product companies.

Within the latter, environmental responsibility is coincident with line responsibility; in other words, the president or plant manager is legally responsible for environmental matters. Although responsibility may be delegated in accordance with certain principles, the plant manager is normally the lowest authorized level. An environmental co-ordinator is also employed in each of the production plants to co-ordinate environmental activities, liaise with official agencies and report to the plant manager. However, operational responsibility is exercised exclusively by the line managers.

Environmental committees have been established at different levels within the product companies. Since one of the committee's functions is to develop action programmes and minimum standards, the levels in question are usually those specified in Fig. 23.1. The environmental committee may be regarded as a forum for consultation between environmental specialists and operations personnel.

Environmental auditing

Environmental auditing – the main purpose of which is to monitor and evaluate implementation of the environmental policy and to provide support for a dedicated environmental protection programme – is the final element of the Group's policy. The relationships between policy, implementation and auditing are illustrated in Fig. 23.2. The auditing remit is extremely wide, extending, in principle, to production as well as products.

Auditing activities were initiated, in autumn 1989, in the form of a pilot programme designed to develop an organization and working procedures. A number of production plants, representative of activities both in Sweden and abroad, and of different types of production, were selected for this purpose. The findings were extremely positive, both in terms of the audits *per se* and the manner in which they were received throughout the organization, and it was decided to audit all production units as a first step.

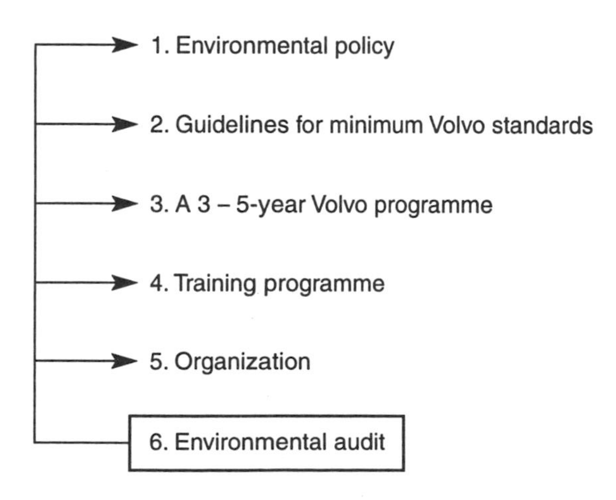

Figure 23.2 Relationships between policy, implementation and auditing

In its programme for 1991–93, Volvo declared environmental auditing to be an integral part of environmental policy implementation, noting that it should be developed in parallel with the Group's environmental activities as a whole.

In more specific terms, the purposes of environmental auditing are:

- to verify compliance with existing legislation and to evaluate the conditions for complying with future regulations;
- to evaluate the conditions for implementing action programmes and attaining minimum standards;
- to provide a basis for the assessment and improvement of management functions and procedures in the environmental area;
- to identify areas and activities which may prove environmentally hazardous in the future.

Although existing legislation forms the primary basis of environmental auditing, it is equally important, in many cases, to take cognisance of future developments. This is particularly important at a time when legislation in the field is becoming ever stricter.

Since the action programmes and minimum standards are the main instruments used for environmental policy implementation, the review of existing programmes and standards is an extremely important aspect of auditing. This is a matter of verifying compliance with company policy and of determining whether the stated objectives are achievable with the resources allocated to them.

The third of the objectives listed above – that of evaluating the related management functions – hardly merits detailed explanation. In a decentralized organization, such as Volvo, it is important that environmental responsibilities be clearly defined and delineated in each individual case, given the fact that many environmental problems are delegated to plant managements, and that responsibility in this area is part of operational line responsibility.

The fourth, and final objective implies that even greater efforts must be made to anticipate the demands of future legislation. This is a matter of identifying, on the basis of research findings, processes or materials which may pose environmental hazards at a future date.

As in most companies, environmental audits are carried out as part of acquisition and disposal procedures. Audits of this nature are normally undertaken by external consultants who are independent of both vendor and purchaser.

Provision is also made in the programme for environmental auditing of suppliers. This has occurred in some instances in which an environmentally hazardous process, such as painting, has been carried out as a sub-contract rather than by Volvo itself. In a case of this type, the aims are to ensure that the environmental impact of the process is not increased and to verify that the supplier complies with the relevant legislation.

Volvo has adopted an environmental auditing organization in which head office is responsible for the auditing programme and direction of the audits, backed by the resources and expertise available throughout the Group at large. In practice, this means that about twenty experts from the various product companies, and from the Technological Development Department, are available to participate in auditing assignments. To meet the requirements of independence and objectivity, these individuals never take part in audits of their own companies.

Performance of the audits is based largely on the ICC Guide to Effective Environmental Auditing and the Swedish-language manual published by the Federation of Swedish Industries.[1, 2]

Management support

Earlier in this chapter, we saw how the vision of industrial activity embodying concern for the environment could be realized in practice as a contribution to long-term, sustainable development. In this respect, the work begun in 1988-89 may be regarded as successful, although arduous.

The strong support of management has contributed significantly to the speed with which the programme has been developed and launched. This is particularly true of the personal involvement of Volvo Chairman Pehr G. Gyllenhammar who, in opening the 1989 managerial seminar, defined the thrust of the company's policy by declaring: 'This is a time for action, not excuses'.[3]

Training activities have played a crucial role in heightening environmental awareness and laying the foundation for initiating the process of change. Education in the main environmental problems, and Volvo's environmental policy, has enabled various categories of personnel to incorporate environmental concern in their own activities.

The system of action programmes and minimum standards has both provided the opportunity and imposed the requirement of formulating environmental protection objectives at local level, for example in a production plant. In this respect, central direction has been minimal, perhaps even too little. However, once the objectives have been formulated, their implementation is based on a solid foundation, both technically and financially. This scenario is also ideally suited to Volvo's highly decentralized type of organization, as reported in a study carried out at the Massachusetts Institute of Technology.[4]

The unreserved support of management has been a major factor in the success of environmental auditing at Volvo, giving the programme a flying start and gaining it respect within the organization. In this context, the company has been careful to emphasize that the results are intended primarily to assist operations personnel in improving the standard of environmental protection. In other words, auditing is not an inspection procedure imposed by Group management, but is a sophisticated tool which affords impartial assessment of environmental protection programmes and highlights the influence of pollution problems on a particular operation.

Overall view

As already described, the environmental policy reflects an overall view of the environmental impact of the company's products 'from cradle-to-grave'. However, methods of translating theory into practice were unavailable when the policy was first formulated in 1989, although designers were requesting the development of practical, environmentally sound product design tools.

At this juncture, Volvo sought the assistance of the Federation of Swedish Industries and the Swedish Environmental Research Institute (IVL) in developing a method of calculating the environmental impact of a product over its complete life cycle.

Known as EPS (Environmental Priority Strategies), the system is designed to:

- systematically provide information for use in evaluating the overall environmental impact of the product;
- describe the environmental effects of energy and raw materials consumption, and of pollutant emissions, during the various phases in the life cycle of the product;
- evaluate, in relative terms, the environmental effects of alternative production processes and design concepts, enabling different process configurations and product designs to be compared.

The purpose of the EPS system is to make environmental impact calculations' visible to all interested parties. The methods used should be transparent, while the environmental impact indices should vary accordingly as new knowledge is added or evaluations modified.[5]

On completion of testing, the full-scale system will be implemented mainly by designers at Volvo Car Corporation and Volvo Truck Corporation, while the Federation of Swedish Industries has undertaken further development work in collaboration with other Swedish industrial manufacturers, including ABB, Electrolux and Stora.

Future developments

Volvo's environmental policy was initiated during a period of economic buoyancy and high profitability. Since then, however, economic conditions have changed dramatically and the effects have been felt particularly in the auto industry. In Volvo's case, this has necessitated severe cutbacks in all sectors, with the imposition of strict financial constraints on investment.

The fact that the policy was implemented largely during the latter period creates an obvious risk that the programme will be curtailed as financial restrictions become tighter. On the other hand, the working flexibility afforded by the company's high degree of decentralization should enable optimum solutions to be developed at local level. The new circumstances represent a transition from capital investment to the aspects of monitoring, optimization and the development of solutions for the future.

Nevertheless, it is unrealistic to rely exclusively on local initiatives in the long term; constant reference to a common Group strategy is essential, and a strategy document outlining the progress which should be achieved by 1995 has been prepared for this purpose. This document does not supersede the environmental policy, but defines the policy guidelines for the period 1993–95 in concrete terms.

The strategy document details a number of specific guidelines under the following headings, based on evaluation of the Group's position in relation to various environmental problems in 1988 and 1992:

- Management
- Production
- Products
- Product utilization (transport)
- Recycling

As an example, the EPS system described above is dealt with under the heading of *Management*, while the document also contains proposals relating to the EC's newly announced Eco-management and Audit scheme, and to the future development of environmental auditing. The most urgent pollution problems are discussed under *Production*, while *Products* is specific to each product company. The section entitled *Product utilization* deals with Volvo's general stance on transport problems and its participation in various development projects. Finally, *Recycling* describes the company's involvement in the recycling of vehicles both at home and abroad.

It is hoped that this initiative will serve to concentrate efforts in a number of particularly important areas, enabling Volvo to maintain its position among the world's automakers as a leader in environmental protection.

Key learning points

It is important to acknowledge that any process of change is influenced by a number of external and internal factors influencing the process.

The experience in two corporations will therefore never be alike. There are, however, some factors which tend to recur and for Volvo the following key learning points can be singled out as particularly important:

- Acknowledge the environmental problems and the contributions from your own activities.
- Ensure full management support.
- Give priority to information and training.
- Define environmental responsibilities as coincident with line responsibilities.
- Allow objectives to be formulated and implemented locally.
- Monitor and evaluate implementation through environmental auditing.
- Integrate environmental strategies in overall corporate strategies whenever they are formulated.

References

1 International Chamber of Commerce. *ICC Guide to Effective Environmental Auditing*, ICC Publishing S.A., London, (1991).

2 Almgren, R. Miljörevision *Environmental Auditing*, Federation of Swedish Industries, Sweden, (1990).

3 Willums, J-O. and Golüke, U. *From Ideas to Action: Business and Sustainable Development*, ICC Publishing S.A. and Ad Notam Gyldendahl, Oslo, (1992).

4 Rothenberg, S. , Maxwell, J. and Marcus, A. 'Issues in the Implementation of Proactive Environmental Strategies', *Business Strategy and the Environment*, vol. 1, no. 4, pp. 1–12, European Research Press, Ltd, Bradford, England, (1993).

5 Ryding, S-O. *Environmental Management Handbook*, IOS Press Amsterdam (1992).

Recommended reading

Schmidheiny, Stephan, with the Business Council for Sustainable Development. *Changing Course: A Global Perspective on Development and the Environment*, MIT Press, Cambridge, Massachusetts (1992).

Smart, Bruce. *Beyond Compliance: A New Industry View of the Environment*, World Resources Institute Washington, DC (1992).

24

The Environmental Manager as a change agent

JOHN LAWRENCE

Environmental Adviser to ICI,
and Visiting Professor at King's College, London

■ The task of the Environmental Manager is to act as the environmental conscience of the company and to ensure that changes occur in various ways in response to the wider expectations of society.

■ To bring about change the Environmental Manager has three tools: providing information, exerting influence and using authority.

■ There are many publications and services that can help: in the UK there is Environmental Data Services' *ENDS Report*, also *Integrated Environmental Management*; in Europe there is *Europe Environment*; and in the USA, *International Environmental Reporter*. (See 'Sources' at end of chapter.)

■ It would be wise to have a qualified lawyer allocated to help, when necessary, with the task of assessing the implications of the law for the company.

■ In parallel with the development of systems and procedures the Environmental Manager has an important task in persuading management of how the world is changing, to accept the need for change, and to get on with the job. Part of this is, of course, to complete the loop of the management system by communicating how change has happened and how performance has improved.

■ The Environmental Manager does not only exert influence at all levels with his/her company, he also has an important role in presenting the face of the company to the outside world.

■ The Environmental Manager has the backing of mandatory company policies and standards that go beyond the law. Sir Denys Henderson, the Chairman of ICI, has stated that 'Good environmental performance is not an option. It is essential if ICI is to continue as a leading international chemical company into the next century.'

Introduction

There are many forms of pressure on the management of a company to modify its practices in favour of a cleaner environment; they come from its own staff, from its customers and possibly its suppliers, from the environmental movement, from the State (in the form of legislation and occasionally tax incentives), and from shareholders. These are the channels through which society tells us that there has been a change in the balance of priorities.

But in the traditional organisation structure – that is, one that existed before the present level of environmental concern – staff at all levels will not necessarily act wisely or consistently in response to these pressures, particularly if the organisation is one in which business decisions are decentralised to profit centres. Managers will have different views about the wisdom of what is suggested; some will be under extreme financial pressure, and the reward system may not recognise the value of a change which improves the environment but which costs money; they may simply not have the mechanisms to know what is the state of current and pending legislation, or what are the technical possibilities within their own units.

To come to terms with this dilemma many companies have appointed *Environmental Advisers* often leading to the development of an advisory functional network. Sometimes the title has been *Environmental Manager*, implying a greater degree of authority. As with other functions in the past, we have seen the parallel development of codes of practice, formal systems and other features of emerging professionalism.

The job of the Environmental Adviser, or Manager, who is put into this position is to assist the company to behave responsibly with respect to the environment, to be consistent across all its operating units, and to achieve an acceptable balance between environmental performance criteria and all the other performance criteria by which it will be judged. This might be summarised as providing the company with a corporate environmental conscience, but it is more than that because it means effecting change. It is in that sense that the Environmental Manager is a 'change agent'.

To bring about that change the Environmental Manager has three tools:

- providing information;
- exerting influence;
- using authority.

Providing information

It would be very regrettable if the only signal to which a company responded was the law, because the law is just one of the ways in which society articulates its collective wish. Nevertheless the law specifies the minimum standard to which the company must operate. Not only is the law complex and often difficult to interpret, and becoming more so in the environmental field, it is constantly developing and changing, and a company is wise to take account of likely changes before making significant decisions. Furthermore the legal requirements of different countries can vary widely, raising the question, to which we shall return, of whether it is acceptable to operate to different standards in different territories.

An important function, therefore, for the Environmental Manager, is to ensure that the company, at all levels, is provided with information as to the law and related regulation, and of likely future developments in all the territories in which it operates. For example, in the United Kingdom this would embrace a good understanding of such things as:

- The Environment Protection Act.
- The regulations regarding sea disposal.
- The regulations implementing the EEC Dangerous Substances in Water Directives.
- The regulations implementing the EEC Transport, Packaging and Labelling Directives.
- The proposed EEC Landfill Directive.

There are many, many others, of course, to say nothing of the variations as they apply (in Britain) to Scotland and Northern Ireland.

Some countries have supplemented this 'command and control' approach to environmental management with the use of economic instruments. Straightforward fines for offending the rules are the simplest of these. A more recent innovation is the incentive type of charge, such as a charge for a discharge to the aquatic environment. For example, France, Holland and Germany have all had such charges, the amount being based on both the quantity and the properties of the particular discharge. Incentive systems that are more fundamental to the structure of the economy may also become widespread. These include the so-called carbon tax and taxes on packaging. It is not intended to discuss the merits of these concepts here, but they are certainly popular with many economists and politicians. This suggests that the Environmental Manager should have considerable economic competence within his armoury to help the company to know how to respond and also, as will be discussed later, to ensure that legislators appreciate the possible impact on the company. Obviously, understanding the law, and the detail of the huge body of regulation, poses immediate problems, especially for multinationals. ICI sells in over 150 countries to about 100,000 customer companies. The range of products is immense, ranging through seed corn for breakfast cereals, plastic material for floppy discs and drink bottles, dyes for fabrics and leathers, paints for industry and the home, heart drugs, superconducting ceramics for advanced medical imaging, and so on; there are more than 15,000 products in all. The Company employs more than 120,000 people at 600 locations in more than 40 countries.

Keeping track of the regulatory environment as it applies to the manufacturing processes and the distribution and use of the products is difficult in itself, but ensuring compliance at all levels is an even greater challenge. Fortunately there are many publications and services that can help, including, in the UK, Environmental Data Services (ENDS), *Integrated Environmental Management* and *Europe Environment*, and in the USA *International Environmental Reporter*. It is surprising to find how many firms do not make use of these.

It would be wise to have a qualified lawyer allocated to help when necessary with the task of assessing the implications of the law for the company. This is particularly important in states where the relationship between the state and the company is becoming, or has already become, litigious. In the United States, for example, a disconcertingly high proportion of the enormous sums spent under the Superfund legislation has gone in the form of lawyers fees. This does nothing directly to improve the environment. It is to be hoped that this will be moderated and that the experience will encourage other countries to adopt less wasteful approaches.

Since improving environmental performance will often involve technical change in products, in processes or in logistics, such as transport systems, the company needs a source of information on those technical options. The Environmental Manager will acquire a broad grasp of this but, in a highly technical company, (s)he should develop contacts in the science and technology departments who will provide specialist knowledge in the fields of the company's operations.

It may well be important to keep track of the outside world's view of the company and its environmental policies and practices. This may be done informally or, in a larger company, by commissioning opinion polls of the public, the media or important customers.

Finally, the Environmental Manager will become a broad source of intelligence about what is going on in different parts of the company, what competitors' and other companies' policies are, and what they are achieving, and how thinking is developing in Government and other external organisations. The information and awareness publications which were referred to above can be of great value here also.

Exerting influence

As in so many things, it is the influence that the Environmental Manager brings to bear as a result of his wisdom, sapiential authority, and powers of persuasion that really determines the rate and extent of the change that he or she is able to bring about. That influence may usefully be employed in a number of directions for it is useless to persuade top management about an overall policy if plant managers have no enthusiasm or wherewithal to carry the policy out. Equally if the plant managers are clear about what is needed, change may be frustrated if the shop floor is not persuaded by the rationale. Of course an authoritarian style can achieve many things, but good environmental performance depends on the willing co-operation of the workforce, and a committed attention to detail, as well as on straightforward good housekeeping. A spill of raw material during the night shift, due to careless unloading by an uncommitted or discontented workforce can easily negate many months of successful environment protection.

Starting at the top it is assumed that the appointment of an Environmental Manager is evidence of the support of top management. The attitude to the environment is determined by the lead given by top management, and so an important task for the Environmental Manager is to propose an environmental policy for publication both within and outside the company. ICI's environmental policy (see Fig. 24.1) is a suitable illustration.

The policy document goes on to indicate the type of action that is necessary to fulfil this policy. It sets out the spirit of the Group's actions and is the framework which governs behaviour with respect to the environment. The policy is mandatory for all business units and employees world wide. But, left at that, the policy might be seen as 'PR', the term being misapplied with a cynical connotation. It will be dismissed as 'motherhood statements', so it should be followed up with realistic targets for change. Again, taking ICI as an example, the Chairman, in an important public address, announced a series of targets including a 50 per cent reduction in waste production over a 5-year period (see Fig. 24.2)

It is ICI's policy to manage all of its activities so as to give benefit to society, ensuring that they meet relevant laws and regulations; that they are acceptable to the community at large; and that their environmental impact is reduced to a practicable minimum.

Figure 24.1 ICI's environmental policy

ICI will require all its new plants to be built to standards that will meet the regulations it can reasonably anticipate in the most environmentally demanding country in which it operates that process. This will normally require the use of the best environmental practice within the industry.

ICI will reduce wastes by 50 per cent by 1995. It will pay special attention to those which are hazardous. In addition, ICI will try to eliminate all off-site disposal of environmentally harmful wastes.

ICI will establish a revltalised and more ambitious energy and resource conservation programme, with special emphasis on reducing environmental effects.

ICI will establish a clear policy and practice on waste recycling.

Figure 24.2 An ICI target for change

Moving down the organisation, and simplifying for the sake of clarity, we come to the managers whose products and plants are the source of environmental impact. Personal contact is, of course, important; but there are bound to be too many of them for the amount of time available in the Environmental Manager's day. The Environmental Manager has to influence them in groups, through the written word or possibly through the medium of video.

To build on the Board's environmental policy and broad improvement targets, ICI has taken further steps to turn the Board's wishes into hard practice. The initial step was to ensure that the Company had the right management systems for environmental protection. Sites and businesses have appointed managers whose tasks are to ensure that their units take proper account of the environmental dimension; managers at all levels have environmental responsibilities built into their job descriptions and are set environmental objectives; information systems have been established.

Within the management system, further levels of supportive material have been provided. The first level is a series of mandatory standards. These are not numerical requirements but rather specifications of management practice requiring, for example, that management at each site should maintain an up-todate assessment of its environmental impact, including all wastes and emissions and their method of treatment or disposal. Similarly other standards deal with the protection of the land and of groundwater, with the control of contractors, with the environmental dimension of acquisitions and divestments, and many more. The existence of these mandatory standards are an important factor in ensuring that units can satisfy the requirements of national and international standards as illustrated by BS5750 and 7750 and ISO 9000.

In a complex multinational, multi-business company like ICI, or even in a diverse national company, it is neither wise nor possible to specify precisely how all this should be implemented for each site and for each product. Broad standards are not enough to produce the best results everywhere. Even given equivalent levels of motivation, there is no guarantee that a manager in location A will arrive at the same solution as one in location B, where local pressure and culture may be very different. What is important, however, is that the company makes uniformly good progress in its businesses across the globe.

At this practical level, the Company provides a range of non-mandatory, advisory documents to help the manager do his job and fulfil the ICI Group Environmental Policy. These documents might be thought of as guidelines or advisory codes of practice, and there may be several to back up any one of the mandatory standards. For instance, there is a guide to carrying out an environmental impact assessment on an existing site, and another for a new site or addition to an existing one; there is a series of guides to assessing the possibility of land being contaminated, giving advice on the approach to managing it if it is. Nor is this system fixed; it is added to and amended as circumstances require.

This quantity of documentation may sound bureaucratic but it is no more than is necessary to provide the busy manager with information when he (or she) accepts the need for change but is uncertain of how to go about it. Perhaps he could work it out for himself, but in a large company it would be madness for each site management to 'invent the wheel'. Indeed all of the work need not be done within the company; national and international trade associations are ways of sharing the load. The promotion of the 'Responsible Care' concept within the chemical industry by the Chemical Industries Association is a good example. Not only has this been effective at the national level, it has also been promoted by local Responsible Care 'cells' operating in regions such as Teesside and Severnside where several companies have plants.

In parallel with the development of systems and procedures, the Environmental Manager has an important task in persuading management of the ways in which the world is changing, and of the necessity to accept the need for local changes, and to get on with the job. Part of this is of course to complete the loop of the management system by communicating how change has happened and how performance has improved. The manager must be visible personally and can, of course, do this by moving around the company svstematically and talking to groups of managers. Another route is to arrange for a regular slot at management courses and conferences, and to ensure that special environment courses are always available for those who need to develop environmental knowledge and skills in depth.

The final step in the development of the 'bureaucracy' is the need for sites and businesses to have their own environmental procedures. There will, of course, be much in common but each site must support its own staff with standards and practices that meet the local culture and local regulations, and which apply to particular products and processes – all coming within the company's overall policy and standards.

Perhaps the most practical level at which the Environmental Manager should exert his influence is on the 'shop floor'. The actual physical practices that determine the company's environmental performance are the actions of its operatives from the shop floor upwards. And the benefit of their commitment does not end there, because our employees are probably our most influential ambassadors to the outside world, through their families and their contacts in schools, clubs and pubs, and through every form of contact in the community at large.

Within ICI a whole range of techniques has been used to inform and educate the workforce about environmental issues in general, about the Company's policies and about the Company's stance on particular issues that concern it.

- There have been regular presentations to gatherings of the staff consultation network up to national level.

- A 'programmed learning' package has been prepared and distributed widely for use by staff and their families.
- A number of businesses have produced environmental 'Issues' broadsheets, highlighting the policies and activities of particular interest.
- Pamphlets have been distributed widely outlining not only the broad policy but also the Company's stance on important issues such as CFCs and the Ozone Layer.
- Time has been made available for employees to attend environmental forums, on the lines of 'Question Time', to exchange views about policy and practice.

The Environmental Manager does not only exert influence at all levels within his/her company, he also has an important role in presenting the face of the company to the outside world. Trade associations are an important channel for such influence because a poor environmental performance by one company within an industry invariably reflects on the whole industry. Of course, the chemical industry is a good example of this phenomenon; many people have a rather imprecise concern about 'chemicals', a concern that is heightened whenever the media feature an accident or a spillage. The committees of the trade associations are therefore a valuable route for the Environmental Manager to compare experiences, and to help his opposite numbers to develop equivalent policies and practices of their own.

The manager also has a role in influencing Government. Legislators do not always have first-hand experience of the practical problems of industry, and it is legitimate for the Environmental Manager to seek to influence legislation and regulation before it is cast in stone; indeed it is happily the practice in many states for industry's views to be sought actively. In the early application of Integrated Pollution Control (IPC) and the development of Her Majesty's Inspectorate of Pollution (HMIP), many important issues of fact, of interpretation, and of behaviour emerged. By active co-operation between companies and the Inspectorate in trial notifications, important decisions were made leading to more rapid and effective environmental improvement.

Using authority

The fact that the title 'Environmental Manager' may imply greater authority than 'Environmental Adviser' was mentioned earlier. This is no more than a reflection of changing attitudes. Society has raised its expectations of environmental performance and this has often given the Environmental Manager the backing of the law. Companies also accept that the environmental dimension of decision-making must be given greater weight. For example, Sir Denys Henderson, Chairman of ICI, has stated that 'Good environmental performance is not an option. It is essential if ICI is to continue as a leading international chemical company into the next century.' So not only does the Environmental Manager have the backing of law, he has the backing of mandatory Company policies and standards that go beyond the law.

Perhaps the most important area of mandatory behaviour is the environmental audit. This is the essential checking process which is the tool for assuring management, up to Board level, that policies are being implemented throughout the Company, and that the standards are being adhered to. Audits are designed to uncover shortcomings and form the basis of improvement plans. They must have the formality of protocols and rules, and must incorporate a continuous system of checking that improvement plans are implemented and achieving results. To be effective, auditors must be properly trained. The lynch-pin of the system is the annual Letter of Assurance that, in ICI, every business, territorial or subsidiary head must provide.

Within this exercise senior executives must describe the management system that ensures that ICI policies and standards are being met within their own areas; they must also describe the improvements that are being sought, and what improvement plans are in place.

As mentioned earlier, the ICI Board has set a challenging target for the reduction of waste. This would be meaningless if it were not followed up by suitable definitions of waste, in itself not a simple exercise, a requirement for all of the units to provide baseline and annual data, and for that data to be collated and published. The Group's Annual Reports

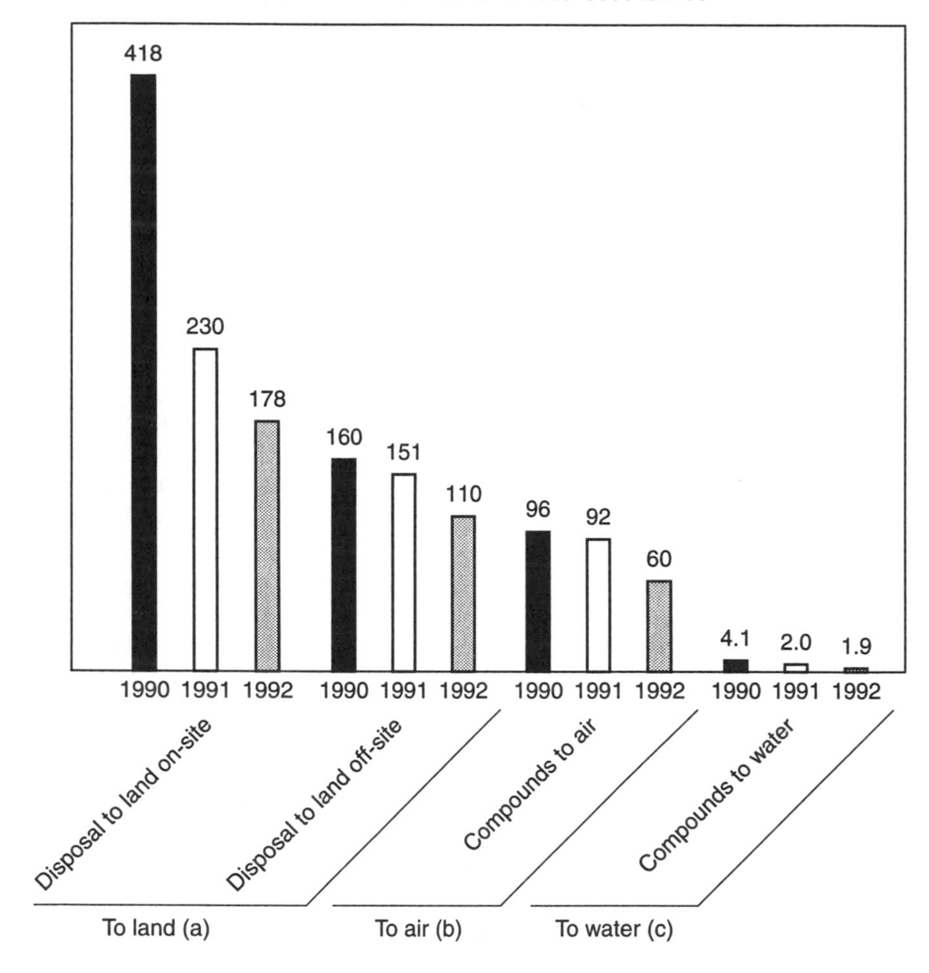

Total wastes 1992 (1990)

Non-hazardous 93.2% (88.8%)

Hazardous 6.8% (11.2%)

Waste considered hazardous

(a) To land: are conventionally defined by national regulations as requiring special treatments or disposal procedures.

(b) To air: are predominantly the volatile organic compounds (i.e., those which evaporate in air at normal temperatures).

(c) To water: those chemicals defined by national and international lists (e.g., EC black list, UK Red list).

Figure 24.3 Quantities of waste for 1991/2, compared with 1990 baseline

for 1991 and 1992 were accompanied by brochures giving the quantities of waste for those years compared with the 1990 baseline. The chart in Fig. 24.3 illustrates the content of these reports.

Other mandatory policies and standards were referred to earlier and these also give the Environmental Manager the base for requiring action. These include the preparation of Environmental Impact Assessments for operating sites and for new projects, environmental aspects of proposals for the acquisition or of the disposal of sites and businesses, reviewing the possibility of contamination of land or groundwater at every site and maintaining a dossier of that information together with data on spills and leaks. There will be others; the notion of cradle-to-grave product stewardship imposes additional disciplines, for example for measuring the environmental impact of their products in use using, where appropriate, the technique of Life Cycle Analysis (LCA), itself not yet a fully developed methodology.

Conclusion

The task of changing the ways in which companies act with respect to the environment is of massive importance. It is a fact that in some areas man now has the capacity to harm his environment on a global scale. It is probably at least as important that society, under the influence of the media and the environmental movement, fears that he may actually be causing that harm in numerous ways. It is a fear that must be respected and resolved. Industry must involve itself by understanding the issues, developing wise policies and practices, and play a proper part in reaching acceptable compromises. The role of the Environmental Manager and, in a large company, his functional associates, has been described. Because the field is developing rapidly, with companies being at different stages, the theme has often been illustrated by reference to the ICI Group.

The theme has been that of the Change Agent. Change is occurring and will certainly continue. Industry can either let it wash over itself or play an active part. Whatever his particular title in the organisation, the Environmental Manager is a key player in helping his/her company to come to terms with that change and remain successful.

Sources

The environmental awareness publications referred to are:

ENDS Report
Published by Environmental Data Services Ltd, Finsbury Business Centre, 40 Bowling Green Lane, London EC1R 0NE

Integrated Environmental Management
Published by Blackwell Scientific Publications, Osney Mead, Oxford OX2 0EL

Europe Environment
Published at 46 Avenue Albert Elizabeth, B 1200 Brussels, Belgium

International Environmental Reporter
Published by BNA International, 17 Dartmouth Street, London SW1H 9BL

The chapter is illustrated by the policies and practices of Imperial Chemical Industries PLC. Further information is available from:

ICI Public Affairs
9 Millbank
London SW1P 3JF

Publications include:

Investing in the Environment, Sir Denys Henderson, 28 November 1990

The ICI Policy on the Environment

The ICI Approach to Safety, Health and Environmental Management

The ICI View on Integrated Pollution Control

Problems and Solutions: the Ozone Layer and CFCs

Environmental Awareness. A learning package for ICI staff

Progress towards Environmental Objectives 1991

Environmental Report 1992: Improvement by Design

Numerous other publications cover the products and practices of the various territories and business within the ICI Group. See also:

Responsible Care

Published by CIA Publications, Kings Buildings, Smith Square, London SW1P 3JJ

Working with suppliers to reduce waste and environmental pollution

J. ANTHONY HILL

Environmental Quality Manager,
Procter & Gamble European Technical Center, Brussels

- The supplier programme is strategic in motivation – ensuring long-term growth by doing what is right and valued by customers.
- You might think that if a supplier has a problem and you question it he/she will want to pass the bill on to you for fixing it. Fortunately for the environment, business is not that simple. Paying attention to your supply base can provide many benefits for your business and for the environment as well.
- Fruitful ways of working together include:
  - working jointly to reduce waste through source reduction, re-use or recycling;
  - using the supplier's expertise to help introduce environmental improvement in your products;
  - improve your environmental management system by learning from them.
- A good approach is to follow Pareto's principle, focusing first on those elements that will provide the biggest return and can be accomplished simply.
- It is not sustainable to pick and choose which bits of the environmental impact of your business you will address; rather you need to strive for continuous improvements in all facets of the environmental impact of your products.

Why work with suppliers at all?

It is tempting not to enquire too deeply about where your supplies come from. You could take the view that most companies are introducing measures to improve the environmental performance of their operations, so there is no real need to become involved in what your suppliers are doing. They will be putting measures in place in any case. You might also think that if a supplier has a problem and you question it he/she will want to pass the bill on to you for fixing it. Fortunately for the environment, business is not that simple. Paying attention to your supply base can provide many benefits for your business, and for the environment as well. Progressive companies look at the environmental impact of their products over the total life cycle, and the source of supplies is part of this. Neglect, on the other hand, can put your business at risk if you become associated with a source of pollution. Your supply source can be shut down. Environmentalists may even consider using your business as a more effective way of gaining public attention to the pollution, instead of campaigning against the supplier. However, the main rationale for a supplier programme is not defensive; it is for the benefits it offers. In this chapter we will look at the many reasons why an environmental purchasing policy makes good business sense, as well as being the 'right thing to do'. We will then discuss ways of implementing a programme.

IT'S A MATTER OF SUSTAINABLE DEVELOPMENT

The 'International Chamber of Commerce (ICC) Business Charter for Sustainable Development', launched in April 1991,[1] and signed by most leading companies, is an internationally recognized statement of the principles for environmental management. The first 10 principles address the internal measures a company needs to adopt. The concept of working with suppliers to reduce waste and environmental pollution is embodied in principle 11. This calls upon companies to promote the adoption of the same standards by their contractors and suppliers. Furthermore, suppliers who have made strategic investments in designing systems to reduce waste and avoid pollution, offer long-term supply.

All companies are faced with meeting an increasing number of environmental requirements just to stay in business. Companies committed to sustained, long-term success go beyond compliance, anticipate change and seek continuous improvement. The ICC charter recognizes this. An environmental purchasing programme is part of this strategy.

THERE ARE BENEFITS FROM CHOOSING ENVIRONMENTALLY-COMMITTED SUPPLIERS AND WORKING WITH THEM

As well as entitling you to the moral high ground, working with your suppliers to improve environmental quality also serves your overall business interests. Suppliers who pay proper attention to waste management and pollution control are also likely to have high standards of service, product quality and competitive costs.

There are a number of fruitful ways in which you can work with these suppliers to tackle environmental issues. Examples will be discussed in more detail later, but basically they fall into one of three categories. You can:

- work jointly to reduce waste through source reduction, re-use or recycling, and often save money;
- use the supplier's expertise to help introduce environmental improvement in your products;
- improve your environmental management system by learning from them.

THE REQUIREMENT TO WORK WITH SUPPLIERS IS APPEARING IN ENVIRONMENTAL MANAGEMENT PROGRAMMES.

Many of the principles of the ICC Charter have been incorporated into emerging eco-audit and management programmes. The UK eco-audit and environmental management system proposal BS7750 includes a requirement[2] for control and verification of all functions, activities and processes which could, if uncontrolled, have a significant environmental impact. This includes, where relevant, raw material sourcing and purchasing. Similarly, the Code of Practice for Product Stewardship in the Chemical Industry 'Responsible Care©' programme is concerned with the environmental impact of a product over its total life cycle. In the USA, the Environmental Protection Agency proposed (1993) an environmental leadership programme to recognize and reward high quality environmental practices.[3] The programme will provide formal recognition to companies that go beyond compliance by incorporating pollution prevention into all operations, *including purchasing*. This vision to go beyond compliance is welcomed by forward-looking companies that are committed to setting the pace for environmental improvement.

INDUSTRIAL COMPETITIVENESS AND PROTECTING THE ENVIRONMENT GO TOGETHER

A Cranfield School of Management survey (1991) found that less than 10% of 200 small companies surveyed had any environmental supplier policy. Now, however, companies are seeing a supplier policy as a useful way of promulgating their own environmental credentials. Setting high standards of environmental management, including monitoring your supply base, will be important elements in your competitiveness, or indeed qualification, as a supplier to others. For example, the EC

Commission Communication on 'Industrial Competitiveness and Protecting the Environment' that was adopted by Industry Ministers in November 1992 states:

> It is not in the interests of small and medium sized enterprises (SMEs) to cut themselves off from developing markets by being unable to meet environmental requirements. Industrial clients are increasingly demanding that their suppliers and sub-contractors also meet equally high levels of environmental performance in order to demonstrate their own environmental commitment.[4] Public Authorities also have a major responsibility when accomplishing their activities as purchasers.

It is reasonable to assume that suppliers to industrial clients (or Public Authorities) will need to demonstrate that they have checked that their suppliers and sub-contractors meet high levels of environmental performance, and so on down the supply chain.

CONSUMERS ARE TAKING THE ENVIRONMENTAL PEDIGREE OF PRODUCTS INTO ACCOUNT IN THEIR PURCHASING DECISIONS.

If your products are sold direct to the public, these may be looked upon as a way of applying pressure on your suppliers. While a consumer product may have no environmental issues *per se* it can be linked with the environmental performance of its suppliers. Consumers and influential green groups have access to a growing amount of information about the environmental performance of companies. In Europe, Friends of the Earth have said that they are compiling a database on performance from all public sources and intend to use it against those they see as transgressors. When direct contacts with suppliers do not get the results environmentalists are looking for, they have used consumer boycotts of products that use the materials under attack.

On the other hand, products that give excellent value *and* care for the environment over their total life cycle will attract the growing number of consumers who want to demonstrate their environmental concerns.

Examples of environmental purchasing

This section examines some approaches being followed by a number of companies. All the examples improved environmental quality *and* made good business sense. They cover three aspects of supplier contact:

- joint work to reduce waste through source reduction, re-use or recycling;
- use of the supplier's expertise to help introduce environmental improvement in the customer's products;
- a comprehensive supplier programme.

REDUCE, RE-USE, RECYCLE

Reduce:

- Procter & Gamble has worked closely with a number of suppliers to develop high-performance packaging systems that use less material. To quote one example, the refill system for compact detergent powders saves 90% of the packaging used for the primary package. This has been successful because consumers save money while doing something to help reduce waste.
- Plysu Containers produce ultra light-weight plastic containers for fresh milk and juice. They have introduced a transport packaging system where the containers are delivered in plastic shrink-wrap modules – the transit packaging being fully recyclable. This cost-effective system reduces the amount of packaging materials used by traditional delivery systems and was developed by working with customers to make it compatible with their high-speed filling systems.
- A UK catering company started measuring the packaging waste from suppliers and gave priority to those supplying goods with the minimum packaging needed to protect the product, and with packaging that could be returned. In doing so the company expects to save on product, delivery and waste disposal costs.

Re-use:

Many customers are looking for returnable delivery systems as a means of reducing their waste. Two examples illustrate the innovation in this area.

- Dow Corning (Europe) developed returnable, foldable Intermediate Bulk Containers (IBC) and now license this system to other suppliers.
- A supplier of agrochemicals is working with a packaging supplier to develop refillable, multiple-use containers. The containers have to be designed so that they can only be refilled with the correct product. This system will replace existing packaging and reduce the number of one-trip packs in future markets.

Recycle:

A large number of companies have introduced voluntary policies to take back materials for recycling. With targets being introduced for waste recovery this trend can only increase. Leading companies have been offering schemes before they are mandated. Listed below are just a few of the many items that suppliers are currently offering to recycle.

- A typical example is the UK 'Save-a-Cup' scheme where a plastics supplier collaborated with a cup manufacturer and the vending machine operator to recycle plastic cups.

- Many suppliers of toner cartridges have Euro-wide recycling services.
- BASF launched fully recyclable video cassettes – and expect to 'shake up the video sector' with this revolutionary product.
- IBM (UK) introduced a scheme to take back obsolete computers in the early 1970s. During 1992, recycling rates of 86% were achieved. The company's latest personal computers now incorporate clip-together technology for ease of disassembly, and code the plastics components to improve identification and future recycling efficiencies.
- ICI Katalco arranges recycling of spent catalysts through its 'Catalyst Care Programme'.
- The automobile industry provides many examples where suppliers are collaborating with manufacturers to develop components that will allow the car to be more easily dismantled at the end of its life.

Lessons to be learned:
The first thing that can be done to reduce waste is to look what is being thrown away. Talk to your suppliers to see whether:

- the secondary packaging you are being sent can be reduced;
- items that are thrown away can be refilled or regenerated;
- any items can be recycled.

The supplier may already be working on recycling programmes or be able to direct you to associations that are running pilot schemes. The important fact is that you can improve your waste performance by encouraging suppliers to work with you to reduce waste. Even if you do not get any revenue from this, you will save on disposal costs.

WORKING IN PARTNERSHIP TO SOLVE PROBLEMS

The hallmark of a sound customer-supplier relationship is the supplier's concern to work with the customer to provide a better service. Suppliers now recognize that this can give them a competitive edge and have started being proactive with their customers about improving the environmental quality of their products.

Dow Plastics advises customers on environmental design – how to design items to use the minimum material possible; designs that facilitate later recycling and with component parts that have approximately equal durability. They have also published an instructive booklet – *Designing Green – a guide.*[6]

BP Chemicals have developed literature about the properties of volatile organic compounds (VOCs) and the environmental impact of different chemicals they supply. They offer advice about abatement techniques and work with customers to develop formulations with the minimum environmental impact.

In the following examples the customer has taken the lead in working with the supplier to achieve environmental quality improvements in purchases.

S. C. Johnson provides a case study 'Catalyzing improved supplier performance' in *Changing Course*.[7] Senior representatives of top supplier organizations were invited to a high-level conference to learn of Johnson's environmental goals, and to come up with ideas to help achieve them. Following the conference, suppliers came up with projects that will lead to a substantial reduction in packaging, and to the phasing out of controversial chemicals.

Procter & Gamble pioneered the development of the incorporation of post-consumer recycled (PCR) plastics in detergent and softener bottles with one of its suppliers in Europe. This entailed technical experts working with this supplier to develop a new process. P&G funded the equipment changes needed, helped set up a supply chain of suitable recycled material, and proved that mixed household plastics waste could be incorporated into many types of bottle, despite a general scepticism in the industry. The technology was not protected but made available to anyone interested. Other suppliers have now developed this capability, and they supply bottles containing recycled plastics.

Lessons to be learned:
It is appropriate to survey what opportunities there are for these partnerships. You can expect suppliers to use their inventiveness or expertise to help you in a number of ways, such as:

- reducing the use of resources through projects such as light-weighting designs or incorporating recycled material in your packages;
- making your products more readily recyclable;
- identifying alternatives to replace controversial chemicals;
- abatement techniques for the materials supplied.

This is a very important aspect of working with the supplier interface to improve environmental performance and gain business benefits. However, companies committed to long-term sustainable development go further by encouraging their suppliers to protect the environment in their operations.

A COMPREHENSIVE SUPPLIER PROGRAMME.

A number of leading companies have a comprehensive environmental purchasing programme as part of their overall aim to achieve environmental excellence. A few examples demonstrate how comprehensive this can be.

The Do-it-Yourself chain B&Q was one of the first companies to introduce, in 1990, a comprehensive purchasing policy. The programme demanded that its suppliers should enter into a contractual agreement to reduce the environmental impact of their products. The impact of materials from different sources was determined, and where an alternative product performed as well and cost the same, the product with the lower environmental impact was preferred. A detailed questionnaire was sent to 450 suppliers to establish the environmental effect of their operations. In many cases the questionnaire was followed up with a visit. Suppliers were encouraged to remedy environmental issues that were identified. If they were unwilling or unable to they were de-listed. B&Q teamed up with the World Wide Fund for Nature (WWF) to devise a scheme to phase out purchase of timber from 'unsustainable' sources.

IBM UK set up a task force within the purchasing organization to enhance environmental standards at its suppliers. Self-assessment guidelines were developed and issued to suppliers. Major suppliers were then asked to complete a detailed environmental questionnaire, and a number were audited. A few were subsequently dropped from the supply base on purely environmental grounds. IBM UK has participated, with the Institute of Purchasing and Supply, management consultants KPMG, and other companies, on a Business in the Environment (BiE) initiative to develop a code of practice on assessing the environmental performance of suppliers.[5]

British Telecom has published a guide for suppliers entitled *Selling to BT*.[8] As well as spelling out the requirements for quality, reliability and competitive pricing, the guide lays out the company's environmental purchasing policy. This is backed up by an Environmental Impact Standard that sets out in a helpful way what a prospective supplier needs to do to satisfy BT's environmental criteria. While the environmental impact of suppliers will be taken into account in awarding tenders, the emphasis is on partnership, inspiring suppliers to explain and develop their own programmes, and fostering an exchange of initiatives and ideas on good environmental practice.

Procter & Gamble works closely with its suppliers of wood pulp for its paper business with respect to their environmental practices. This includes reviewing the supplier's forestry practices and the environmental performance of pulp mills. In addition, P&G has learned from its suppliers and other professional experts in the field about the 'state-of-the-art' developments in these areas. When it is appropriate, and there is an agreed-upon environmental benefit to be gained, P&G will work closely with suppliers to make improvements. An example of such an action occurred in the late 1980s when the company-led hygiene product industry moved to elemental chlorine-free (ECF) fluff pulps. This occurred after environmental scientists in Scandinavia discovered that the use of elemental chlorine during bleaching could pose potential local environmental risks because of discharge of untreated effluent, due to the unintended production of highly chlorinated organic compounds.

Lessons to be learned:
These programmes can represent a significant undertaking of resources. B&Q, with a budget of £500,000 for the first year employed 6 consultants to evaluate the questionnaires and follow up with site visits. However, the environmental impact of this approach by major players can be dramatic and worthwhile – CFCs were replaced in most consumer products, and in electronic cleaning applications, considerably faster than was required by global protocols.

Small and medium-sized enterprises (SMEs) will need to look at the resources they can afford. However, it is a worthwhile undertaking and does have a direct impact on overall environmental performance. There should be savings and it is important for long-term business success. Furthermore, there is increasing commercial pressure from bigger customers on their suppliers to be able to demonstrate their environmental impacts. Help is available for smaller enterprises.

- Government bodies and/or trade associations may be able to give advice on the environmental issues that are important for your supply base.
- External alliances may be able to provide guidance. For example, WWF has been active in advising a number of timber users supporting sustainable forestry practices.
- As already mentioned, Business in the Environment is developing a code of practice.
- Friends of the Earth have advised local authorities about the environmental impact of the products they use. That is not to say the advice will not sometimes be without controversy. Customers should listen to their suppliers as well as advisers.

A good approach is to follow Pareto's principle, focusing first on those elements that will provide the biggest return and can be accomplished simply. The policy commitment is a strategic decision – it's for the long term. Do not expect immediate results. The following ground rules are suggested:

- Decide your policy and share it with your suppliers.
- Get their energy, as good suppliers, working for you.
- Initially make contacts and pose general questions. Establish that you can go back for more information when appropriate.

- Never ask for information you don't know you need – it's a waste of your suppliers' time and undermines your credibility when they ask you what you did with it.
- Keep abreast of environmental issues and think how they may impact on your suppliers.

Formulating a purchasing policy and programme

An environmental purchasing policy has to be part of a total environmental quality management programme which is actively supported by top management. Focusing on your supplier's environmental policy is only worthwhile when it comes out of an integrated approach and a corporate commitment to environmental quality.

POLICY STATEMENT

The elements of a purchasing policy need to balance and include:

- a requirement for competitive and fair pricing;
- meeting quality standards;
- compliance with human and environmental safety requirements with respect to transport, delivery and documentation;
- security of supply;
- meeting agreed delivery schedules;
- exercising sound environmental practices in the provision of goods and services.

In formulating the environmental component of a purchasing policy, the buyer will rely on advice and measures from a variety of specialists in the organization. Many companies set up a team of experts to advise the buyer and to draw up a programme. The buyer must remain in overall control of the contact with suppliers while including environmental considerations in his/her decision-making.

The environmental purchasing policy is likely to start with a statement reiterating the purchasing company's commitment to improving the environmental quality of its products, packaging and operations. The company positively encourages its suppliers to have an environmental policy and programmes. These supplier programmes will include procedures, such as the auditing of their operations, aimed at minimizing the environmental burden of their processes. EQ performance of the supplier will be one of the factors taken into account when allocating business. Where a choice exists, the purchaser will prefer suppliers whose products cause the least environmental impact during production and use. Assessment of environmental impact must be based on science, not emotion. The tone should be one of partnership – working together to improve the environment. However, suppliers with a chronic record of pollution that they are unable to or unwilling to put right will be disqualified.

Procter & Gamble, for example, encourages its suppliers to have an environmental quality policy and to further improve their EQ performance through clear objectives with specific, quantifiable goals and strategies. EQ performance will be one of the factors taken into account when allocating business.

Some other companies have very detailed purchasing policy statements. For example, British Telecom issues the following guidance to its suppliers in *'Selling to BT – a winning partnership'*.[8]

> BT's buyers are committed to:
> (1) Purchasing goods and services which can be manufactured, used and disposed of in an environmentally responsible way.
> (2) Meeting, and where appropriate exceeding, the standards required by environmental legislation.
> (3) Specifying and purchasing items which can be recycled and reused.
> (4) Specifying and purchasing items which can be operated in an energy efficient manner.
> (5) Selecting suppliers who are committed to sustainable environmental improvement.
> (6) Requiring all our suppliers to identify any harmful processes and materials they currently use and where feasible securing their commitment to a phased elimination.
> (7) Enabling disposal of goods in ways that minimize their environmental impact and, progressively, requiring suppliers to take responsibility for disposal.

Another approach is to define environmental requirements that suppliers must meet in material specifications or in annexes to contracts.

THE PROGRAMME – APPLYING THE TOTAL QUALITY CIRCLE: PLAN, DO, CHECK, ACT, AND START AGAIN

Setting priorities (Plan)

Having formulated and agreed a purchasing policy, the first step in implementing the programme is to prioritize the approach to suppliers. This requires assessing the biggest environmental impacts. A Life Cycle Analysis (LCA) study, if available, may quantify these. If not, then research and some judgement will be needed. Are you purchasing from industries with known water or air pollution problems? Are there supplies that create large quantities of solid waste in their manufacture?

Are there opportunities for a re-use or recycling alliance with particular suppliers? Are there supplies that cause you or your customers particular problems of disposal that could be solved by working with your supplier? If no particular issues stand out then rank supplies in terms of tonnage purchased.

The most practical approach is to confine initial enquiries to the policies of direct suppliers. There may be exceptions where an environmental burden down the supply chain needs addressing at the outset. However, generally, it is best to focus on where you can get the fastest return.

Putting a programme into practice (Do)

The appropriate programme will depend upon circumstances, but some steps can get you off to a quick start. You will have an immediate impact just by stating your interest. A brief statement about your policy, together with a general questionnaire directed to the target group, will not be too heavy on you or your supplier's resources, and provides a fast screening. You are likely to find that suppliers who have made the biggest commitments in this area will have material prepared and be keen to tell you what they are doing.

This brief initial contact is fast, flags your concern for the environment, and screens out the suppliers that do not need follow-up. It also provides a benchmark against which to base your assessment of the others.

Reviewing the data (Check)

The initial responses will be reviewed either by the co-ordinator or by a purchasing environmental steering team. Hopefully, the majority of suppliers will have convincing evidence of policies and programmes. Follow-up with some of these suppliers would be confined to exploring opportunities to work together. Or you could evaluate their programmes to benchmark you own.

For suppliers without any policies or programmes, the next step is to draw up a follow-up list on the basis of the biggest potential environmental impacts. This entails going into more detail about the issues that are likely to arise with these suppliers with respect to waste or pollution. Alternatively, you may have come across data about pollution issues with certain suppliers that you had not been aware of. Sources of this information can, for example, be environmental news services, environmental organizations or consultants specializing in this type of search.

Follow up (Act)

Based on the evaluation of risks it is now time to establish a second round of contacts. This can proceed via a detailed questionnaire, or a site visit or both. Some companies use consultants for an independent environmental audit.

If, after this second review, there appear to be serious pollution issues, share the findings with the supplier and involve him in trying to solve the problem against an agreed timetable. If you do not have enough influence with the supplier individually, see if other customers share your concern. Work in partnership with the supplier, although he needs to understand that it is his problem, not yours. Eventually, the problem must be fixed or the supplier changed. This message is not intended to be threatening, but there are basic company values and they cannot be compromised.

And start again (Plan, and so on . . .)

The cycle is then repeated, with progress on a number of fronts – prioritized according to the size of environmental impact and opportunity. The areas to address are:

- extension of the enquiry to the suppliers not yet contacted;
- new suppliers;
- new issues arising, or new standards of pollution control attainable, with the initial supply base.

THE CUSTOMER/SUPPLIER RELATIONSHIP

The importance of working with the supplier to achieve environmental quality improvement has been stressed. The following guidelines are recommended for a fruitful customer/supplier relationship.

- Treat suppliers courteously without compromising other suppliers.
- Operate with integrity. Do not mislead with dubious or false statements. Share what you know that is in the public domain and might be helpful to them.
- Always be fair and factual.
- Avoid any form of favoured treatment.
- Treat proprietary information in strict confidence.
- Don't mix up environment and commercial negotiations.
- Exercise the highest standard of personal conduct. For example, do not accept gifts or entertainment.

Your suppliers will soon notice if you are less serious about the environment than they are. That extends to all contacts. Sloppy layout of letters and questionnaires conveys the message that the task is routine rather than committed. Confine general questions to company policies and programmes. Don't ask general questions like 'what are the environmental effects of your products?' and provide 3 lines for reply; it conveys a lack of seriousness and will be ignored or answered superficially. Specific questions should relate directly to your purchases. However, if you learn of a serious environmental incident at a supplier, even though it did not involve your products, that would be grounds for auditing the operation and reviewing sourcing.

Conclusions

Industrial competitiveness and protecting the environment are inextricably linked. The commitment of the business has to be total, with top management support. It is not sustainable to pick and choose which bits of the environmental impact of your business you will address. An environmental purchasing programme can bring benefits in waste reduction and cost savings. However, the essential reason for it is that a supplier programme is part of your total commitment to sustainable development through environmental excellence. This means working on continuous improvement of the environmental impact of your products in all ways.

Recommendations:

The following steps are recommended for incorporating purchasing policy within Total Environment Quality Management:

(1) Formulate and write down what your environmental purchasing policy is. Get it agreed and supported by company senior management. Tell your suppliers about it and what you expect of them.

(2) The programme should have the following elements:

- Prioritize which suppliers you will contact.
- Work through your purchasing organization – be seen as part of a total company approach. However, keep commercial discussions and environmental discussions separate.
- When first contacting suppliers tell them what your environmental purchasing policy is and how it is part of your total environmental quality approach.
- Determine what your suppliers are doing to manage their environmental impact. Collect information about their programmes, their involvement in environmental activities and any environmental publications. Ask only for what you know you will use. Compare programmes with your own and with those of your other suppliers.

- Encourage continuous improvement by your suppliers.
- Visit key suppliers to establish standards you can expect. Exchange views and share learnings.
- Keep abreast of environmental issues in the media to give early warning of supply issues that could affect your business. Consider subscribing to specialist news services.
- For areas of high concern find out what data is in the public domain. The supplier expects you to be informed and fair minded about issues. Base contacts on science and fact, not scare stories. If appropriate, ask that your audit specialist be allowed to look at the supplier's facilities. Be prepared to make a confidentiality agreement with the supplier to protect his proprietary interests, but don't compromise your commitment to avoid pollution.
- Encourage technical meetings to solve environmental problems you have with a supplier's product.
- Look out for opportunities to work with your suppliers on projects to reduce waste.
- Participate in industry technical groups with suppliers to pool data about shared environmental concerns.
- Discuss problem areas with suppliers and ask them to remedy issues within a reasonable period of time. If encouragement and discussions do not solve a pollution problem, then disqualify a supplier, leaving the door open to resupply when the problem is fixed.
- Have a clearance procedure for new suppliers.

References

1 Bright, Peter. Introduction to the 'ICC Business Charter for Sustainable Development', WICEM II Official Report, published in *Environment Strategy Europe 1991,* Campden Publishing Limited, London (1991).
2 British Standard Specification proposal for environmental management systems BS7750, Annex A.4, BSI publication, London (1993), (1994).
3 US Federal Register, Washington DC, 15 January 1993.
4 Communication (SEC (92) 1986), 'Industrial Competitiveness and Environmental Protection', EC Commission, Brussels, 24 November 1992.
5 Environmental Data Services Ltd (ENDS), report, *Buying into the Environment*, May 1993, p. 22. (Available from ENDS, Finsbury Businesss Centre, 40 Bowling Green Lane, London EC1R 0NE.)
6 Dow Europe S.A., CH-8810 Horgen, Switzerland.

7 Schmidheiny, Stephan, with the Business Council for Sustainable Development. *Changing Course: A Global Perspective on Development and the Environment*, MIT Press, Cambridge, Massachusetts (1992), Case 12.4, p. 217.
8 *Selling to BT – a winning partnership,* British Telecommunications plc, London (1991).

Appendix

The Business Charter for Sustainable Development

■

There is widespread recognition today that environmental protection must be among the highest priorities of every business.

To help business around the world improve its environmental performance, the International Chamber of Commerce established a task force of business representatives to create a Business Charter for Sustainable Development. It comprises sixteen principles for environmental management which, for business, is a vitally important aspect of sustainable development.

It was formally launched in April 1991 at the Second World Industry Conference on the Environment. The objective is that the widest range of enterprises commit themselves to improving their environmental performance in accordance with these principles, to having in place management practices to effect such improvement, to measuring their progress, and to reporting this progress as appropriate internally and externally.

The sixteen principles are as follows:

1. *Corporate priority*
   To recognise environmental management as among the highest corporate priorities and as a key determinant to sustainable development; to establish policies, programmes and practices for conducting operations in an environmentally sound manner.

2. *Integrated management*
   To integrate these policies, programmes and practices fully into each business as an essential element of management in all its functions.

3. *Process of improvement*
   To continue to improve corporate policies, programmes and environmental performance, taking into account technical developments, scientific understanding, consumer needs and community expectations, with legal regulations as a starting point; and to apply the same environmental criteria internationally.

4. *Employee education*
   To educate, train and motivate employees to conduct their activities in an environmentally-responsible manner.

5. *Prior assessment*
To assess environmental impacts before starting a new activity or project and before decommissioning a facility or leaving a site.

6. *Products and services*
To develop and provide products or services that have no undue environmental impact and are safe in their intended use, that are efficient in their consumption of energy and natural resources, and that can be recycled, re-used, or disposed safely.

7. *Customer advice*
To advise and, where relevant, educate, customers, distributors and the public in the safe use, transportation, storage and disposal of products provided; also to apply similar considerations to the provision of services.

8. *Facilities and operations*
To develop, design and operate facilities and conduct activities taking into consideration the efficient use of renewable resources, the minimisation of adverse environmental impact and waste generation, and the safe and responsible disposal of residual wastes.

9. *Research*
To conduct or support research on the environmental impacts of raw materials, products, processes, emissions and wastes associated with the enterprise, and on the means of minimising such adverse impacts.

10. *Precautionary approach*
To modify the manufacture, marketing or use of products or services or the conduct of activities, consistent with scientific and technical understanding; to prevent serious or irreversible environmental degradation.

11. *Contractors and suppliers*
To promote the adoption of these principles by contractors acting on behalf of the enterprise, encouraging and, where appropriate, requiring improvements in practices to make them consistent with those of the enterprise; and to encourage the wider adoption of these principles by suppliers.

12. *Emergency preparedness*
To develop and maintain, where significant hazards exist, emergency preparedness plans in conjunction with the local services, relevant authorities and the local community, recognising potential transboundary impacts.

13. *Transfer of technology*
    To contribute to the transfer of environmentally sound technology and management methods throughout the industrial and public sectors.

14. *Contributing to the common effort*
    To contribute to the development of public policy and to business, governmental and intergovernmental programmes and educational initiatives that will enhance environmental awareness and protection.

15. *Openness to concerns*
    To foster openness and dialogue with employees and the public, anticipating and responding to their concerns about the potential hazards and impacts of operations, products, wastes or services, including those of transboundary or global significance.

16. *Compliance and reporting*
    To measure environmental performance; to conduct regular environmental audits and assessments of compliance with company requirements, legal requirements and these principles; and periodically to provide appropriate information to the Board of Directors, shareholders, employees, the authorities and the public.

An up-to-date list of companies who have signed the charter is available from ICC offices.

Glossary

■

The definitions presented in this glossary have been compiled from various sources. Some are taken from *Caring for the Earth – A Strategy for Sustainable Living* produced jointly by IUCN, UNEP and WWF. Other definitions have been supplied by chapter authors or abridged from *The Green Book* – Stephen Pope, Mike Appleton, Elizabeth-Anne Wheal, Hodder and Stoughton and *Into the 21st Century – a handbook for a sustainable future*, Brian Burrows, Alan Mayne, Paul Newbury, Adamantine Press.

Acid rain

Sulphur dioxide (SO_2) and nitrogen oxides (NO_x) are the main pollutants which cause acid rain. A colourless gas, $S0_2$ is emitted principally by power stations, commercial installations when burning coal and oil, and by metal smelters. NO_x – three compounds of nitrogen and oxygen ($N0$, N_20, NO_2) – is produced during the burning of coal, oil and petroleum. It is also emitted by stationary sources such as power stations, and by motor vehicles.

Acid rain results from SO_2 or NO_x combining with atmospheric water droplets. 'Dry' fallout of the same pollutants also takes place. Acid rain is a classic form of trans-boundary pollution and its effects are widespread: for example it occurs over much of western Europe and Scandinavia. It is also a growing problem in developing countries where rapid industrialisation is taking place.

Acid rain damages many ecosystems, particularly freshwater and forests. It also leads to loss of soil fertility and it damages human-made structures.

Biodegradability

Any substance or item that is capable of being broken down into harmless constituents, such as water, carbon dioxide, or individual elements by the action of living organisms is biodegradable. Organisms, including bacteria, in detritus ecosystems are particularly involved in such breakdown. However, biodegradable substances or items are not always broken down into their constituents when released into the environment. This is frequently the case where the environment is swamped by excessive amounts of materials, or where local conditions – for example, lack of sufficient oxygen in municipal refuse disposal sites – prevent biodegradation.

Biogeochemical cycles

The flow of elements such as calcium, carbon, nitrogen, phosphorus, potassium or sulphur, through the physical and biological components of the biosphere. The operation of biogeochemical cycles is strongly dependent on the actions of living organisms. At different stages in the cycles, these elements are combined with other elements in a variety of chemical forms. At some stages in biogeochemical cycles, some elements accumulate in large amounts, the sizes of which are far in excess of annual flows into or out of those stages. Those stages are often referred to as environmental sinks.

Biological diversity or biodiversity

The totality of genes, species and ecosystems in a region or in the world. It includes genetic diversity (the variation in the genetic composition of individuals within or among species); species diversity (the variety and frequency of different species); and ecosystem diversity (the variety and frequency of different ecosystems).

Biomass
A general term that refers to the total mass of living organisms in a defined area or belonging to a particular species. It sometimes includes both living and dead material. It is also used to refer to the production and use of biomass as an energy resource.

Biosphere
The thin covering of the planet that contains and sustains life. The biosphere (life) includes the hydrosphere (water), atmosphere (air) and lithosphere (rock, the crust of the earth) as well as ecosystems and living organisms.

Carbon cycle
The flow of carbon through the biosphere and its various components. Carbon dioxide in the atmosphere is absorbed by green plants and 'fixed' through the process of photosynthesis into simple sugars. These are used as a source of energy and as building blocks for the growth of plants, which as primary producers ultimately provide the food for all other living organisms. CO_2 is released by all organisms through the process of respiration, and through the decay of dead organic matter.

As well as being present in the atmosphere, CO_2 is also present in solution in water, and the oceans contain a huge store of carbon far in excess of that in the atmosphere. The concentrations of CO_2 in the world's oceans and the atmosphere ultimately tend towards a balanced equilibrium. This, however, has been disrupted by geological processes in the past, and is now being disrupted by the large inputs of CO_2 into the atmosphere arising from the combustion of fossil fuels during the last 150 years.

Carrying capacity
The capacity of an ecosystem to support healthy organisms while maintaining its productivity, adaptability and capability of renewal. Continued use of an ecosystem in excess of its carrying capacity will ultimately result in ecosystem degradation.

Cleaner production
Cleaner production is an operational approach to the development of production and consumption. It is based on addressing all phases of the life cycle of a product or process in order to prevent or minimise short and long-term risks to human health and to the environment.

Climate change
A change in climate resulting primarily from human activities, and their direct or indirect effects on the climate system which comprises the atmosphere, biosphere, oceans, water resources, soils and geological processes. Human activities affect these components and their interactions through releases of gases such as carbon dioxide, CFCs, nitrogen oxides, sulphur dioxide and water vapour directly into the atmosphere; through changes to ecosystems within the biosphere such as deforestation, desertification, or the draining of wetlands; and through the knock-on effects that these and similar changes cause to other components.

Critical loads approach
A critical load is the estimated exposure to one or more pollutants above which significant harmful effects on specified sensitive elements of the environment occur. Critical loads vary geographically according to the nature and sensitivity of ecosystems to specific pollutants. Below a critical load, the environment is able to accommodate an activity, or rate of activity, over time without unacceptable impact. This ability is finite and is calculated on the basis of the physical, chemical and biological characteristics of any site.

Cyclical industrial systems
Our future on the Earth can be likened to that of astronauts on board an enclosed spaceship far from Earth, in which survival depends on recycling wastes into freshwater and food, on re-use of materials and on the maintenance of vital life-support functions through this recycling. There is no possibility of expansion or growth beyond what is there already. Providing for equivalent practices in industrial societies requires the adoption of Cyclical Industrial Systems.

Cyclical Industrial Systems are those where the industrial system in its entirety, rather than just its individual component processes, has ways of recycling or utilising a high proportion of wastes for further productive industrial activity, and where this re-use or recycling is equally well developed to the production systems of linear industrial systems. An example of a prototype cyclical industrial system is the disassembly on an industrial scale of cars and electronic goods such as is being developed in Germany. Product stewardship is highly developed in cyclical industrial systems and is a key aspect of business development. As yet no industrial system has developed to a state anywhere near that which would approach a cyclical industrial system.

Deforestation
Deforestation is the permanent destruction and loss of forests. More than 50 per cent of tropical forests that existed in 1900 have been destroyed through logging and agricultural expansion. Temperate forests suffered the same fate for the same reasons, and are now under renewed pressure from expanded logging operations: virgin forests in Canada and Russia are particularly affected. Some forest cover in temperate regions has been replaced by forestry plantations, although these lack the diversity – and particularly the genetic diversity – of natural forests.

Reduction in the number of trees has repercussions for the Earth's overall ability to sustain life because the Earth's atmosphere, water cycle and climate is destabilised by forest loss.

Desertification
Desertification is the ultimate stage in the process of unnatural land degradation, when the fertility of its soil has been completely and irreversibly destroyed. Causes include erosion, salinisation or pollution of the soil, and are often linked with overgrazing or intensive agriculture.

Detritus ecosystem
Detritus ecosystems comprise organisms that live or feed on dead and decaying organic matter. They break it down over time into simple components that are recycled within the soil and immediate surrounding ecosystem, within adjacent ecosystems or throughout the biosphere, via biogeochemical cycles. Worms, fungi and bacteria are examples of such organisms. Their action in detritus ecosys-

tems accelerates the process of decay to more than ten times the rate of purely physical decay. Without detritus ecosystems, dead material would accumulate and would lock up nutrients, preventing them from being recycled. This would limit the potential for renewed growth dependent on those nutrients.

Duty of care
The Duty of Care is a legal approach in pollution control and environmental legislation applying to people who handle or control environmentally harmful substances or processes. It includes the duty to prevent anyone else contravening legislation; to prevent escape of potentially damaging substances and wastes; and to transfer potentially damaging substances and wastes to individuals or organisations which satisfy legislation for dealing with such materials. Duty of care also covers responsibility to ensure that legal and technical requirements are fulfilled.

Eco-efficiency
The refocusing of goals and assumptions that drive corporate activity in such a way that they play a major role in reducing the environmental and human impacts of competitive and successful companies. Improvements in technological 'hardware' – the development and use of new equipment – and 'software' – the management practices of business – are central to eco-efficiency.

Increasingly efficient use of raw materials is a feature of industrial activity, due largely to improved technology and processes. It is driven by pressures to reduce costs and enhance competitive positioning, especially among market leaders. Reducing inputs of raw materials in relation to profit generation also leads to a reduction in waste and pollution. However, such gains in efficient use of energy and materials, while valuable in themselves, can ultimately be outpaced by growth in the scale of economic and industrial activity. The environment responds to absolute levels of use and consumption, and cumulative impacts, rather than to relative improvements in materials use and pollution intensities.

Eco-labelling
The labelling of products deemed to have lesser impacts on the environment than alternatives, taking into account all stages of production, use and final disposal. Canada, Germany and Japan have eco-labelling schemes, and the European Community set up a programme that currently applies only to manufactured goods. In order to use eco-labels, producers and suppliers must comply with certain criteria. For example, WWF is working closely with a variety of companies that buy and sell timber products and who have made a commitment to phasing out the use of all unsustainable wood and wood products. WWF is also promoting an independent labelling scheme under the auspices of the Forest Stewardship Council. A voluntary body, the FSC, will be responsibe for approving and accrediting the certifying agencies which, in turn, will monitor and approve individual timber producers.

Ecological diversity
The variety and frequency of different ecosystems.

Ecological footprint
The overall environmental impact of a country, region, community or individual on both the immediate territory and areas elsewhere in the world from which they obtain resources, or which they pollute, directly or indirectly. To understand

the concept of an ecological footprint, imagine a farmer with one cow. The cow needs about half a hectare of grassland to feed it during a year, and the farmer either supplies this on the farm, or keeps the cow in a stall, buying in feed from elsewhere. That feed has to be grown somewhere, so the cow still consumes the produce of half a hectare of grassland, or the equivalent, whether it is kept in a field or a stall.

All human and corporate activities leave a footprint. The aim is to minimise it, by reducing consumption of raw materials and the generation of pollution and wastes, and so for us all to tread more lightly on the planet.

Ecological processes
Ecological processes sustain the Earth's life-support systems. They involve continuous actions or series of actions that are governed or strongly influenced by one or more ecosystems. They sustain the productivity, adaptability, diversity and capacity for renewal of the biosphere as a whole and of its components, including ecosystems and living organisms.

Ecology
The branch of science concerned with the relation of living things, including human beings, to each other and to their living and non-living surroundings, among them those created by humans. It is concerned with the study of the distribution and abundance of plants, animals and other living organisms and the factors that influence this and their interactions.

Ecosystem
A dynamic system of plants, animals and other living organisms together with the non-living components of their environment. Natural ecosystems are those which (apart from global influences) are unaffected by human activity. Other ecosystems are modified by human activity to varying extents and include those that are built or cultivated. Ecosystems that have been so damaged that they are unlikely to recover without help, are termed degraded ecosystems.

Energy flow
The flow of energy through ecosystems and the biosphere. Energy (apart from geothermal and nuclear energy) is ultimately derived from sunlight through photosynthesis by green plants, and through direct warming of the Earth's surface.

Environmental assessment
A process that assesses and predicts the environmental impact of a proposal – for example, a road scheme or new development – that is subject to a decision by a competent national authority. It identifies alternatives and presents its findings in such a way that decision-makers can be informed of what needs to be done. Environmental assessment is equally applicable to policy proposals.

Environmental audit
An inspection system that assesses the environmental effects of a company's activities, products and suppliers. It covers specific audits of health, safety, waste prevention and other matters and focuses on environmental issues of key concern – the organisation's impact on ozone depletion, pollution control, contamination of land or water, noise and odour pollution and waste minimisation, for example. It also takes into account the environmental performance of suppliers of raw materials, goods and services.

Environmental investing
Environmental investing is practiced by funds which seek to invest in companies providing environmental technologies and services. Environmental companies are defined in terms of the products and/or services they provide or the proportion of turnover from a particular product or service. A sound environmental performance track record is not a necessary criterion for investment.

Environmental management
Management that enables an organisation to establish an environ-mental policy and objectives, comply with them and demonstrate them to the outside world. The policy must be relevant to the organisation's activities, products and services and their environmental effects. It must also be understood, implemented and maintained at all staff levels.

Environmental management system
This covers the organisational structure, responsibilities, ways and means of implementing environmental management. It ensures that the activities of an organisation, and their effects, conform with environmental policy and associated objectives and targets. It includes the preparation and implementation of a documented system of procedures and instructions providing the basis for a programme of continuous environmental improvement.

Environmental sinks
Those parts of the living or non-living environment where large accumulations of particular elements or materials build up over time, and to which particular elements or materials flow in biogeochemical cycles.

Ethical investing
An ethical fund is defined by the Ethical Investment Research Service (UK) as any fund which restricts its choice of shares according to at least one ethical criterion and whose portfolio is published. Many ethical funds include environmental factors and exclude, for example, companies engaged in trade in tropical hardwoods or the manufacture of CFCs.

Gaia hypothesis
The Gaia hypothesis proposes that the global environment is maintained within the narrow 'window of opportunity' necessary for the existence of life, by a self-balancing feedback mechanism. This functions through the interactions of the world's biological and physical systems.

The Gaia hypothesis highlights continual interactions between the biological and physical worlds. The atmosphere – a physical resource – has been generated by biological activity. So have coal, chalk, certain types of metal-rich ores and many other types of rock. Fertile soil is created by the interaction between mineral rocks and physical weathering, plants and soil-dwelling organisms. Interactions of this sort are pervasive and essential to the functioning of biogeochemical cycles, and to the health of the planet.

Genetic diversity
The variety and frequency of different genes and of genetic stocks.

Global warming
The term used to describe the rise in the Earth's overall temperature caused by human activities. Fossil fuel burning, deforestation and other practices have combined to produce more and more greenhouse gases, which absorb infra-red radiation emitted from the Earth's surface and heat up the atmosphere in the process. As a result, average temperatures around the world have risen by about

0.5°C since the beginning of the century. Carbon dioxide (CO_2) emissions from human activity are estimated to have caused around 50 per cent of this increase; a further 25 per cent is contributed by CFCs, and another 15 per cent by methane.

Green consumer
A green consumer is a person who buys goods or services which are (or claim to be) less harmful to the environment and/or human health than alternatives. Increased environmental awareness of consumers in the developed world, linked with important issues of public concern such as global warming or standards of food safety, has had a significant impact on retailing in many industrialised countries.

Green investing
This is a similar term to socially responsible investment (SRI) covering a wide range of styles and types of investment vehicles. What distinguishes it from SRI is the greater emphasis it places on environmental considerations. Green investment takes as its guide the principles of sustainable development as defined and expanded upon at the United Nations Conference on Environment and Development at Rio de Janeiro in June 1992.

Industrial systems
Dynamic networks of industries and the associated environments, infrastructure, social and economic organisation, communities and institutions with which they interact. Industrial systems are dependent on these interactions for their inputs of materials, energy and labour, for their markets and for the disposal, recycling or re-use of waste materials.

Integrated pollution control (IPC)
A method of handling pollutants and connected aspects of environmental problems. IPC seeks to reduce impacts of pollution on the environment to an overall minimum and emphasises 'at source' rather than 'end of pipe' pollution control measures. IPC recognises that if pollution control is dealt with by separate arrangements covering emissions of pollutants to land, air or water, there is a high risk of reducing one problem and causing another. For example, technology used in power stations to remove sulphur dioxide (SO_2) reduces acid emissions but creates additional solid and liquid wastes.

Internalisation of costs
Internalisation of costs is a crucial step towards reducing environmental damage from economic activity.

Adverse impacts on the environment or social groups often arise from activities of business or other organisations – where these impacts are not prevented, fully mitigated, or compensated, they are termed externalities or external costs. Internalisation of costs refers to measures taken to ensure that external costs are paid for by those that cause them, and are taken into account in their decision making.

One of the major concepts supporting the internalisation of costs is the Polluter Pays Principle, requiring that polluters pay for damage caused. Internalisation can also be supported by fiscal measures (such as charges and taxes), regulations or legal remedies (such as strict liability for environmental damage).

Internalising environmental and social costs ensures that those potentially causing adverse impacts take the full cosequences of their activities into account in economic decision making. This influences them to favour environmentally-sounder production and consumption choices over 'cheap and dirty' ones.

Life-cycle assessment (LCA)

The evaluation of a particular material or activity from generation to final disposal. This is also known as 'cradle-to-grave' or 'earth-to-earth' analysis. The underlying concept is to address environmental degradation before it occurs, rather than wait until problems have been created.

By understanding the fundamental systems that impact on the environment, LCA provides an aid to the design of new and less harmful business methods. Life-cycle assessment accounts for all inputs to, and outputs from, a system and an understanding of the system itself.

Limits to growth

Economics and ecology both recognise that the resources of the Earth such as land, fossil fuels and mineral reserves are finite – as is the capacity of ecosystems to regenerate, to detoxify pollutants or recycle wastes. As the demand made on these resources grows, scarcities arise and place limits to growth.

'Limits to Growth' is also the title of the best known of the studies that have modelled and reported on the world situation and future trends. These studies have greatly contributed to a global view of world and human problems, and have made it clear that there are physical, psychological and social limits to growth, even though they may not be known precisely or fully understood. It is therefore important to change direction now before it is too late, in accordance with the Precautionary Principle.

These models have clarified interactions between many problem areas previously treated separately, and have shown that a long-term perspective is required for the handling of world problems.

Linear industrial systems

These convert raw materials into wastes via a series of different processes and industrial activities. These wastes include the final disposal of goods or wastes arising either directly from production processes or from the use and disposal of products and services. While within linear industrial systems, some processes may be 'closed loops' – for example, the recycling of water at a particular site. Across their entire set of industrial processes, linear industrial systems have only poorly developed means of recycling or utilising wastes in further productive industrial activities.

Monocultures

The cultivation of just one crop or variety of crop over a large area. Monocultures are cultivated ecosystems of extremely low biodiversity at genetic and species level, and of ecosystem diversity. Monocultures are frequently dependent on the heavy application of agrochemicals, including fertilisers and pesticides, and exclude other possible land uses. Pesticides are used to control pests and disease organisms, which in the absence of predators or other natural measures of control found in natural and more diverse ecosystems, can cause serious problems. Fertilisers are frequently required in monocultures to compensate for the absence of natural processes, present in more diverse ecosystems, for maintaining soil fertility..

Non-renewable resources
Resources which, once used, can never be renewed. Fossil fuel energy such as coal, oil or minerals are examples.

No regrets policy
This concept applies in areas where there is uncertainty as to future developments. It identifies policy actions designed to address unexpected changes, but where – even if there is no anticipated change – the policy is desirable in itself. Energy efficiency is an example.

No Regrets Policies are related to 'Adaptive Strategies' which aim to preserve flexibility to respond and adapt to change, thereby meeting increasing human needs such as population levels and per capita consumption rise. They also aim to preserve and widen options for the future. It is important to avoid irreversible commitment to institutions and infrastructure that locks development onto an unchanging path. In industrial terms, adaptive strategies mean reducing waste and pollution, considering life cycle impacts of products and services, and considering the role and spectrum of industrial activity in delivery of national and regional socio-economic and environmental goals. Industrial development that is inflexible in responding to changing human needs, and which promotes resource consumption rather than resource efficiency and conservation, is not adaptive and is unsustainable in the long term.

Ozone layer
A thin layer of ozone gas (O_3) in the upper atmosphere at altitudes between 25 and 35 kilometres that forms a protective screen against harmful solar radiation. In 1984, a 30 per cent reduction in ozone levels above Antarctica was detected and the first 'hole' in the ozone layer was recorded. In 1987, 46 nations signed the Montreal Protocol of the Vienna Convention on the Protection of the Ozone Layer.

It has since been generally agreed that anything less than an immediate and international halt to CFC use would result in accelerated depletion for another 50 to 100 years, because of the long atmospheric life of CFCs. By March 1989, the Montreal Protocol signatories had agreed to eliminate, by the end of the century, the production of the most damaging CFCs and halons. Less environmentally harmful substitutes are now commercially available.

Polluter pays principle
The principle that those who cause pollution should bear the costs not only of damage caused by pollution, but also of measures necessary to reduce pollution.

Pollution prevention pays (PPP)
This concept is based on the fact that pollution prevention reduces costs incurred by an enterprise, and that integrated action to prevent pollution often leads to cost savings generally. PPP originates from the pollution reduction strategy pioneered by the 3M Corporation in the United States.

Precautionary principle/precautionary approach.
Where there are threats of serious or irreversible damage, lack of full scientific certainty shall not be used as a reason for postponing cost-effective measures to prevent environmental degradation.

If there is any uncertainty about the effect that an action may have, the environment must be given the benefit of the doubt. Prevention is better than cure, so releases should be prevented even before there is evidence of damage. The precautionary principle is part of a policy of risk prevention to reduce emission levels of all human-introduced substances.

Protected area
An area dedicated primarily to protection and enjoyment of natural or cultural heritage, to maintenance of biodiversity, and/or to maintenance of the Earth's life-support systems, within their environments.

Rectification at source
The prevention of environmental or other problems through the fundamental redesign of processes, operating facilities and techniques, products, working practices, etc, so as inherently to prevent or minimise production of effluent and wastes which would otherwise impact on the environment. This contrasts with measures implemented to treat effluents and wastes immediately prior to their discharge into the environment – an approach often referred to as 'end of pipe' (from the practice of building treatment plants at the end of discharge pipes).

Recycling
The process by which waste or used materials are put back into productive use. Efficiently operated recycling systems can reduce pollution problems caused by waste disposal. In addition, some recycling schemes may provide alternative sources of energy, and all help conserve energy and natural resources.

Renewable resources
A resource which can be harvested or extracted regularly without diminishing its yield. All biological resources are renewable if used sustainably, as are some physical resources such as power derived from wind, water flow or waves. In many cases, however, potentially renewable biological resources are harvested in excess of their capacity to regenerate. This leads to declines in yield over time. In agriculture, this decline may be compensated by the use of fertilisers and pesticides which are not in themselves renewable resources, and which through their use cause other problems.

Socially responsible investment (SRI)
A term used to embrace all forms of investing which include non-financial criteria as part of the investment management brief. It includes community banking, funds screened ethically or investments in environmentally-responsible enterprises, or institutional funds which take the position as shareholders to encourage higher standards of corporate social and environmental responsibility.

Sustainable development
Sustainable development is defined in the Brundtland Report as 'development that meets the needs of the present without compromising the ability of future generations to meet their own needs'. The aim of sustainable development is 'improving the quality of human life within the carrying capacity of supporting ecosystems' – a definition used in the joint IUCN, UNEP and WWF report entitled 'Caring for the Earth: A Strategy for Sustainable Living'. Sustainable development emphasises the need for a balanced relationship between environmental, social and economic factors.

Any industrial or other development will only be sustainable when it sustains the communities and environments on which it depends both now and in the future.

Sustainable resource use (SRU)
The wise and controlled use of natural resources, including wildlife, so that it always remains within the limits of environmental capacity and can renew itself. SRU is one component of sustainable development and must meet the needs of the present generation, particularly the poor, without compromising the ability of future generations to meet their own needs.

Water cycle
The flow of water through the biosphere and its various physical and biological components.

Bibliography

■

Advisory Commitee on Business and the Environment, *Report of the Financial Sector Working Group*, Department of Trade and Industry and Department of the Environment, London (1993).

Adams, Richard, Carruthers, Jane and Hamil, Sean. *Changing Corporate Values*, Kogan Page, London (1991).

Allaby, Micheal (ed.). *Thinking Greener: An Anthology of Essential Ecological Writing*, Barrie & Jenkins, London (1989).

Association of Environmental Consultancies. *Qualified Environmental Auditors: Code of Practice and Registration Procedures*, AEC, Washington DC (1991).

British Standards Institute. *BS7750: Specification for Environmental Management Systems*, BSI, London (1992).

Brown, Lester, *et al. State of the World 1993*, Earthscan, London (1993).

Brown, Lester, Kane, Hal and Ayres, Ed. *Vital Signs: The Trends that are Shaping our Future 1993/94*, Earthscan, London (1993).

Burrows, Brian, Mayne, Alan and Newbury, Paul. *Into the Twenty-first Century: A Handbook for a Sustainable Future*, Adamantine, London (1991).

Business in the Environment. *A Measure of Commitment: Guidelines for Measuring Environmental Performance*, London (1992).

Business in the Environment. *Buying into the Environment: Guidelines for Integrating the Environment into Purchasing and Supply*, London (1993).

Cairncross, Frances *Costing the Earth*, The Economist Books and Harvard Business School Press, Boston, Massachusetts (1991).

Charter, Martin (ed.). *Greener Marketing: A Responsible Approach to Business*, Greenleaf Publishing, Sheffield (1992).

Clark, Mary E. *Ariadne's Thread: The Search for New Modes of Thinking*, Macmillan, London (1989).

Columbia Journal of World Business (Fall/Winter 1992) 'Focus Issue: Corporate Environmentalism', Columbia Business School, USA.

Commission for the European Communities. *Proposal for a Council Regulation (EEC) allowing voluntary participation by companies in the industrial sector in a community Eco-Management and Audit scheme*. S218/98 Env. 64 Com. (91) 459, Final, Brussels (1993).

Commoner, Barry. *Making Peace with the Planet*, Victor Gollancz, London (1990).

Daly, Herman. *Steady-state Economics*, Second edition, Earthscan, London (1992).

Daly, Herman and Cobb, John. *For the Common Good: Redirecting the Economy towards Community, the Environment and a Sustainable Future*, Greenprint, London (1990).

Davis, John. *Greening Business: Managing for Sustainable Development*, Blackwell, Oxford (1991).

Deloitte Touche Tohmatsu International. *Coming Clean: Corporate Environmental Reporting*, DTTI, London (1993).

Department of the Environment. *This Common Inheritance: Britain's Environmental Strategy*, HMSO, London (1990).

Department of Trade and Industry and Department of the Environment Advisory Committee on Business and the Environment. *Third Progress Report*, London (1993).

Ehrlich, Paul and Ehrlich, Anne. *The Population Explosion*, Hutchinson, London (1990).

Ehrlich, Paul and Ehrlich, Anne. *Healing the Planet: Strategies for Resolving the Environmental Crisis*, Addison-Wesley, Reading, Massachusetts (1991).

Elkington, John and Knight, Peter. *The Green Business Guide: How to Take up and Profit from the Environmental Challenge*, Victor Gollancz, London (1991).

Elkington, John and Hailes, Julia. *The Green Consumer Guide* (1988) and the *The Green Consumer's Supermarket Shopping Guide* (1989), Gollancz, London.

Elkington, J. and Dimmock, A. *The Corporate Environmentalists: Selling Sustainable Development, But Can They Deliver?* A Report on the Greenworld Survey, SustainAbility, London (1991).

Fava, J. A. *et al* (eds). *A Technical Framework for Life-Cycle Assessment*, Society of Environmental Toxicology and Chemistry (SETAC) and SETAC Foundation for Environmental Education, Pensacola, Florida (1991).

Fava, J. A. *et al* (eds). *A Conceptual Framework for Life-Cycle Impact Assessment*, Society for Environmental Toxicology and Chemistry (SETAC) and SETAC Foundation for Environmental Education, Pensacola, Florida (1993).

Global Environmental Management Initiative (GEMI). *Corporate Quality/Environmental Management: The First Conference*, GEMI, Washington DC (1991).

Good, Ben. *Industry and the Environment: A Strategic Overview*, Centre for Exploitation of Science and Technology, London (1991).

Goodland, Robert, Daly, Herman, El Serafy, Salah and von Droste, Bernd. *Environmentally Sustainable Economic Development: Building on Brundtland*, UNESCO, Paris (1991).

Gore, Al. *Earth in the Balance: Forging a New Common Purpose*, Earthscan, London (1992).

Gray, R. H. *The Greening of Accountancy: The Profession after Pearce*, London, ACCA (1990).

Grayson, L., *Environmental Auditing; A Guide to Best Practice in the UK and Europe*, The British Library, London (1992).

Grubb, Michael, Koch, Matthias, Munson, Abby, Sullivan, Francis and Thompson, Koy. *The Earth Summit Agreements: A Guide and Assessment*, The Royal Institute for International Affairs, Earthscan, London (1993).

Handler, Thomas (ed.). *Regulating the European Environment*, Second edition Baker & McKenzie, London (1994).

Harper, Malcolm. *Small Business in the Third World*, Intermediate Technology Publications, London (1984).

Holden Meehan. *An Independant Guide to Ethical and Green Investment Funds*, Fourth edition, Holden Meehan, Bristol (1993 & 1994).

Hundred Group of Finance Directors, *Statement of Good Practice: Environmental Reporting in Annual Reports*, HGFD, London (1992).

Hutchinson, Colin. *Vitality and Renewal: A Manager's Guide to the Twenty-first Century*, Adamantine, London (1994).

International Institute for Sustainable Development. *Business Strategy for Sustainable Development*, IISD, Winnipeg (1992).

IUCN, UNEP and WWF. *Caring for the Earth: A Strategy for Sustainable Living*, Gland, Switzerland (1991).

Jacobs, Michael. *The Green Economy: Environment, Sustainable Development and the Politics of the Future*, Pluto Press, London (1991).

Johnson, Stanley (introduction and commentary). *The Earth Summit*, Graham & Trotman, London (1993).

Kennedy, Paul. *Preparing for the Twenty-first Century*, HarperCollins, London (1993).

King, Alexander and Schneider, Bertrand. *The First Global Revolution*, Simon & Schuster, london (1991).

Leggett, Jeremy (ed.). *Global Warming*, Oxford University Press, Oxford (1990).

Lovins, Hunter, Lovins, Amory and Zuckerman, Seth. *Energy Unbound*, Sierra Club Books, USA (1986).

MacNeill, Jim, Winsemius, Pieter and Yakushiji, Taizo. *Beyond Interdependence: The Meshing of the World's Economy and the Earth's Ecology*, Oxford University Press, Oxford (1991).

Meadows, Donella H., Meadows, Dennis L. and Randers, Jorgen. *Beyond the Limits: Global Collapse or a Sustainable Future?* Earthscan Publications, London (1992).

OECD. *The State of the Environment,* Paris (1991).

Parikh, J., Parikh K., Gokarn, S., Painuly J. P., Saha, B. and Shukla V. *Consumption Patterns: The Driving Force of Environmental Stress*, Indira Gandhi Institute of Development Research, Bombay India (1991).

Pearce, David, Markandya, Anil and Barbier, Edward B. *Blueprint for a Green Economy* Earthscan, London (1989).

Regester, Michael. *Crisis Management: How to Turn a Crisis into an Opportunity*, Hutchinson, London (1987).

Renaux, G. *EC Waste Policy*, Club de Bruxelles, Brussels (1993).

Schmidheiny, Stephan, with the Business Council for Sustainable Development. *Changing Course: A Global Perspective on Development and the Environment*, MIT Press, Cambridge, Massachusetts (1992).

Smart, Bruce. *Beyond Compliance: A New Industry View of the Environment,* World Resources Institute, Washington DC (1992).

Stead, W. Edward and Stead, Jean Garner. *Management for a Small Planet: Strategic Decision Making and the Environment*, Sage Publications, Newbury Park, California (1992).

Tromans, Stephen and Grant, Malcolm (ed.). *Encyclopedia of Environmental Law*, Sweet & Maxwell, London (1993).

Vaughan, Dion and Mickle, Craig. *Environmental Profiles of European Business*, the Royal Institute of International Affairs, Earthscan, London (1993).

Wells, Phil and Jetter, Mandy. *The Global Consumer: Best Buys to Help the Third World*, Victor Gollancz, London (1991).

Williams, Jonathan. *Environmental Opportunities: Building Advantage out of Uncertainty*, Centre for Exploitation of Science and Technology, London (1992).

Willums, J-0. and Golüke, U. *From Ideas to Action: Business and Sustainable Development*, ICC report on the Greening of Enterprise. ICC Publishing, Oslo (1992).

World Commission on Environment and Development. *Our Common Future*, Oxford University Press, Oxford (1987).

Zadek, Simon and Evans, Richard. *Auditing the Market: A Practical Approach to Social Auditing*, published jointly by Traidcraft Exchange, Kingsway, Gateshead, Tyne & Wear NE11 0NE, and New Economics Foundation, London (1993).

Contact addresses

■

INTERNATIONAL

Business Council for Sustainable Development (BCSD)
World Trade Centre Building
3rd Floor, Route de l'Aeroport 10
CH-1215 Geneva 15, Case Postale 35
Switzerland
Tel: +41 22 788 3202
Fax: +41 22 788 3211

International Chamber of Commerce
14-15 Belgrave Square
London SW1X 8PS
UK
Tel: +44 71 823 2811
Fax: +44 71 235 5447

International Institute for Sustainable Development
161 Portage Avenue East
6th Floor, Winnipeg
Manitoba R3B 0YA
Canada
Tel: +1 204 958 7700
Fax: +1 204 958 7710

International Network for Environmental Management
Helgrund 92, D-2000
Wedel
Holstein
Germany
Tel: +49 41 03 84 019
Fax: +49 41 03 13 699

World Industry Council for the Environment (WICE)
40 Cours Albert 1er
75006 Paris
France
Tel: +33 1 4953 2891
Fax: +33 1 4953 2889

EUROPE

Bundesweiter Arbeitskreis fur umweltbewusstes Management (BAUM)
Lohnsteinstrasse 35
A-2380, Perchtoldsorf
Austria
Tel: +43 222 860614
Fax: +43 222 8653893

FIFEGA
Banernmarket 22
A-1010 Vienna
Austria
Tel: +43 1 535 467012

Erhvemlivets Ledelsesforum for Miljofremme
Brodrene Hartmann A/S
203 Klampenborgvej
DK-2800 Lyngby
Denmark
Tel: +45 45 875030
Fax: +45 45 873321

Confederation of Finnish Industries
Etelaranta 10
SF-00130 Helsinki
Finland
Tel: +358 0 180 9250
Fax: +358 0 180 9209

Association francaise des enterprises pour l'environment
5 Esplanade
Charles de Gaulle
92733 Nanterre Cedex
France
Tel: +33 1 4724 6455
Fax: +33 1 4724 6177

Organisation pour le respect de l'environment dans l'enterprise
BP 296 - R8
F-67000 Strasbourg
France
Tel: +33 88 323 443
Fax: +33 88 235 652

Bundesdeutscher Arbeitskreis fur umweltbewusstes Management (BAUM)
Christian Forster - Str 19
2000 Hamburg 20
Germany
Tel: +49 40 81 0101
Fax: +49 40 49 2138

Gesellschaft fur Abfallvermeidung und Sekundarrohstoffgewinnung
MBH Adenauerallee 73
5300 Bonn 1
Germany
Tel: +49 228 228020
Fax: +49 228 2280213

Irish Productivity Centre
IPC House
35-39 Shelbourne Road
IRL-Dublin 4
Ireland
Tel: +353 168 6244
Fax: +353 166 86525

Centrum voor Engergiebesparin en Schone Technologi
Oude Delft 180
NL-2611 HH Delft
Netherlands
Tel: +31 15 150150
Fax: +31 15 150151

Drustvo Poslovodnih Delavcev Slovenije
c/o Manager, Dunajska 106
61000 Ljubljana
Slovenia
Tel: +38 61 181169
Fax: +38 61 183106

Naringslivets Miljoforum
Svenska BAUM
Artillerigaten 38
S-1145 Stockholm
Sweden
Tel: +46 8 6620388
Fax: +46 8 6621496

Eco-Rating International
Fabrikweg 2
8306 Zurich Bruttisellen
Switzerland
Tel: +41 1 808 5620,

Schweizerisch Vereiningung fur okologisch bewusste Unternehmensfuhrung
PO Box 9
CH-9001 St Gallen
Switzerland
Tel: +41 71 286302
Fax: +41 71 286306

Advisory Committee on Business and the Environment
c/o Environment Unit
Department of Trade and Industry
151 Buckingham Palace Road
London SW1W 9SS
UK
Tel: +44 71 2151042

Association for Management Education and Development
(The AMED Sustainable Development Network)
14/15 Belgrave Square
London SW1X 8P5
UK
Tel: +44 71 235 3505

Business in the Environment
8 Stratton St.
London W1X 5FD
UK
Tel: +44 71 629 1600

Business in the Community
8 Stratton St.
London W1X 5FD
UK
Tel: +44 71 629 1600

Centre for Business in Scotland
58-59 Timber Bush
Edinburgh EH6 6QH
UK
Tel: +44 31 555 5334
Fax: +44 31 555 5217

The CERES (Valdez) Principles
c/o Franklin Research and
Development Corporation
The Chartered Association of Certified
Accountants
29 Lincoln's Inn Fields
London WC2 3EE
UK
Tel: +44 71 2426855

Ecumenical Committee for Corporate Responsibility (ECCR)
11 Burnham Wood
Fareham
Hants. PO16 7UD
UK
Tel: +44 329 239390

Ethical Consumer Research Association
16 Nicholas Street
Manchester M1 4EJ
UK
Tel: +44 61 2371630

Ethical Investment Research & Information Service (EIRIS)
504 Bondway Business Centre
71 Bondway
London SW8 1SQ
UK
Tel: +44 71 735 1351

International Institute for Corporate Environmentalism
Barleythorpe
Darham
Leicestershire LE15 7ED
UK
Tel: +44 572 723711
Fax: +44 572 757657

NPI
48 Gracechurch St.
London EC3P 3HH
UK
Tel: +44 71 623 4200

Pensions and Investment Research Consultants
Challoner House
19-21 Clerkenwell Close
London EC1R 0AA
UK
Tel: +44 71 972 9060

Royal Society for the Encouragement of Arts, Manufactures and Commerce (RSA)
8 John Adam Street
London WC2N 6EZ
UK
Tel: +44 71 930 5115
Fax: +44 71 839 5805

UK Social Investment Forum
Room 23, Vassalli House
20 Central Road
Leeds LS1 6DB
UK
Tel: +44 532 429600

North America

Canadian Chamber of Commerce
Focus 2000 Project
1160-55 Metcalfe
Ottawa
Ontario K1P 6N4
Canada
Tel: +1 613 238 4000
Fax: +1 613 238 7643

The Council on Economic Priorities
30 Irving Place
New York, NY 10003
USA
Tel: +1 212 4201133

Environmental Business Council
Choate, Hall & Stewart
Exchange Place
53 State Street
Boston, MA02109
USA
Tel: +1 617 227 5020
Fax: +1 617 227 7566

Franklin Research and Development Corporation
711 Atlantic Avenue
Boston
Mass 02111
USA
Tel: +1 617 423 6655

Global Environmental Management Initiative
1828 L Street
NW Suite 711
Washington DC, DC 20036
USA
Tel: +1 202 296 7449
Fax: +1 202 296 7442

Investor Responsiblity Research Centre
Suite 600
1755 Massachusetts
Washington DC 20036
USA
Tel: +1 202 234 7500

Kinder Lindenberg & Domini
129 Mount Auburn Street
Cambridge
Mass. 02138
USA
Tel: +1 617 5477479

Presidents Commission on Environmental Quality
3M Company
3M Center
St Paul, MN55144
USA
Tel: +1 612 733 1110
Fax: +1 612 733 9973

South America

Asociacion para Desarrollo dela Gestion Ambiental
Maria del Carmen Longa Virasoro
Juramento 3030, 29B
AR-1428 Buenos Aires
Argentina
Tel: +54 1 312 4775
Fax: +54 1 755 9497

Sociedade para o Incentivo ao gerenciamento Ambeintal
Rio Chamber of Commerce
Rue da Candelaria
9/Sala 707, BR-20091
Rio de Janeiro,
Brazil
Tel: +55 21 253 8232
Fax: +55 21 253 6236

Middle East

Society of Industry for Ecology
PO Box 68
70650 Yavne
Israel
Tel: +972 8 433 777
Fax: +972 8 439 901

Asia

Private Sector Committee on the Environment
1 Queens Road Central
Hong Kong
Tel: +852 822 4993
Fax: +852 845 0113

Indonesian Environmental Management and Information Centre
Environment Building
Jalan Kramat IV Nr. 8
Jakarta Pusat 10420
Indonesia
Tel: +62 21 357449
Fax: +62 21 3101656

Eco-Life Centre
Mitani Building
4th Floor
2-13-6 Nishi-Shinbashi, Manto-ku
Tokyo 105,
Japan
Tel: +81 3 358 08221
Fax: +81 3 358 08265

BCSD in Malaysia
11th Floor
Exchange Square, Off Jalan
Semantan
Damansara Heights
50490 Kuala Lumpur,
Malaysia
Tel: +60 3 2561606
Fax: +60 3 2561641

Environment Management and Research Association of Malaysia
38A Jln SS21/58
Damansara Utama
47400 Petaling Jaya,
Malaysia
Tel: +60 3 7177588
Fax: +60 3 7177596

Philippine Business for the Environment
PO Box 12228, Ortigas Centre
Post Office 1600
Philippines
Tel: +63 2 631 3138
Fax: +63 2 631 5714

Natural Resource and Environment Programme
Thailand Development Research
Institute
Rajapark Building
163 Asoke Road
Bangkok 10110,
Thailand
Tel: +66 2 258 902729
Fax: +66 2 258 9046

Australasia

Environment Management Industry Association of Australia
Cumbrae-Steward Building
University of Queensland, GLD 4072
Australia
Tel: +61 7 3653800
Fax: +61 7 3653787

Environment Access, Inc
PO Box 5067
Moray Place
Dunedin
New Zealand
Tel: +64 34 774047

Africa

Industrial Environmental Forum of South Africa
ESKOM
Location 01T40, PO Box 1091
RSA-2000 Johannesburg
South Africa
Tel: +27 11 800 5401
Fax: +27 11 800 4360

Environmental Forum of Zimbabwe
PO Box BW 294
Harare
Zimbabwe
Tel: +263 4 739 822
Fax: +263 4 739 820

Index

■